T0272673

GROWTH FOR GOOD

GROWTH

for

GOOD

Reshaping Capitalism to Save Humanity

from Climate Catastrophe

ALESSIO TERZI

HARVARD UNIVERSITY PRESS

Cambridge, Massachusetts   London, England

2022

Library of Congress Cataloging-in-Publication Data

Names: Terzi, Alessio, author.
Title: Growth for good : reshaping capitalism to save humanity
from climate catastrophe / Alessio Terzi.
Description: Cambridge, Massachusetts : Harvard University Press, 2022. |
Includes bibliographical references and index.
Identifiers: LCCN 2021051705 | ISBN 9780674258426 (cloth)
Subjects: LCSH: Economic development—Environmental aspects. | Global
warming—Economic aspects. | Green movement. | Fossil fuels—
Economic aspects. | Capitalism.
Classification: LCC HD75.6 .T47 2022 | DDC 333.7—dc23/eng/20211208
LC record available at https://lccn.loc.gov/2021051705

To my parents, Fiorella and Mario, who taught me the fine balance between taking responsibility for one's destiny, and at the same time honoring the hard work and personal sacrifices of those who came before us.

Contents

Preface

Like many good stories, this one starts around a dinner table, specifically on a Roman terrace. The date was 2007. I was just turning twenty and had recently started an undergraduate program in economics in Milan. The hosts were a couple of old family friends—both consummate intellectuals from a left-leaning tradition—and I had been given the chance to tag along with my parents. Respecting the standard Italian dinner script with clockwork precision, between the first and the second course the conversation veered toward politics, the dire state of the economy, and what the government needed to do to get the country out of its impasse.

By the time of the digestif, I had pulled out my basic economics toolkit and was parroting the conventional policy recommendations I had heard so many times in university classes. "Italy needs to implement structural reforms that will boost economic growth" was my main contribution to the debate. At this point, and to my shock, Giorgio—the host—turned my world upside-down by proclaiming, "We need to *stop* growing! As my friend Serge always says, we need to *de*grow!" At the time, I had no idea who Serge was. Only years later would I discover that Giorgio's reference was to the French intellectual Serge Latouche, one of the masterminds behind the *decroissance,* or degrowth, movement. On that note, moved by some paternal protective spirit, my father declared the end of the discussion with the Italian code word *eh vabbé,* and we left shortly thereafter.

Nonetheless, Giorgio's words echoed in my mind for years, laying the seed of doubt and therefore of inquiry into what I otherwise saw as a given, both in my economics classroom and in the public debate. Do we actually need economic growth? And, to begin with, what *is* growth? Since then, I have devoted most of my academic and professional career to exploring these questions, and issues related to economic growth more broadly, in both theory and practice. In many ways, this book is the culmination of these reflections, and the manual I wish I could have read early on in my economics undergraduate course.

In the 1980s, economist Robert Lucas was prodded by the extreme differences in countries' growth rates to pursue a general theory for economic development. He wondered: Could government actions in a slow-growth nation allow it to match the growth rates of a fast-growing one? If not, what in the first country's "nature" prevented that? "The consequences for human welfare involved in questions like these are simply staggering," he mused in a 1985 Marshall Lecture at Cambridge University. "Once one starts to think about them, it is hard to think about anything else." My own passion for understanding the nature, drivers, and consequences of economic growth has even deeper roots. Coming from a family of scientists, I strongly recognize that to contribute to improving society is to stand on the shoulders of past discoverers and benefit from past accumulation of knowledge.

As it turns out, decrypting the secrets of the accumulation of wealth allows us to understand much else about the world. Questions about growth are deeply interwoven with some of the most important perennial puzzles across social sciences. They link to the theories of value that kept so many classical economists busy from the eighteenth century onward, including Adam Smith, David Ricardo, and Karl Marx. Digging even deeper into history, they link to the interplay between the quest for happiness and the accumulation of material possessions, power, and profit that philosophers have reflected on at least since 380 BCE, when Plato's *Gorgias* narrated a dispute on the topic between Socrates and Callicles. Social tensions within countries, and therefore large parts of national politics, are ultimately clashes over distributions of proceeds of economic growth and the power imbalances that these generate.

Likewise, the resources made available by economic growth, accumulated over time, can help explain the power relations between countries. The turn to industrialization and the rise of the West, considered by some to be the most consequential developments of the last five hundred years, are a story of unprecedented acceleration in economic growth that changed the course of global history. The recent rise of China can be read along similar lines. Geopolitical power and economic growth go together. Naturally, there is another side to the same coin—meaning the continued poverty of many nations and mass migration—which also demands explanations of why economic growth and material progress have not taken off to a similar extent elsewhere in the world. Finally, thinking about economic growth does not matter only to understanding the past or the present. As we will see, growth is tightly linked to scientific and technological progress, what humans are and will be capable of doing—in other words, it matters to humanity's future.

In light of all this, it will be evident that the ideas, paradoxes, and principles contained in this book have been brewing in my mind for now over a decade. At the same time, 2020 marked a turning point. The year before, upon rejoining EU institutions after my PhD, I had started working on economic questions related to the European Green Deal—the flagship project of the freshly minted Von der Leyen Commission to make Europe the first climate-neutral continent in the world. It was, however, the Covid-19 pandemic that prodded me toward further work related to economic growth— namely, to investigate its relationship to changes in climate, the environment, and nature more broadly. As economies were brought to a grinding halt by lockdown measures aimed at containing the virus, the general feeling was that nature started to heal after decades of abuse at the hands of an industrial society. In 2020, global CO_2 emissions dropped at the fastest clip since World War II, suggesting to many that actively shrinking an economy could perhaps be part of the strategy to combat climate change. This hope was encouraged as images circulated on social media of animals flourishing with the retreat of human activity. The inauthenticity of some of these notwithstanding (dolphins were not swimming in Venice's canals), seeing them led me to wonder: Is it the economy *per se* that is at odds with our planet's well-being, or is it simply the *current type* of economy? More formally, Covid-19 pushed me to deeper exploration of "green growth." This is the widely accepted idea that economic growth need not cause climate and environmental change—the two realms can be decoupled, and rapidly.

As I engaged with the challenge of understanding how capitalism, economic growth, and climate change relate to each other, I felt it was not only my own ideas that needed to be pieced together. Everyone in the hard and social sciences seemed to be working on climate change, using the tools of their various disciplines, but doing so in silos and engaging only superficially with findings from other fields. Throughout history, the growth and increasing complexity of economic activity have gone hand in hand with greater specialization, and the work of producing knowledge has not proved immune to this tendency. To be sure, specialization brings clear advantages in terms of depth and rigor, but if there is one problem that calls for broad interdisciplinarity because it affects practically all aspects of human life, it is climate change. Avoiding climate catastrophe demands an all-hands-on-deck approach, with all disciplines, all sectors, all agents in society contributing to better understanding and response. For me, locked down in my Brussels apartment, watching the world's experience of Covid-19 was a starkly clarifying lesson in why, when a problem threatens humanity so broadly, effective solutions must draw on many realms of expertise. This is the spirit in which I weave together various strands of knowledge in the pages to come. My hope is that readers will not balk at what might feel like excessive discipline hopping, but will recognize the good-faith effort to give them as rich and complete a picture as possible.

Climate change is too important an issue to allow ideology to blind us to truth. I have worked hard to free myself from preconceptions, taking as much as possible an open-minded approach, even to theories normally belittled by mainstream economics. This book carefully considers the points raised by those who criticize capitalism, its relation to nature, and therefore the mainstream strategy—green growth—being pursued to deal with climate change. Many of these critiques of capitalism have already permeated the collective imagination, whether consciously or surreptitiously, and in combination with the growing sense of urgency to act against climate catastrophe have created fertile ground for systemic changes that, in my view, would take us to places where nobody wants to go.

Finding the societal and economic model that allows humanity to avoid climate catastrophe is just too important a quest to leave to the pages of technical manuals and debates among experts. I have tried here to explore the issues in a way that is accessible to anyone with a keen interest in the subject, whether or not they have any prior knowledge of economics. Technical and ancillary ruminations are consigned to the endnotes, for the keenest (and bravest) of readers.

Will this prove to have been a useful endeavor? By enriching the debate can this book contribute to prompt climate action? Will the 2020s be remembered as the decade when a visible reconciliation began between economic and natural processes—when faith in progress was rekindled? Most importantly, will it turn out that the policy recipe I outline is sufficient to head off catastrophic climate change, while also avoiding an equally destructive global scramble over limited resources? I can only quote Italian poet Alessandro Manzoni, who wrote upon Napoleon's death two hundred years ago, "Ai posteri l'ardua sentenza": only posterity will judge.

At the very least, I hope this book will be useful to any readers whose dinner conversation turns to debate over economic growth, perhaps prompted by discussion of the climate emergency. May you have a deeper understanding to draw on than I had when confronted by Giorgio's provocative statement fifteen years ago.

GROWTH FOR GOOD

INTRODUCTION

The Origins of Discontent

Few of us can stand prosperity. Another man's, I mean.
—MARK TWAIN

The economic system now dominant in the world—falling under the catchall term of *capitalism*—is failing us. Or, at least, that is the perception in many quarters, to varying degrees. The geography of discontent paints a paradoxical picture: it is the West that, having served as the cradle of capitalism, has now grown most disenchanted with the system behind its unmatched prosperity. Meanwhile, Asia, and in particular China, has shifted to staunch support of capitalism (or at least state capitalism), after decades of ideological battles against it.[1]

The grievances against capitalism today make for a long list, but generally align with two broad indictments. First, capitalism seems to work for the few, not the many, and therefore to accelerate the concentration of wealth and power. Second, as a system powered by fossil fuels, it has put humanity on a path that, if left unaltered, will lead to climate disaster within a matter of years. Caught up in these condemnations is economic growth: If its benefits are not spreading broadly through society, and it is harming the planet, who needs it? Would ending the pursuit of it alleviate materialistic tensions within and between countries, and take much of the pressure off nature?

To answer these questions, we will take a multidisciplinary approach to understanding the nature of economic growth and the origins of the imperative to keep growing. To identify where problems truly lie, we will put a finer point on the vague, ill-defined concept of capitalism to explore the

1

mechanics of a system in which production is organized for profit using voluntary labor and mostly privately owned capital, and market prices serve to align supply with demand. Informed by a variety of fields, including economics, history, sociology, anthropology, social psychology, cognitive sciences, and evolutionary biology, we will develop a better understanding of why economic growth has been taken up as a "religion of the modern world."[2] It will also become apparent that now is not the first time that this fundamental tenet of the modern economy has been questioned. Detractors of growth have historical and philosophical roots in the work of such towering intellectuals as Thomas Malthus, Jean-Jacques Rousseau, John Stuart Mill, and even John Maynard Keynes. We will see how growth is linked to the concepts of progress, well-being, liberal democracy, science and innovation, consumerism, and capitalism, and how the latter can actually become a precious ally in the effort to rapidly green our economies. In brief, we will see how growth can once more be a force for good when the power of capitalism is harnessed to join the fight against impending climate catastrophe.

A particular focus in these pages on rich countries makes sense for two prominent reasons. First, most of the calls to shelve economic growth originate from thinkers in advanced economies, and it is there that these ideas are spreading most widely. Second, these calls specifically target rich countries, where the environmental footprint per capita is highest. Poorer nations, it is generally recognized, will still need economic growth just to provide for the basic needs of their people. At the same time, this book's analyses of growth hold wider relevance for the world. The inherent mechanics of capitalism, and the interplay with nature, are the same regardless of geography, and obviously the need to address climate change is a global problem. Toward the end of the book especially, as we explore geopolitical interactions between countries and their implications for climate action and growth, the focus will move further beyond rich countries.

In a sentence, the argument I will make is that capitalism can once again become a force for good, but that this will not happen without resolute policy action steadily supported by citizens. The latter phrase is essential to emphasize; this is no Panglossian claim that we can adopt a laissez-faire attitude toward the climate challenge and see a looming catastrophe magically wash away. Both sets of grievances mentioned above, the socioeconomic and the environmental, contain entirely legitimate points that should compel us to change the direction in which capitalism is currently going.

The list of grievances with capitalism

Let us consider the two broad causes of discontent with the current economic system. First, in his recent magnum opus on the future of capitalism, inequality guru Branko Milanovic documents the sheer extent to which power and wealth have concentrated over the past few decades.[3] *Capitalism, Alone* shows how this trend has played out in the capitalist champion par excellence: the United States. We will inspect some data and trends from America before considering how they can be generalized to the larger set of advanced capitalist economies.

The top 5 percent of US households—those with incomes above $250,000 in 2018—command a growing share of national income, from 16 percent in 1968 to 23 percent in 2018, making the United States, at least among the G7 countries, the income-inequality leader.[4] As one striking part of the trend, we can look at the pay of top executives in corporate America. The Economic Policy Institute calculates that CEO compensation grew by a staggering 940 percent from 1978 to 2018, while the pay of typical workers went up by a meager 12 percent. The ratio of average CEO pay to average worker pay rose from 20:1 in 1965 to 58:1 in 1989 and to 278:1 in 2018.

Daniel Cohen's work to factor in costs of living finds that 90 percent of the US population has gained no actual purchasing power over the past thirty years.[5] As time has gone by, widening income divergences have led to greater concentrations of wealth, a trend dissected by another magnum opus, Thomas Piketty's *Capital in the Twenty-First Century*. In the United States, the top 0.1 percent holds roughly a fifth of recorded household wealth—as much as the bottom 90 percent has. This was not always the case under American capitalism. After World War II, the wealth share of the bottom 90 percent increased for decades; it started declining only in the mid-1980s.

While the degree of income inequality seen today in America and some other English-speaking countries (including the United Kingdom, Canada, Ireland, and Australia) is staggering, even smaller levels of wealth concentration create inequality of opportunity and hence injustice. This is what economist Alan Krueger, former advisor to President Barack Obama, dubbed the *Great Gatsby curve,* showing a strong connection between one generation's concentration of wealth and the next generation's economic mobility—the likelihood that those on lower rungs will move up the ladder.[6] Indeed, even in France, where income inequality has not grown much according to some analyses, one study calculates that at current rates of mobility it would take six generations on average, or 180 years, for the offspring of a family at the bottom 10 percent to reach incomes at the level at the overall society's mean. Economic mobility was a reality for many people born of low-educated parents between 1955 and 1975, but it has stagnated for those born after 1975, implying that the social elevator is effectively broken.[7]

Adding to the prevailing feeling that these income and wealth gains are unfair, or at least not simply driven by greater personal talent or harder work, is the scale of tax evasion in many countries. By some estimates, even in low-inequality, high-transparency Scandinavia, the wealthiest 0.01 percent could collectively be evading as much as 25 percent of its tax obligation thanks to savvy exploitation of loopholes in tax codes and jurisdiction hopping.[8]

Adding insult to injury, while most people would say that with great power should come great responsibility, the ultrawealthy channel much of their money in directions seen by the rest of society as at best frivolous, at worst destructive. While society's wealthiest tend to donate large sums to philanthropies, in the tradition of the Rockefellers and the Carnegies, many can also be seen amassing collections of luxury cars, colossal yachts, and private jets, and engaging in outrageously lavish entertainment.[9]

Second, with regard to climate impact, the facts on the table are well known but worth reiterating. Over two hundred years of escalating production and consumption have caused a sharp rise in CO_2 concentration in the atmosphere, which in turn intensifies Earth's greenhouse effect. The impact of rising temperatures is not entirely known, but climate scientists have predicted rising sea levels, more frequent extreme weather events, more droughts and floods, and the collapse of ecosystems.[10] The beginnings of such devastation are already evident today, from Cyclone Idai in southern Africa in 2019—it was one of the worst storms ever recorded in the Southern Hemisphere and killed thirteen hundred people—to Australia's ravaging wildfires in 2020, and the freezing cold spell that paralyzed Texas in 2021.

In the face of all this evidence, one might well conclude that to increase national income—to achieve economic growth, that is—should not be an objective for policy and society. It would seem that most people do not benefit from it, that it puts more wealth in the hands of the few (who squander much of it), and that it leaves those irresponsible few only further entrenched in positions of power. On top of all this, it harms the very planet on which human prosperity and well-being depend.

Very likely, the Covid-19 pandemic and its economic fallout have deepened these sentiments, for at least three reasons. First, as economies were brought to a grinding halt, people saw nature thriving. The concentration of pollutants in the air temporarily dropped, CO_2 emissions saw a larger year-over-year decline than has happened since World War II, and water flowed clean again, as seen in international media images of Venice's murky canals turned transparent. Even as the pandemic ends (one hopes) and lockdowns become distant memories, unlikely to be repeated in the near future, the impression that economic growth is inevitably interlocked with climate change and environmental degradation will remain in people's minds. Second, because the most vulnerable groups in society—women, disadvantaged minorities, those with least savings and most precarious earnings, workers in lower-qualification jobs that cannot be performed remotely—were hardest hit, pandemic lockdowns exacerbated inequalities, stoking further discontent. Third, the policies enacted in response to the pandemic will create heavy burdens in years to come. As governments in advanced economies scrambled to contain the economic fallout, they accepted bigger annual deficits, borrowed on financial markets, and sharply increased public debts. In societies where people already argued over the fairness of taxation and burden sharing, the added pressure of new debts to pay off will push resentments to new levels.

The recessionals

There is one generation that has legitimate reason to be especially dissatisfied with our current economic system, and that is the Millennials. A widespread feeling among these young adults, born between 1981 and 1996, is that they did what they were supposed to do and yet they are not reaping the promised benefits. Indeed, Pew Research shows that Millennials are the most educated generation in US history: as of 2020, as they reached twenty-four to thirty-nine years old, around forty percent had a bachelor's degree or higher. Contrast this with roughly a quarter of Baby Boomers and about thirty percent of Gen Xers at the same point.

Graduating into the 2008–2009 Great Recession made it harder to find a job, however, and most importantly left what economists call "wage scars"—the lasting damage that a lower starting salary has on earnings throughout one's career. By some estimates, every percentage-point bump in unemployment costs a cohort of new graduates 7 percent in earnings at the start of their careers, and 2 percent in earnings nearly two decades later.[11] Millennials are living through the lowest-growth generational period in US economic history (Figure I.1). This was true even before the economic shock of the Covid-19 pandemic, and the disadvantage will only be exacerbated by it. No wonder a headline in the *Financial Times* renames them the "Recessionals," and the *Washington Post* calls Millennials "the unluckiest generation in US history."[12]

Lower earnings—combined with costs of higher education that have grown by 70 percent since the turn of the century, far outpacing the overall inflation rate—leave Millennial borrowers with an average of $33,000 in student debt.[13] Another development is that home-buying is harder for this generation. For the first time since the Great Depression, a majority of young adults (under thirty) live with their parents.[14]

Millennials, along with the following "Gen Z" cohort, are also distinguished from earlier generations by their much stronger environmental consciousness.[15] For example, 70 percent of US adults aged eighteen to thirty-four say they worry about global warming, compared to 56 percent of those aged fifty-five or older. Crucially, this trend seems to be cutting across the left-right divide.

Against this background, one might only expect Millennials to be disenchanted with the economic system they live in, which delivers neither sufficient growth and prosperity nor rapid-enough progress on climate change. Unsurprisingly, survey evidence shows that among members of the Millennial and Gen Z cohorts, only half have a favorable opinion of capitalism, compared to 63 percent of Baby Boomers. This growing resentment and appeal for radical alternatives led the *Economist* to dedicate a recent cover story to rising "Millennial socialism."[16] All of a sudden, giving up on "fairy tales of eternal economic growth," as suggested by environmental activist Greta Thunberg in a speech at the United Nations, might not seem so outlandish.[17] In all fairness, we should not be shocked to hear such sentiments, given that mainstream economists have for the most part stopped explaining to wider audiences the purpose of economic growth— perhaps thinking it is obvious to everyone that pursuing post-growth views is economically illiterate and politically infeasible.[18] Yet, at the cur-

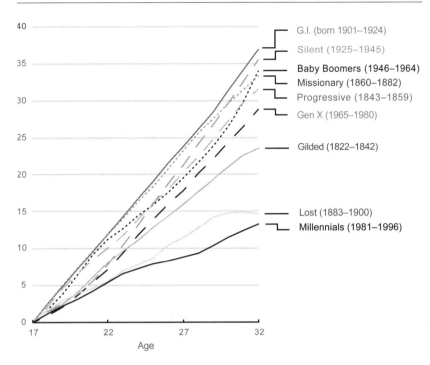

FIGURE I.1: GDP growth experienced by generations after entering the workforce

Data sources: Maddison Project (Bolt and Van Zanden 2020); IMF WEO (2020).
Notes: Chart shows cumulative real GDP per capita growth during each generation's first fifteen years in the workforce, starting at eighteen years of age, averaged across birth years. Data adjusted for inflation and population. Averages for Millennials are based on fewer years due to data availability. Estimates of GDP per capita in 2019 and 2020 are from the International Monetary Fund's October 2020 World Economic Outlook survey.

rent juncture, the quest for solutions to climate change can leave no stone unturned, and economic growth can therefore be legitimately challenged. Chapter 1 will consider critiques of growth advanced by a variety of scholars on the grounds of both its disputed usefulness to human well-being and the hazards it poses to nature.

If I emphasize Millennials it is not because they are my generation and I feel some special attachment, but rather because, as of 2019, they are the largest adult generation in the United States. Between their voting choices and the fact that, in coming years, they will increasingly move into leadership positions, their perceptions and value system will shape political debates and policies for decades to come.

Most of the facts mentioned up to now come from America. Their implications, however, although with some nuances, are relevant to the wider set of rich, capitalist countries. While university tuition, for example, is generally lower in Europe than in the United States, it is still true that European Millennials share with their American counterparts a debt-to-asset ratio greater than past generations. In Europe, Millennials have lived their working lives while the continent was going through a state of quasi-perma-crisis, lurching from the Great Recession to the eurozone crisis to the Covid-19 pandemic—all within a decade or so. A 2018 study by the International Monetary Fund shows the younger generations bearing the brunt of the economic crisis in Europe.[19] Especially for those in Italy, Spain, Greece, and other Southern European countries, the result has been a mix of precarious, underpaid, and undesired-part-time employment, long-term unemployment, and inactivity. The only way to avoid this malaise is emigration in search of jobs.

Among Millennials on both sides of the Atlantic, the feeling is that the social contract—with its implicit promise of ever-increasing socioeconomic improvement over previous generations—has been broken. The result? Trust in the political elite has waned, as has satisfaction with the economy, and younger generations in many nations are looking for radical alternatives.[20]

Big, structural change

Beyond radical Millennials, the idea that our economic system is not on the right track has by now become mainstream. In 2019, the *Financial Times,* not the classic anti-capitalist outlet, launched a call for reform ideas, as capitalism was "in need of a reset."[21] Milanovic, too, even as he proclaims the superiority of the "system that rules the world," acknowledges the important limitations of the current setting, pointing especially to concentrations of wealth and power and the damage done by hyper-consumerism.

Mainstream economists have added much in recent years to the literature on how to reform capitalism.[22] Policy options now being considered include what some might regard as radical plans, such as providing universal basic incomes, sharply increasing wealth taxes, breaking up tech giants, taking the fight to tax havens, capping CEO pay, and replacing narrow measures of gross domestic product (GDP) with more broadly defined assessments of well-being. In the United States, we see growing receptivity to major shifts (which from a European perspective do not seem all that radical) such as free access to healthcare and higher education—and some politicians advocating further for cancellations of student debt and otherwise trumpeting campaign slogans that sound like "big, structural change."[23]

It is beyond the scope of this book to analyze closely the pros and cons of these various measures—over which much ink has been spilled already, and much more will be in coming years. Zooming out and taking the long view, however, we are only seeing in this reform dialectic the standard, positive, corrective forces that have always operated at the intersection of democracy and capitalism and always sustained their combined success.[24] In a compelling recent book, political scientist Francesco Boldizzoni traces the history of two centuries of mistaken predictions of the end of capitalism.[25] Again and again, detractors failed to understand capitalism's incredible adaptability, and the system's ability to change proved key to its durability. Perhaps the most notorious historical example—the one that disproved Karl Marx's forecast of doom—was German Chancellor Otto von Bismarck's innovation in 1889. His establishment of the first old-age security system based on a payroll-financed public pension was an extremely effective political tactic to stifle growing support for socialism. That policy, which then spread to the rest of Europe, is widely seen as the origin of welfare capitalism, and credited with extending the benefits of material progress more tangibly to the working class. We have witnessed the same dynamic repeatedly: when the economic system runs out of control, public pressure grows, democracy channels that discontent into political processes, and eventually there is correction. To state this so simply is not to downplay the painful operational hurdles of a process that has always been and will always be messy. It is far from easy to assemble the required citizen pressure, political leadership, election results, consensus-building capacities, pushback against entrenched vested interests, and so on. Yet, ultimately, change is brought about. In short, as "radical" proposals continue to arise, and in all likelihood some of those listed above are implemented in coming years, we should not be surprised when they not only make the system more equitable but prove to be compatible with capitalism.

Can the same be predicted on the subject of this book's prime concern—the tensions among disenchantments with current capitalism, pushes toward environmental sustainability, and calls to abandon economic growth? Climate change will be in all likelihood the defining challenge for humanity for decades to come, and no one seriously discussing the future prospects of capitalism can shy away from its interaction with nature. But while anyone drawing up a radical reform agenda might consider it a natural fit, abandoning economic growth does not belong on the wish list of reforms. You may sympathize with minimalism, abhor material accumulation and clutter, or embrace the anti-flying social movement known as Flygskam. You may loathe the omnipresent cars choking the streets, despise the vast islands of plastic floating in the oceans, or love small local shops versus large chain department stores. None of this is cause for suppressing economic growth. Your vision of a better world might clash with the reality of today but it is perfectly compatible with capitalism.

Meanwhile, for reasons explained in Chapter 2, halting growth would mean felling one of the fundamental pillars of our economic system; rather than a reform of capitalism, it would bring about its demise. This alone signals a need for caution. Too many academics, natural scientists, and members of the general public genuinely concerned about climate change and the environment are flirting with post-growth ideas, I suspect, without fully realizing their wide-reaching economic and political implications. To them is devoted the first part of this book.

To be sure, for some, the idea of tearing down the whole economic system we live in might seem perfectly reasonable. As an *Economist* article put it recently, "many voters will feel that the social contract has been so badly breached that they would rather rip it up altogether."[26] Environmental activists might view capitalism as so inevitably intertwined with resource extraction and climate change that only its demise could preserve the planet. To them is devoted the second part of this book, in which tearing down the current economic system is shown to be neither inevitable nor desirable.

What a post-capitalistic world might look like is anybody's guess. While some believe that it would usher in a utopian society in which people live in peaceful harmony with each other and with nature, there are significant reasons to predict a much grimmer outcome—and to doubt that people will spontaneously embrace a "less is more" philosophy en masse. If this reality were to be ignored and a model of simple living imposed top down, it is likely that the organization of society would rapidly produce misery. Sold as a dream of liberation from the iron cage of capitalism and consumerism, the utopian society would paradoxically turn dystopian. In spite of the initial aspiration of its ideologues, it would be exposed to the risk of becoming illiberal, inward-looking, and conflictual, and would stoke worse tensions within and among countries.

Under any circumstances, post-growth or not, the only credible way of avoiding a climate catastrophe is to accelerate the development and widespread adoption of "green" innovations. To accept this fact is to understand that abandoning capitalism would take us further away from our climate goals, by throwing sand in the gears of an innovation machinery that is unmatched in human history. Innovation and economic growth are, in fact, inextricable. This is no call for despair. With the right policy actions, capitalism can be enlisted as an essential force in the fight against climate change, and humanity need not devolve into violent scrambles over scarce resources.

It is possible to envision a credible decarbonization strategy—meaning one that creates strong incentives, at the individual, group, and national levels—to meet ambitious climate targets. That strategy will need to respect six basic principles, as summed up in Chapter 8: growth is imperative; the price mechanism shapes supply and demand; government action accelerates change; social cohesion must be maintained; early adopters pave the way; and international cooperation can achieve loose coordination.

Decarbonizing the global economy at the necessary speed will take an effort by literally all actors in society: not only governments and green-innovation firms, but all businesses, every part of the financial sector, and, crucially, every individual. No one should be under the impression that this is going to be easy, but we cannot be deterred by the difficulty. To borrow from President John F. Kennedy's galvanizing announcement, seventy years ago, that the United States would send astronauts to the moon, our resolve today to save the planet should "serve to organize and measure the best of our energies and skills, because that challenge is one that we are willing to accept, one we are unwilling to postpone, and one which we intend to win."[27]

I

Economic Growth and Capitalism

THE LIMITS OF ECONOMIC GROWTH

Anyone who believes exponential growth can go on forever in
a finite world is either a madman or an economist.
—KENNETH BOULDING, 1973

As recent trends such as widening inequality and environmental awareness
provoke startling questions about the very desirability of economic growth,
it might seem that this fundamental tenet of our economic system is under
attack for the first time. To the contrary, recent critiques are just the latest
expression of an old school of thought. It saw its last peak in popularity in
the 1960s and 1970s, and has been simmering ever since. In that era, blatant cases of pollution and environmental destruction—not least, a decade
of nuclear bomb tests at Bikini Atoll following the Second World War—
caused environmental activism to gather steam in America. These sentiments were reinforced by the first full picture of Earth, known as *Earth-
rise*. Taken during the Apollo 8 mission on Christmas Eve of 1968, this
image, as Andrew McAfee puts it, "helped us see that the human condition is inseparable from the state of the planet that we live on."[1] On April 22,
1970, the United States celebrated its first Earth Day.

On the academic side, an MIT team was inspired to create a computer
simulation of the world, illustrating the interlinkages between global population, affluence, and the environment. The resulting report, *The Limits
to Growth*, was issued in 1972 by the Club of Rome, founded only four years
earlier by a group of thirty philanthropists, intellectuals, and members of
civil society looking for long-term, holistic solutions to humanity's most
pressing problems. In it, biophysicist Donella Meadows and her coauthors
argued that exponential economic growth could not go on forever and had
to be brought to a halt quickly to avoid civilizational collapse. The modern

degrowth movement was born. Soon, the 1973 oil crisis helped capture the public's attention—the report had warned that raw materials would run out, precipitating the collapse of civilization, and to some it now seemed (even though this crisis was due to geopolitical tensions) that there must be something to that prediction.[2] Just over a decade later, on the other side of the Atlantic, French philosopher Serge Latouche eloquently linked degrowth to an anti-capitalist narrative and became perhaps the best-known European voice in favor of *la decroissance*.[3]

Subsequently, the movement declined to the point of being considered a fringe cult at the borders of economics, but its stalwarts continued thinking deeply about policies that could finally achieve growth curtailment, and how to make a success of it. Occasionally rekindling some interest by the public, but ostracized by mainstream economics, the degrowth band remained vibrant by establishing its own discipline—ecological economics— complete with focused academic journals and self-standing conferences. This chapter presents the degrowth adherents' perspective, objectively laying out the key planks of their criticisms of society, of the usefulness of economic growth, and of the current mainstream approach to climate mitigation. It will be the work of later chapters to offer a rebuttal, arguing that continued economic growth is natural and imperative, and that, all things considered, pursuing green growth is the only credible strategy for staving off climate catastrophe.

Degrowth in all but name

Given that degrowth is formally confined to a small set of radical academics and activists, one might wonder why a full chapter should be devoted to the movement's arguments.[4] The first answer is that, while most readers might never have encountered the term *degrowth,* many of the arguments made by the movement are actually pervasive in current popular debates on the rights and wrongs of our economic system, especially in relation to climate change. You might, for example, have heard the slogan "You cannot have infinite growth on a finite planet." Or the view that "GDP is not what matters—we should instead focus on happiness and well-being." The fact that these critiques of the current economic system sound commonsensical is a testament to just how fully they have permeated the collective imagination.

The second reason to take time to understand the arguments of degrowth scholars is that, over the past fifty years, they have been doing the intellectual heavy lifting of taking the complaints lodged more generally against economic systems that pursue growth and assembling these into a coherent vision for a post-growth world. Their work therefore provides a useful starting point for reflecting on the origins and nature of economic growth itself, before going on to ponder its relationship with capitalism and climate change—a relationship that has only recently strayed onto the mainstream economics radar screen. Learning the arguments of the founding fathers and mothers of the movement also allows us to recognize that the ideas of recent degrowth proponents, their claims to novelty notwithstanding, are essentially identical to those past ideas. This may be why the snappy slogans posted on social media and quoted in articles about degrowth and the environment are almost always dug up from long ago (the epigraph of this chapter being a prime example). Unoriginality is not the issue here. Rather, the repetition suggests that the degrowth critique, framed as a matter of economy versus nature, may be one of those "internal contradictions of capitalism" that Marx spoke of—a tension that is ever-present, and indeed, in his view, so unresolvable that it must eventually cause the system to collapse. Again, the important function of this chapter is to explain where degrowth thinkers are coming from; even if by the end of this book you are persuaded, one hopes, against their policy recommendations, the concerns they raise are more often than not entirely legitimate and demand a response. This is especially true now, as our economic system seems to be underserving many people and threatening the very planet. To be credible at all, the strategy this book offers to head off climate catastrophe (or any other strategy for doing so) must address these legitimate concerns.

Another merit of the degrowth school is that it draws on multiple disciplines—from economics and sociology all the way to anthropology, philosophy, ecology, and design—as it imagines an alternative society for the post-growth world.[5] The downside of the movement's sprawl over multiple disciplines is that there are several high priests in the degrowth church, and the common heading often conflates their quite different worldviews. This makes it challenging to discern the fundamental principles of degrowth. Still, to stick to the religious metaphor, *The Limits to Growth* can be seen as the equivalent of the timeless Old Testament in the degrowth community, and as for the New Testament, that would probably be Tim Jackson's best-selling *Prosperity without Growth*. Many other texts exist and deserve close attention, but in what follows we will turn often to these two fundamental ones as we consider the building blocks of the degrowth vision.[6]

Slogan 1: "Decoupling is a myth"

Historically, the acceleration of economic growth since the Industrial Revolution of the eighteenth century went hand in hand with the soaring greenhouse gas emissions that are now recognized as the primary cause of global warming. This is because carbon dioxide (CO_2), by far the largest component of greenhouse gases, is released during the combustion of fossil fuels such as coal, oil, and natural gas. While the notion that human activity could modify the planet's climate has been disputed for decades, particularly in the United States, there is now a strong consensus among climate scientists that this is so. The question, then, is whether these seemingly interrelated trends—growing economic activity and growing pressure on the atmosphere—can effectively be decoupled. Is it, in short, possible that economic growth can go on while CO_2 emissions are simultaneously reduced? Advocates of degrowth, while certainly allowing that production processes can be less polluting, reject the possibility of a complete separation (or, in the field's jargon, "absolute decoupling") of the two phenomena. To support their view they use statistics like those charted in Figure 1.1, showing CO_2 emissions for advanced (OECD) economies and the world as a whole.

First, the chart shows global CO_2 emissions still rising. If the goal is to slow down climate change, this trend alone stands as prima facie evidence that the current strategy, based on decoupling and improving energy efficiency, fails on the one metric that matters. The status quo is a road to disaster and must be profoundly modified. Second, degrowth advocates would argue that the trend shown for the advanced economies is deceptive. It is true that these countries' carbon emissions have largely stabilized or even edged down a notch since the mid-2000s—but this is mainly because, responding in part to decades of tighter environmental regulation, they have pushed production of the most polluting items to the developing world. Yet these products are promptly imported back into the rich world to feed consumerism. Indeed, Figure 1.1 shows a widening gap be-

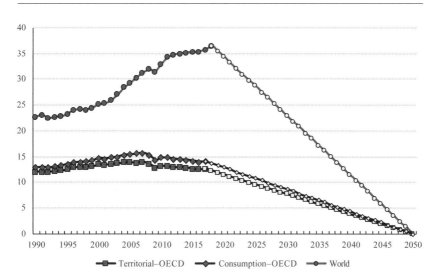

FIGURE 1.1: The path to zero absolute CO_2 emissions

Data source: Global Carbon Project.

Notes: Emissions shown in billions of tonnes. Solid markers indicate real data; unfilled markers approximate a linear trend needed to reach zero absolute emissions by 2050.

tween emissions as calculated based on the production (territorial) versus the consumption of goods. While the effect of this accounting trick might not seem very consequential in the aggregate, for some countries, like Switzerland, it does completely transform the picture. Swiss CO_2 emissions based on production flattened, while those based on consumption went up. For degrowth proponents, this closer scrutiny reveals that decoupling is a myth.

Slogan 2: "Technology will not save us"

Humankind, when challenged by complex problems, has a penchant for innovation and technology solutions. For example, when the Covid-19 pandemic hit, public attention immediately focused on scientific efforts to develop a vaccine. The belief was that, once effective inoculation was produced and rolled out, lifestyles could promptly go back to normal. Degrowth advocates see climate change as a problem in a different league, and consider any high hopes placed on technology to be misplaced.

First, despite all the renewable energy sources (including solar, wind, hydro, and geothermal) being brought online, and all the regulatory actions in richer countries to compel greater energy efficiency, encourage shared mobility, and so on, CO_2 emissions have, at best, dropped very little. This paltry change is particularly disappointing when we contrast it with what climate scientists are saying is necessary. To give a rough sense of the sheer scale of the green transition needed, Figure 1.1 includes a linear trend line approaching zero emissions by 2050. Of course, this may overstate the challenge; after all, climate experts do not say that we must achieve zero emissions by 2050 to avoid climate catastrophe but rather "climate neutrality" by then. This allows for the possibility that the account could be balanced by carbon capture and storage technologies or other solutions that effectively take CO_2 out of the atmosphere.[7] Such technologies, however, are only now being tested and have yet to prove successful on a large scale.

This brings us to the second part of the critique—that we cannot bet the future of the planet on technologies that do not yet exist. Given historical trends, technological development and deployment would need to accelerate at an incredible rate to ensure rapid-enough reduction of greenhouse gas emissions, and even faster if economic growth continues unabated. Even the least pessimistic voices in the degrowth movement would say that, while technology can be part of the solution, it will not save humanity from climate change.

Jevons's paradox, or the "rebound effect"

We are all by now familiar with the pattern that, when a new model of car, television, refrigerator, heating system, or other consumer product is released, it generally rates as more energy-efficient than ones of the previous generation. How can it be, then, that even as this process has been going on for decades, and the upper tiers of efficiency scales have stretched into extreme ratings like AAAA+++, aggregate CO_2 emissions have not descended accordingly?

This is a paradox that has historical roots. In 1865, William Stanley Jevons observed that, as technology improved the efficiency of coal use, the surprising outcome was an increase in the consumption of coal in a wide range of industries. Jevons concluded that, contrary to intuition, technological progress could not be relied upon to reduce fuel consumption.[8] An increase in the efficiency of an appliance reduces the energy needed to perform a specific task, but simultaneously makes it cheaper to use said appliance more extensively. If your new car guzzles less gasoline, then for the same price of a full tank, you can now drive around more. If you replace the light bulbs in your house with more energy-efficient and long-lasting LEDs, you might worry much less about turning off the lights. If you run an airline, and you can fly planes with less fuel, you might promote cheaper weekend-getaway fares to get customers traveling more. This links to the degrowth argument that, in the words of Tim Jackson, "simplistic assumptions that capitalism's propensity for efficiency will allow us to stabilize the climate or protect against resource scarcity are nothing short of delusional."[9] The extent to which energy-efficiency improvements gave rise to a broad-based rebound effect in CO_2 emissions over the past decades remains a disputed empirical question.[10] At any rate, degrowth proponents use Jevons's paradox to argue that today's prevailing value system, whereby "more is always better," is a recipe for disaster.

Slogan 3: "You can't produce stuff out of thin air"

To appreciate another pillar of the degrowth worldview, consider a thought experiment. Assume for a moment that renewable energy sources could indeed be rolled out at an accelerating pace, to an extent that CO_2 emissions could be brought down to safe levels. This would mean solar panels everywhere, vast wind parks both on land and offshore, complete conversion of all planes, cars, trucks, and buses to electric or hydrogen, and retrofitting of virtually every commercial and noncommercial building. Still, in the words of systems ecologist Charles Hall, "If we were to replace traditional non-renewable energy with renewables, which seems desirable to us in the long run, it would require the use of energy-intensive technology for their construction and maintenance."[11] Moreover, most of these things would need to be produced out of raw materials including steel, cement, rare earths, copper, and so on. Even assuming these could be extracted with minimal carbon emissions, the overall process would exert pressure on the environment.[12] By trying to fix one problem—notably, climate change—

humanity might still fall short on other dimensions of environmental preservation, and exceed one or more of the other planetary boundaries (to use environmental scientists' term) within which society's activities must remain.[13] This brings us to the argument the degrowth school considers a checkmate statement: You can't have infinite growth on a finite planet.

Malthus reloaded

This first set of propositions leads degrowth adherents to the overall conclusion that consumption, production, and ultimately income are inescapably tied to greenhouse gas emissions and environmental degradation. Infinite growth is not possible on a finite planet and therefore its pursuit should be set aside, or else we face an inevitable civilizational collapse. If all this sounds dismal, consider that a similar concern for material limits is what led economics to be dubbed the "dismal science" in the first place. Near the end of the eighteenth century, an English demographer, economist, and cleric named Thomas Robert Malthus observed growing food production and increasing well-being, and predicted that these would have the effect of expanding the human population. His *Essay on the Principle of Population* (1798) postulates that, because population growth is exponential and therefore exceeds the more linear expansion of resources available to feed people, famine and disease must follow. Such large-scale reductions of the population serve to restore equilibrium. The growth predicament he described became known as the "Malthusian trap." Perhaps due to his clerical background, Malthus interpreted the trap as a punishment from God, and an invitation to fight primordial instincts and voluntarily follow more conservative sexual behaviors.

While this dismal pattern had held true to a large extent for centuries, Malthus was unlucky enough to publish his book precisely at the inception of the Industrial Revolution, which unlocked a period of unprecedented prosperity. Global population increased almost sevenfold in just over two hundred years, from roughly one billion in 1800. Thanks to technology, humanity escaped the Malthusian trap, at least temporarily. To degrowth advocates, however, climate change and planetary boundaries are a return of the Malthusian trap, and with a vengeance. While innovation (in the form of industrialization) allowed us to escape it that time around, today's renewable technologies are not proving successful in decoupling economic growth from greenhouse gases. To degrowth advocates, it seems clear that we face a stark choice: we can either curtail economic growth voluntarily, or else face Malthusian-style catastrophe.

Illustrating its capacity to connect across disciplines, the degrowth movement combines this first set of environmental critiques with a second major concern—that there is too much "obsession with economic growth." Considerations of the size of our economy take a central role for politicians and policymakers, to the point that elections are won or lost on this battleground. This special attention to the size of the economy and specifically to GDP, the objection goes, is wrongheaded for a variety of reasons.

Slogan 4: "GDP measures everything, except what we really care about"

The expansion of gross domestic product was always supposed to be a means to an end rather than a societal goal in and of itself. As Thomas Jefferson once wrote, "The care of human life and happiness . . . is the first and only object of good government."[14] In the 1930s, the inventor of GDP measurement himself, Nobel laureate Simon Kuznets, recommended not focusing on it too much, as "the welfare of a nation can scarcely be inferred from a measure of national income."[15] In the same spirit, degrowth preachers fear that, as a society, we fail to see the forest as we focus on the trees. We have been deluded by a culture that equates happiness with consumption, trapped in an iron cage of hyper-consumerism. The truth is that, as Tim Jackson puts it, "prosperity . . . transcends material concerns. It resides in the quality of our lives and in the health and happiness of our families. . . . Consuming less, voluntarily, can improve subjective well-being, completely contrary to the conventional model."[16] GDP's detractors point out that it does not even differentiate between types of consumption that normatively should have different weights. Consumption for basic needs—say, access to running water—does not count more than consumption for wants, such as buying a new garden irrigation system for a posh Bel Air villa.[17]

The list of grievances aired against the single most well-known economic metric in our societies extends well past its failure to see different shades of consumption.[18] First, GDP says nothing about inequality: national income might go up but be entirely pocketed by the richest. Second, it does not factor in the damages from our eating into natural resources and polluting the planet. When a forest is chopped down and the wood is sold, GDP goes up, despite the facts that climate change is aggravated and our children will have less nature around them. Third, GDP does not account for non-remunerated activities carried out at home, such as caring for children or elderly parents—work, by the way, which still falls disproportionately to women. In short, the degrowth argument goes, GDP measures everything except what we actually care about. This is the message Robert Kennedy delivered in 1968 in a speech worth quoting:

> Gross national product does not allow for the health of our children, the quality of their education or the joy of their play. It does not include the beauty of our poetry or the strength of our marriages, the intelligence of our public debate or the integrity of our public officials. It measures neither our wit nor our courage, neither our wisdom nor our learning, neither our compassion nor our devotion to our country, it measures everything in short, except that which makes life worthwhile. And it can tell us everything about America except why we are proud that we are Americans.[19]

Degrowth proponents are therefore unanimous in suggesting we should move beyond GDP as the main target variable of economic policy. To some, this means actively shrinking the size of our economy. Others take a more nuanced view. In her 2017 best seller *Doughnut Economics,* Kate Raworth (a self-styled "renegade economist") proposes that we become agnostic about GDP growth, and instead move on to focus on a variety of indicators of "things we care about," such as social equity, gender equality, housing, food, and health, while ensuring we are operating within planetary boundaries. She presents this array visually in the shape of a doughnut, laying out the "safe and just operating space for humanity." At first sight, this fresh economic thinking sounds perfectly aligned with the various governments and international organizations now looking at ways to enhance the scope of their monitoring of the economy, under a "GDP and beyond" agenda. For instance, since 2018, the government of New Zealand has been assessing the impact of its budget measures against a set of indicators covering social, human, and natural considerations. The OECD and the European Union are exploring scoreboards that could account for various dimensions of well-being.

What might seem like an innocuous linguistic nuance between "beyond GDP" and "GDP and beyond," however, carries extraordinary implications for policymaking and the future of our society. We will return to this topic in pages to come.

Slogan 5: "Less is more"

A trope of the degrowth line of argument is that GDP might very well be important in the early stages of development, to meet a population's basic needs for food, shelter, sanitation, and education—unsurprisingly, on average, people living in poorer countries do report lower levels of happiness than those in rich countries. After a certain (low) threshold, however, this relationship breaks down: more GDP per capita does not go hand in hand with higher self-reported happiness.[20] This is known as the Easterlin paradox, after the economist who first detected this empirical pattern in the early 1970s. Later writing an update to his original analysis, Richard Easterlin showed that between 1973 and 2004, real income per capita almost doubled in the United States, but self-reported happiness remained effectively constant. The same is true for other wealthy nations in Europe and beyond. Degrowth advocates take this to mean that one should not conflate the desirable development of poor countries with the economic growth of rich, industrialized nations, where extra GDP growth is useless, even harmful. As Tim Jackson frames it, "more isn't always better," and the truth may be closer to the title that anthropologist Jason Hickel chose for his book, *Less Is More.*

Individualism and the infinite desire for more

We have so far focused on practical critiques of economic growth and GDP, but there are others more philosophical and subtle. These center on whether the quest for economic growth, or the infinite desire for more, is inherent in human nature. It is not, the degrowth argument goes: it is just a social construct, a product of capitalist institutions that influence us to approach the world in this way. This discussion can be traced back even to the father of modern economics. Adam Smith, in *The Wealth of Nations*, notoriously explains that as every individual "endeavors as much as he can both to employ his capital in the support of domestic industry, and so to direct that industry that its produce may be of the greatest value, every individual necessarily labors to render the annual revenue of the society as great as he can." The individual, Smith concludes, is "led by an invisible hand to promote an end which was no part of his intention."[21]

Naturally, Karl Marx passionately rejected the idea that individualism and the pursuit of more are inherent to humans and inevitably give rise to a societal growth imperative, but more recent scholarship is also having a go at it. Historian Rutger Bregman, for example, uses a number of anthropological case studies in his book *Humankind* to argue that ancestral societies lived peacefully in a self-regulating fashion, and it was only when the socially constructed idea of private property took hold that violence and inequality entered the picture. (Bregman's larger argument is that, deep down, "most people are pretty decent.") This line of thinking attacks a central tenet of capitalism, that while the human drive for self-interest should be channeled, it cannot be eradicated. And indeed, Bregman does not shy away from that fight. The implication of his work, he proclaims, is that "to stand up for human goodness is to take a stand against the powers that be."[22] This view is echoed by anthropologist James Suzman in his magnum opus, *Work*. Reading the history of humankind through the lens of work, Suzman points to the advent of agriculture as a development that put humanity on a path to being enslaved by institutions that force us to work and accumulate.[23] Economist and sociologist Juliet Schor argues along similar lines that we have been tricked by the mirage of perennial economic growth into overworking ourselves. This is particularly striking given that even the illustrious economist John Maynard Keynes, writing in 1930, thought that growing wealth should require of people only fifteen-hour work weeks within a hundred years.[24]

In the degrowth view, once the obsession with economic growth is brushed aside, different ways are possible. Suzman highlights the hunter-gatherers of the Ju/'hoansi, an African tribe that had lived isolated since the dawn of our species. They were well-fed, egalitarian, and content, and had plenty of leisure, rarely working more than fifteen hours in a week. Tim Jackson looks to *ecovillages*—consciously formed by groups of people committed to sustainability and content with "sufficiency." The environmental activists of the Extinction Rebellion are exploring similar community arrangements.

There are other philosophies and movements, too, that embrace the kind of "voluntary simplicity" that Mahatma Gandhi advocated. Consider the rise of mindfulness communities, such as Plum Village, founded by Vietnamese monk Thich Nhat Hanh in the south of France. In North America, the Amish way of living stands in contrast to the infinite desire for more. Kate Raworth takes inspiration from the past example of the Cree in Manitoba, Canada, who deliberately embraced the principle of sufficiency.[25] Degrowth adherents draw on all such examples to support their three-part argument: It is a myth that humans are selfish, individualistic, and endlessly greedy. Capitalist institutions made us that way. And because institutions are social creations, they can and should be changed.

The natural death of growth

The last arrow in the degrowth quiver is the argument that, whether we like it or not, economic growth is on an inexorable downward path. Its arrival at its natural end is only a matter of time. Across the decades since the Second World War, global GDP per capita grew 3.6 percent in the 1960s, 2.1 percent in the 1970s, and only 1.2 percent in the 1980s. It has remained below 2 percent since then. In advanced economies, growth has remained particularly sluggish since the Great Recession of 2008–2009—a fact that led former US Treasury secretary Larry Summers to speculate in 2013 that we were experiencing "secular stagnation." In *The Rise and Fall of American Growth*, economist Robert Gordon postulates that productivity growth is on a structural downward trend because the inventions with truly life-altering power are all behind us. Between 1870 and 1970, our world was transformed by electric lighting, indoor plumbing, motor vehicles, the telegraph, air travel, and so on. The impact of these simply cannot be matched by subsequent inventions, as revolutionary as computers, the internet, and smart devices may seem to be.

In desperate, last-ditch efforts to revive growth, governments in advanced economies are increasing public debt to levels never observed outside of war periods. This was true even before the Covid-19 crisis. In the eyes of Tim Jackson and Kate Raworth, the deregulation and financialization observed since the 1980s in the United States and United Kingdom are just short-term fixes being applied along with others to prop up the system—and have only made growth more unstable and increased the frequency of financial crises and recessions. Anthropologist David Graeber blames the misleading religion of economic growth—in combination with the ongoing march of labor-replacing automation—for the rise of the "bullshit job," which he defines as "a form of paid employment that is so completely pointless, unnecessary, or pernicious that even the employee cannot justify its existence."[26] In many ways, then, the degrowth argument points out that growth is abandoning us. Rather than prolong our toxic attachment to it, we should reflect on how society could be different in a growthless environment. We should restructure our institutions to be more resilient. And we should re-define what we mean by prosperity.

————————

Agree with it or not, the degrowth movement has built, through one so-cioeconomic critique after another, a consistent narrative of the world. It can be summarized as follows. The growth of national incomes is inextri-cably tied to the rise of greenhouse gas emissions, and will remain so over the foreseeable future. National income is mismeasured and, to the extent the measurement does reflect reality, it reveals that GDP growth does not equate to happiness. And growth, doomed in any case, is already in the throes of its inevitable decline. Therefore, we should ditch economic growth as a target and shift to an economy and mindset of mere sufficiency, taking the path suggested by inspiring (often Eastern) philosophies, alternative cul-tures, and subsistence communities. At least this is what we in advanced economies should do, while appreciating that poor countries still need to grow further to provide for their peoples—and indeed creating additional space for them to do so. Only with global redistribution of wealth can a credible fight be taken to climate change that will not result in overall de-creased happiness—and that should even boost happiness, as humanity is unshackled from superfluous consumption and production.

It is a narrative that, although it sprang from a different crisis perceived over fifty years ago, speaks to a lot of concerns currently on people's minds, especially in the Millennial and Gen Z generations. Perhaps unsurprisingly, in recent years, this worldview has made conspicuous gains in mainstream media attention.[27] Yet even in some unexpected quarters, elements of the degrowth creed are resonating, eliciting support, for example, from eminent energy experts like Vaclav Smil, advisors to the United Nations, directors at accountancy giant KPMG, physics Nobelists, a former US secretary of energy, the president of Ireland, the European Environment Agency, and His Holiness Pope Francis.[28]

Clearly, abandoning economic growth would have broad and deep im-plications for the way our society is organized and for the capitalist system. What might these include? Here, consensus within the degrowth church breaks down. For some, including early Latouche and recent Hickel, de-growth represents a mortal blow to a system that can only survive on per-petual expansion. For others, such as Meadows, free markets will remain vital, and a degrowth agenda need not entail more than some important

tweaks to the current system.[29] For Tim Jackson, the question is almost ir-
relevant. Having painted a picture in his book of an ideal post-growth
world, he concludes: "Is it still capitalism? Does it really matter? For those
for whom it does matter, perhaps we could just paraphrase *Star Trek*'s Spock
and agree that 'it's capitalism, Jim. But not as we know it.'"[30]

Yet the impact of abandoning economic growth cannot be so easily waved
away. In Chapter 2, we will explore the real-world ramifications of embracing what might come across as a pleasant alternative societal vision
and commonsensical policy agenda. As we'll see, the reverberations extend
further than many can imagine.

2

GROWTH AND THE MECHANICS
OF CAPITALISM

Economic growth has become the secular religion
of advancing industrial societies.
—DANIEL BELL, 1976

What is economic growth? To answer this question might feel like an intuitive, almost trivial task. Humans have known the cumulation of wealth probably since the dawn of time, and at least since the earliest days were chronicled in the Book of the Genesis.[1] It is in the Old Testament, too, that the famous "wealth of nations" gets its first mention.[2] Today, talk of economic growth is constantly around us; in newspapers, television shows, political debates, and presentations by corporate managers and business strategists, inevitably the g-word is invoked. At the same time, it would seem that answering the question must involve real complexity, as a battle for ideas has centered on it since the times of Adam Smith and Karl Marx, and an entire field of study, growth economics, has grown out of it.

Before jumping into that fray, it is worth spending a few words on definitions which will turn out to be useful as we proceed. In its most basic form, the *economy* refers to the set of activities within society aimed at producing goods or services to exchange and to fulfill people's needs and desires. Experiencing *economic growth* therefore means producing more or better goods or services in aggregate than in the past. Whatever it is that you produce, whether goods or services, you will use *inputs*—some combination of labor, capital, land, raw materials, and energy. To clean apartments, custodians use their time (labor), plus vacuum cleaners (capital), detergents (materials), and electricity (energy). Accountants, lawyers, physicians, and ski instructors all draw on inputs, as well.

Achieving any higher level or quality of production requires either increasing the amount of inputs (*extensive growth*) or learning a way to do more with what you have (*intensive growth*). Let's say you are a woodworker, you make kitchen tables for a living, and you want to produce more. You might hire someone to give you a hand (increasing labor input) or you might buy machines to expedite the cutting, gluing, painting, and chiseling (increasing capital input). Alternatively, you might figure out new ways of using your current tools and machines to produce more without adding inputs. You might even find a new way of producing more tables that also reduces your raw material input—perhaps by collecting back old tables from your customers and using new techniques to recycle the wood (making you part of the *circular economy*). Producers across society make such decisions constantly, and the aggregate of their actions is what a country's economic growth is all about.

The measurement of gross domestic product (GDP) has the aim of trying to capture this aggregate—the size of an entire economy—in local currency terms.[3] It has its shortcomings since, at the end of the day, it is simply a statistical artifact that relies on assumptions and is subject to limitations. Nonetheless, it is useful because, while *you* might make kitchen tables, other people produce green tea, software apps, and yoga classes. All use different sets of inputs to produce different outputs. To aggregate all these units and make them comparable we give each a monetary value, which is typically the price paid for it (its *market price*). Aside from allowing the aggregate size of the economy to be gauged, pricing everything with a single metric allows people within the economy to make comparisons, establishing a basis for decisions given that resources must be allocated across otherwise hardly comparable things. Yet many things that surely matter, such as hugs from family members, acts of kindness by neighbors, and time spent with friends, are not factored into GDP. Because they are not transacted, they are not part of the economy, and because they are not on the market, they are not captured by GDP. This explains the common sardonic quip that "economists are people who know the price of everything, and the value of nothing."

Economics is the study of allocation of *scarce resources*.[4] For something to show up on the radar screen of the discipline, it needs to be a resource—that is, somebody must attach some value to it—but crucially it also needs to be in limited supply, meaning not everyone can have as much as they want of it. When property rights are well defined and protected, the very fact that something has a price—meaning people are willing to pay for it—usually suggests that these two conditions are fulfilled. The oxygen we breathe surely fulfills the first condition, being fundamental for human survival, but it has not been scarce (at least so far). As such, oxygen per se does not enter the economy or GDP, even if it is "a thing that truly matters."[5] The same type of thinking applies broadly to nature which, as strange as it may now sound, has for a long time been considered to be unlimited.

Does GDP have some blind spots with respect to things people inherently care about, such as the joy of walking through an urban park, spending hours with grandpa listening to stories of his youth, and playing frisbee with the dog? Yes, perhaps, but GDP was never intended to be an indicator of happiness or well-being. (Chapter 3 will explore why this is less of a concern than "beyond GDP" advocates portray it to be.) Is GDP perfect and a measure that cannot be improved? Absolutely not. For instance, given that our increasingly digital world provides plenty of zero-price services, from social media to GPS to endless reference information, it is likely that current GDP computations are painting a distorted picture of the economy's size.[6] Well aware of all this, skillful statisticians worldwide are constantly working to adapt GDP to the changing nature of economies, in part by incorporating the value of new digital services and of nature.[7]

Shooting the GDP messenger

Because GDP has some blind spots, and focusing on it amounts to constantly checking the pulse of economic growth that is currently harming our planet, some scholars, such as Kate Raworth and Tim Jackson, have argued we should become "agnostic" about GDP and its direction of change. In a recent topography of the degrowth movement published by *Nature Communication,* this view is presented as pragmatic or "reformist," because it allows that degrowth can take place within the current, capitalist economic system.[8] Curiously, this agnostic position seems to reflect the belief that, if we just stopped measuring it, economic growth would magically go away. An old philosophical riddle goes: If a tree falls in the forest, but nobody is there to hear it, does it make a sound? The agnostics seem to offer a similar puzzle: If we live in an economic system, but there is no statistical tool to measure its changing size, does it grow?

A degree of confusion springs from the perception, right or wrong, that "politicians and the media are obsessed with growth." Stretching this premise, one could charge that economic growth is *only* a political obsession, perhaps fanned by narrow-minded economists constantly whispering in the ears of power, drawing leaders' attention to the single metric of GDP. In Chapter 3 we will look at why politicians are so fixated with growth itself, but it is useful to note that economic growth surely did not start in the 1930s, when Simon Kuznets first developed the GDP measure. The objective then was to gain a better sense of the depth of the Great Depression, so that President Franklin Delano Roosevelt could calibrate the large fiscal stimulus of the New Deal. We can stop measuring it, but economic growth will remain an in-built feature of our system, there are many reasons to believe, if not of all thriving civilizations. This suggests that, to the extent a society thrives, growth continues—unless policy measures are actively taken to bring the machine to a grinding halt and keep it from restarting. GDP is like the speedometer on a ship: you can stop looking at it as you navigate at sea, but the ship will continue to plunge ahead unless you reverse the engines.

Beyond the agnostic take on GDP, the questions posed by the "reformists" are legitimate and worth exploring: Is it possible to bring growth to a halt within capitalism, or does the system suffer from a *growth imperative*?[9] Is degrowth compatible with the way our society is currently organized? There are several ways to contest the claim that we can simply pull the growth brake without compromising the entire system. The counterargument can be assembled bottom-up by starting at the microeconomic level, looking at individual growth incentives for people and firms, or top-down, by starting at a macro level, examining what happens to a country when growth dries up. Here, we will address the topic at different levels of abstraction, from more immediate and intuitive to more profound and subtle considerations.

Obviously, there can be years in which growth turns negative or stagnant in capitalist economies without systemic consequences. To say otherwise would be nonsensical: even now, the world is slowly emerging from a deep recession due to the coronavirus pandemic. Yet the quest for growth is a central feature of our economic system. In capitalism, incentives are organized society-wide in a way that constantly kindles a growth impetus. This turns out to be the inseparable other side of the coin—the first side being a feature of capitalism that is highly cherished even by several degrowth advocates, its constant push for efficiency and innovation.[10] If growth, or the yearning for it, were to be suppressed, the entire system would be destabilized.

In what follows, we will focus on the concept of zero growth, otherwise known as *steady-state* capitalism. This focus makes sense given that even those degrowth advocates calling for the economy to shrink see that as only an intermediate, necessary step on the way to a lower, sustainable level at which to stabilize. For the time being, we will concentrate on the stabilization of growth, leaving the transition dynamics for a later chapter.

Institutional structures

One cannot discuss clamping down on growth in a vacuum. Democratic capitalism is inherently complex machinery, and new layers of complexity have been added year after year. Among the various features and structures of our economic system are some that crucially depend on economic growth. The prime example is the welfare state.[11] Think of unemployment benefit schemes, old-age pensions, and healthcare.[12] All of these institutions are luxury goods that can be paid for and sustained thanks to continued economic growth.[13] Across advanced economies, on average, government spending on health and social protection accounts for almost 20 percent of GDP, a much larger proportion than any other budget line items. Add to this that population growth is flattening, or even shrinking, while life expectancy grows. The effect of these trends is that the welfare state takes up a growing share of societal resources (or GDP), as growth rates for these expenditure lines are higher than those for the whole economy.[14]

A second but related point has to do with public and private debt. Even before the Covid-19 shock, total public and private debt combined was north of 300 percent of GDP for practically all G7 economies: the United States (318 percent), Britain (310 percent), France (351 percent), Canada (356 percent), Italy (301 percent), and Japan (444 percent). Thrifty Germany (215 percent) was a notable exception. These debt levels are generally considered sustainable because of an assumption that these economies will keep expanding. This conviction alone keeps a massive debt sell-off from materializing. By the same token, abandoning growth would likely result in a cascade of household, corporate, and potentially sovereign debt defaults. Defaults would then make it harder for governments to borrow from financial markets, should there be the need to later on. If and when an unexpected shock arrived, crisis management and any sort of (Keynesian) fiscal stimulus would have fewer resources to deploy. The Covid-19 pandemic provided a good example: when the crisis hit, governments in rich countries borrowed massively to sustain incomes while their economies went on lockdown. Governments in developing countries, with more limited access to financial markets, could not do so to the same extent and their economies were more exposed to the fallout.

In a steady-state environment, access to financial markets would also be harder because governments would have less to borrow against—that is, less expectation of a larger economy in the future, or else a larger taxable base.[15] The reason that governments can borrow more easily, in general, than private companies is that lenders know that a sovereign state has taxing powers and, should push come to shove, can find a way to repay its debt. Of course, taxation is highly unpopular, so governments prefer to repay their debts with the larger tax revenues that come with economic growth. Or, if a nation cannot grow out of its debts, another alternative is to finance its crisis-management countercyclical spending with subsequent austerity, probably involving more belt-tightening than was seen in Europe after 2009.[16] Still another route is to print money, generating high levels of inflation. Yet these latter options are just as unpopular as higher taxes, and are known to disproportionately harm the poor.[17]

For reasons that will be outlined in Chapter 7, the era of climate change is highly likely to intensify the need for redistribution and welfare spending. We know that curbing emissions—for example, by means of a basic carbon tax—creates a burden that falls disproportionately on the poor. This became evident in 2018 when the Yellow Vest movement launched its protests in France in response to a gasoline tax hike. Furthermore, climate science warns us that shocks from extreme weather conditions, droughts, and flooding will be increasingly frequent in the future. As the rich migrate to less exposed regions, and buy additional private insurance, the poor will be less able to do so. Summing all of this up, having only a steady-state economy in a time of climate change would leave governments with less borrowing power, expose the poor to more harm, and generate higher cyclical volatility for countries and individual citizens alike.[18]

There is more to fear, too, since economic growth tends, heuristically, to be highly intertwined with employment levels, at least in the short term. When growth falters, unemployment goes up (Figure 2.1). Perhaps glorifying this basic empirical regularity too much, economists refer to it as

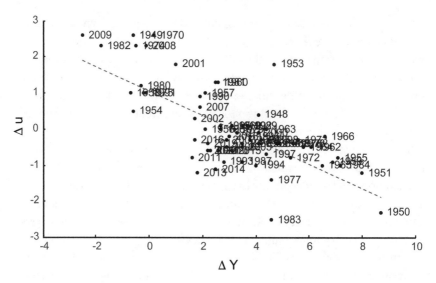

FIGURE 2.1: Okun's law in the United States, 1948–2019

Data source: Ball et al. (2017).

Notes: Horizontal axis shows annual growth rates; vertical axis shows changes in the unemployment rate as tracked by the US Bureau of Economic Analysis and the US Bureau of Labor Statistics.

Okun's Law, after Arthur Okun, who chaired the US Council of Economic Advisers in the Lyndon Johnson administration and documented it in the 1960s. This is why we so often hear the words *growth* and *jobs* used together.[19] Okun's Law implies that a degrowth agenda would increase the need for unemployment benefits and other forms of income support, while also leaving governments less able to meet it. The same would be true for any form of universal basic income, should this be rolled out in the future.

A related point, made in a powerful speech by former Indian central banker Raghuram Rajan, is that new technologies and global competition have had a tendency to destroy jobs made up of repetitive and routine tasks, normally performed by low-education, low-income workers.[20] In this context, economic growth performs a further crucial function: it creates replacement jobs to provide ongoing employment for those for whom substantial new training is simply too onerous.

Going through these arguments, we get a sense of a paradoxical situation. Large parts of the degrowth critique come from a progressive or "Millennial socialist" angle, and are accompanied by calls for tougher stands against inequalities, more support for the poor, and a stronger role for the welfare state more broadly. We have just seen, however, that an economy without growth would leave governments with *less* financial firepower to fight such challenges, even as it expanded their scale—leaving the much-hated alternatives of austerity or iniquitous inflation to pay for them. It comes as no surprise, then, that degrowth advocates have gravitated to the pipe dream of Modern Monetary Theory as an escape route from this policy conundrum. This theory purports to bust the "deficit myth" (to invoke the title of Stephanie Kelton's book) by promoting money-printing as a sustainable and unlimited way of financing government, magically without creating inflation.[21]

Recall that the welfare system was first created back in the nineteenth century precisely to shore up capitalism against some of its most perverse wealth-concentration dynamics, and therefore to contain the insurgence of social unrest. Wouldn't diminishing the welfare state's firepower at a time of pressing need do the opposite? We can credibly suppose that it would further destabilize the whole economic system and that, as Heather Boushey argues, the increase in inequality would also stifle competition and innovation.[22]

This does not necessarily mean that degrowth must imply the complete end of the welfare state. It is possible that people would not object to more meager pensions or even less medical care. In line with what one political scientist calls "the environmental paradox of the welfare state," if people could be broadly persuaded that "less is more," they could live happily with lower levels of support.[23] According to the degrowth narrative, the fact that life in the post-growth utopia would be less stressful, thanks to a better work-life balance, would in part make up for lost healthcare spending. There might therefore be ways to reinvent a smaller and different type of welfare system that could work in a steady-state economy. Still, it should be evident that abandoning growth would cut through several of the institutional features that characterize our *current* economic system. As such, it would force a rather wide-reaching, structural rethink of our economy, much deeper than is reflected in the current public discourse on the topic.

The price mechanism

So far in this chapter, arguments have been made that our capitalist system in the way it is currently organized is incompatible with degrowth. At a deeper level, we are about to discover that the quest for growth is a corner-stone of capitalism itself, and abandoning it would not call for just some reorganizing. It would strike a mortal blow to the entire system, requiring a complete societal rethink. To see this, we must descend into the mechanics of capitalism.

In *Capitalism, Alone,* Milanovic builds on Max Weber to craft a compact definition of capitalism. It is "production organized for profit using legally free wage labor and mostly privately-owned capital, with decentralized coordination." Decentralization requires an automatic system that allocates scarce resources through a series of recursive individual decisions. This is the *price mechanism,* and it is the central feature at the heart of capitalism. It is the mechanism by which demand and supply equal each other at all times: changes in prices are the valve that adjusts the process. The price mechanism is in essence Adam Smith's famous "invisible hand." If demand for avocados, for example, rises higher than supply, perhaps as a result of some culinary fad, the price will go up. That higher price will attract more farmers to shift from producing other crops to producing avocados, thus increasing supply, and this in turn will progressively bring the price back down. Note that this works only where there is a degree of competition in the market and also entry possibility. If the entire supply of something comes from a monopolist and cannot be affected by other producers, the one-to-one relationship between quantity demanded and price breaks down. In many cases, the monopolist has an incentive to undersupply a good to keep its price artificially high. This underscores a wider point, that for the price mechanism to operate smoothly a set of rules (or "institutions") must be in place, including a free and open competitive environment, well-defined and protected property rights, and a functioning legal system to support contract enforcement.

When good institutions are in place, the price mechanism can serve three important coordinating functions:

Signaling. Prices draw resources where they are needed, rising and falling to reflect this pattern.

Transmission of preferences. Through changes in prices, consumers send decentralized signals to producers about their changing needs and preferences.

Rationing. If demanded quantity for a good is higher than the available supplied quantity, prices adjust until the two balance out.

Within this mechanism we can also see that, if consumers have a perceived unmet need, producers are encouraged to provide more of it by their *quest for profit* (or, if you prefer, their "quest for more"). This is why Milanovic's definition includes "production organized for profit." But obviously, in a degrowth world, this cannot be allowed to happen because, at a societal level, quests for more profits fuel economic growth. If degrowth is the goal, we cannot let the price mechanism do its trick.

Up to now, we have considered producers as a bloc. But, of course, in a market economy, producing firms compete with each other for customers. As all seek to gain an edge, they make investments aimed either at increasing efficiency to reduce costs and prices or at innovating to create new and better products. This is the essential process behind economic growth. Once a firm manages to gain an edge—think Apple when the first iPhone came out, or Tesla with its early electric cars—it has an incentive to try and sell as much as possible, matching demand. As all private-sector managers know, the alternative to growth is not to coast along and maintain a status quo, but rather to lose market share to other competitors and begin a descent toward bankruptcy. This leads to the business mantra "innovate or die," and many fates are decided at the nexus of competition, investment, and innovation.[24] Economists refer to this dynamic among firms as "Schumpeterian creative destruction," because it was extensively studied by economist Joseph Schumpeter, who observed in 1942 that "the capitalist economy is not and cannot be stationary."[25] Karl Marx, and also the Communist thinker Rosa Luxembourg, had also identified this competitive dynamic as the force behind firms' resolve to accumulate capital and invest, and therefore the growth imperative. In Marx's words, "the development of capitalist production makes it constantly necessary to keep increasing the amount of the capital laid out in a given industrial undertaking, and competition makes the immanent laws of capitalist production to be felt by each individual capitalist, as external coercive laws. It compels him to keep constantly extending his capital, in order to preserve it, but extend it he cannot, except by means of progressive accumulation."[26]

It therefore becomes evident that the quest for growth, or growth impetus, is an integral part of competition, which in turn is necessary for the smooth functioning of the price mechanism. Note that competition and the quest for profit maximization also drive the private sector's cost-minimization efforts, which lead to the increasing efficiency of processes, or "capitalism's quest for efficiency," to quote Tim Jackson. You cannot strip growth out of the system but at the same time preserve its efficiency-seeking properties.

Financial markets likewise make use of the price mechanism. Adjusted for risk factors, financial resources flow to producers that manage to fulfill the desires of consumers, or are expected to do so. This happens, again, in a decentralized way, signaled by the realization or expectation of profits high enough for firms to reward shareholders with dividends and comfortably repay creditors. The same is true in talent markets, where human capital flows to firms whose growth and profits allow them to offer higher wages and benefits. This is why so many bright young graduates are eager to join the likes of Apple and Google.[27] Firms that meet consumers' demands grow and are rewarded, allowing them to make investments and innovate, and therefore to grow further. In a dynamic, capitalist system, societal resources, in the form of talented employees (labor) and financial means (capital), are constantly reallocated through decentralized decision-making to propel successful, growing companies forward.

Many people, as they think about capitalism, have their own definitions in mind, and often these include a minimal role for government in the economy, limited to simply setting the rules of the game. It should be evident by now that this is not a defining characteristic at the core of capitalism. Different varieties of capitalism display different degrees of government intervention on a spectrum from the interventionist Scandinavian model to the minimalist Anglo-Saxon one, but the price mechanism is always there.[28] This is why Milanovic's definition of capitalism is purposely loose, mentioning that capital, or the means of production more broadly, should be *mostly* (not completely) privately owned.

It should also be clear that embracing a degrowth policy agenda, and therefore dismantling the price mechanism that is the very cornerstone of our economic system, would require a complete societal rethink, beginning with a new mechanism to allocate resources. This is surely not impossible and has happened before, including in recent history. For example, Soviet-style communism abolished decentralized coordination and the price mechanism, opting instead to allocate resources through five-year plans and top-down decisions on what goods were needed at what time and in what quantities. One should be clear, however, about the fact that degrowth entails radical restructuring of society—a truth that is often missed in popular discourse.

There is one way to reconcile zero growth with the price mechanism and capitalism. We have all heard that quintessential capitalist mantra "the customer is always right." If consumers were to radically change their usual preferences and renounce their desire for products of new kinds and in greater quantities and varieties, then supply—meaning profit-maximizing producers—would adjust. This would work if all consumers bought into the "less is more" mantra (the likelihood of which we shall examine later). Nobel laureate Robert Solow's statement that there is "nothing intrinsic in the system that says it cannot exist happily in a stationary state" should be read through this realization that growth, in the end, is merely the result of people's preference for more.[29]

Let's take stock of what we have learned so far. First, the current, capitalist organization of our economic system is not compatible with a steady-state economy, at least given current consumer preferences. Second, beyond the current institutions, and therefore cutting across different varieties of capitalism, whether Anglo-Saxon, Japanese, or Scandinavian, degrowth is incompatible with the price mechanism and therefore with capitalism, *tout court*. This does not prove (yet) that degrowth is a bad idea; it only reveals it to be an idea that undermines the entire way our current economic system is organized. As its advocates push for it, then, they cannot simply tack it on as an extra item in a list of policy suggestions to reduce inequalities or tackle climate change. They must be ready to present a blueprint for a new society.

Innovation as growth

There is, finally, a third level at which we can challenge degrowth and its relation to capitalism. Going to this depth requires a historical understanding of the economic model that allowed nations—first in Europe, starting around the end of the eighteenth century, and later "the rest" outside the West—to expand without precedent.[30] In *A Culture of Growth*, Joel Mokyr masterfully lays out the origins of our modern economy. A consummate economic historian, Mokyr shows that, contrary to conventional wisdom, what happened around the Industrial Revolution is not so much that a specific invention came on line, be it iron manufacturing, steam power, or improvement in transport. Rather, there was a paradigm shift from what Mokyr calls "Smithian growth," which had characterized proto-capitalism as described by Adam Smith, to what followed from then on: sustained, innovation-led "Schumpeterian growth."[31]

Before the Industrial Revolution, growth was mostly driven by commerce, more effective markets improving the allocation of resources, and the division of labor, as described by Smith in his famous observation of how work was organized in a pin factory.[32] This "Smithian model" generated some growth and, occasionally, when significant inventions came along, such as heavy plows, mechanical clocks, spectacles, and windmills, the economy experienced a temporary boost. But while a given innovation might cause something of a revolution in a specific sector, its impact on aggregate growth would eventually peter out. What happened, first in England and then progressively in the rest of Europe, was that this model of sporadic innovation switched to one of sustained and cumulative innovation. Mokyr argues that this transition was the result of a cultural transformation with origins in the sixteenth century, when some meta-innovations took hold. These included the scientific method.[33]

Dubbed "the knowledge machine" by philosopher Michael Strevens, the scientific method led to the triumph of empirical evidence and the demise of the view that ancient knowledge, as epitomized by the Catholic Church's official philosopher Aristotle, was sacred and could not be questioned.[34] This machine gained power from another meta-innovation, the printing press, which sharply reduced the cost of books, democratized access to knowledge, spread awareness of new inventions and technologies, and effectively promoted open science. Crucially, the wide availability of books, thanks to the tireless work of early publishing houses including the Plantin Press in Antwerp, meant that scientific and technical achievements would not be forgotten. The technological regress that had been common in past civilizations was ruled out, and solid foundations for cumulative knowledge gains were laid. Finally, the great voyages after 1500, made possible by improvements in navigation, increased European curiosity about nature and underscored the benefits associated with technology and practical knowledge. All these elements led to the establishment of a strong and widespread belief in progress, which we now typically associate with the Enlightenment, and which has been a cornerstone of Western culture ever since.[35]

The important link between innovation and economic growth was spotted in the early stages of modern growth economics. Seminal work by Robert Solow in 1956 identified changes in *total factor productivity*—

economists' term for how efficiently production inputs are used—as the major factor determining the rate of growth of an economy in equilibrium. In Solow's model, however, total factor productivity was not central to the mechanics of growth, but rather a residual element. Anything that could not be explained by extensive growth had to fall into the category of *efficiency gains,* meaning technological improvements. Starting in the 1980s, a group of economists, including Paul Romer, Philippe Aghion, and Peter Howitt, sparked a Copernican revolution in growth economics by insisting that innovation must stand front and center in an economic model of growth, and could not be treated as a residual or side thought. This led to *endogenous growth models* in which the links between new ideas, technology, and growth were laid out mathematically, built around the concept of Schumpeterian creative destruction.[36] Thus the possibility was revealed that economic growth could effectively be sustained forever—relying as it does on human creativity and ingenuity, of which there is no end.[37] Like a plumbing system, capitalism simply channels human curiosity and the quest for useful knowledge, rewarding these with profits and societal resources. Science and technology were shown to be the ultimate engine of growth.[38]

This brings us to another seeming paradox. These days it feels as if, on one side, we have science—for example, as offered by the UN's Intergovernmental Panel on Climate Change—and on the other side, we have economists and their "obsession with economic growth," and these two sides are clashing over climate mitigation. There was a similar misperception at the height of the coronavirus pandemic, when to some it looked as if epidemiologists were clashing with economists over the desirability of lockdown measures. Taking the long-term view, science and technology are what has pulled humanity out of starvation and propelled it to the top of the planet's food chain. Science and technology allowed population growth to expand exponentially in the last two hundred years, fending off diseases and expanding food production. And since the eighteenth century, their acceleration and accumulation has been the propellant of economic growth.

What does all this imply for degrowth? At the very least, it implies that arresting growth will require dismantling current societal incentives for innovation, or alternatively, proposing a convincing way to decouple innovation from economic growth.[39] This poses a conundrum especially for moderate degrowth advocates, who tend to argue that innovation must continue, particularly in the green sphere. Given that degrowth would tear capitalism apart, a new system that incentivizes, channels, and distributes innovation just as efficiently across society would need to be devised from scratch. Note that this is what the Soviet Union attempted, and in the end it was perhaps its most blatant failure. Dispensing with decentralized coordination mechanisms and monetary incentives and relying instead on other forces—the power of ideology in good times, and the threat of punishment in bad times—to produce knowledge and innovation and channel these to productive ends simply did not work. Capitalism did the job better. Therefore, when we hear someone say that we should ditch economic growth, but at the same time foster innovation, we must be sure to ask the question: Just how this will happen? And specifically, how will credible incentives be set up across society?

The belief in progress

As we approach this chapter's close, it is worth going back to the question with which it began: What is growth? If capitalism is a way to organize production, then it can be seen as the plumbing of the economy. Growth, in this metaphor, is the water pressure that enables water to flow along the pipes, pumped by the engine of sustained innovation.

New pipes and hardware can to a large extent be added, shifted, upgraded, or removed without affecting the tenability of a water management system. And indeed, features have been added to capitalism throughout history. This chapter's examples and narrative present a plain-vanilla model of capitalism, but real economies are much more complex. Not everything responds to pure market-price mechanics. Typically, there is a strong role for government in providing a set of services, ranging from defense to basic education. In some instances, this involves direct ownership of service providers such as postal services, railway network operators, and water utility companies. Moreover, there are philanthropies, NGOs, social impact investors, corporate social responsibility initiatives, and plenty of companies pursuing goals beyond bottom-line profits. Together these correct for some of the problems that result from too-narrow applications of capitalism and too-direct responses to the price mechanism. But they are not completely independent from the logic of the system in which they are nested.

To understand this, imagine you have encountered the abhorrent images of huge islands of plastic waste floating in our oceans. Moved to take action, you decide to found a company to produce water bottles from bioplastic, which is biodegradable. Your product appeals strongly to environmentally conscious consumers and, as they buy more, your profits expand. This allows you to hire more workers, produce at higher volume, and explore new methods to cut costs. As a promising investment opportunity, your company catches the eye of sustainability-oriented portfolio managers—so-called green finance—and your project is funded at a more ambitious level. You can make new investments and expand the set of biodegradable products you offer. The story could continue from there, and actual case examples could just as easily be produced, but the point should be evident. The origin of economic growth is not necessarily selfish interest, narrowly defined as a desire to make as much money as possible. You can start with a Millennial-minded desire to have an impact, and realize it will take resources to do that. If your idea is worthy and society sees value in what you do, then resources will flow to your activity, in the forms of both financial and human capital, and give you the wherewithal to make progress toward your goal.

Examples like this help us see that economic growth is intertwined at a deep level with the very concept of progress, both personal and societal. Commitment to progress is, at its heart, a conviction that the future can be better than the past, and that this depends, at least to some degree, on human action. By means of science, innovation, business creation, and government intervention, people collectively and cumulatively improve society. Progress is the principle according to which humanity takes on responsibility for its destiny. This, according to Mokyr, is the mindset shift that made possible the Industrial Revolution. This was clear to the Enlightenment thinkers who, in the word of economist Daniel Cohen, "made progress a moral value."[40] Commitment to progress has continued to shape Western cultural beliefs up through today.[41] This does not imply a Panglossian credulity that all technology is good or offers immediate benefits, but rather a belief that it will lead to benefits overall and in the long run.

We can therefore conclude that one of the main forces feeding economic growth is the belief in progress itself. It is progress that should be seen as the secular religion of the modern world, rather than economic growth, as a quote early in this chapter claimed. As Mokyr writes, "A critical cultural belief that drives economic growth and complements the belief in the virtuousness of technology is a belief in progress, and specifically in economic progress."[42] Thus, if degrowth advocates want to shift cultural values so thoroughly that they are compatible with steady-state capitalism, they will need to bring down this three-hundred-year-old Enlightenment tenet of Western culture.

To be sure, the very concept of progress has been questioned before. At the very same time that Enlightenment thinkers were embracing it, eighteenth-century philosophers such as Jean-Jacques Rousseau in France and Giambattista Vico in Italy were disputing it, inspired by a nostalgia for a simpler, rural age. Their school of thought came to be called *primitivism*. This phenomenon should be interpreted in light of the great explorations that started in the 1500s and brought Europeans in contact with new cultures, such as those of Native Americans. Major intellectual debates were spurred by these contacts, specifically about the extent to which these civilizations and their organization and traditions provided Europeans with a view into their own past.

Living before the French Revolution under the *Ancien Régime,* Rousseau was impressed by the freedom that seemed to characterize Native Americans, the equality among individuals of few material possessions, and the limited hierarchy in their society. He concluded that the apex of humanity must have predated the very invention of agriculture, which accompanied the creation of permanent settlements and private property.[43] This view was antithetical to the idea of progress, and echoed more classical notions such as that of a past Golden Age, which Renaissance thinkers perceived in Ancient Greece and Rome, or religious concepts such as the prelapsarian Garden of Eden.

Considering the arguments of degrowth advocates, we can see that their philosophical roots go much further back than the 1970s and the environmentalist push for the first Earth Day. They are rather to be found in the primitivism of the eighteenth century. The reason that these critical views remained confined to an intellectual niche at the time, and failed to spread to the wider society, is that the benefits of scientific and economic progress were increasingly evident and experienced firsthand by a growing share of individuals. In the early decades of the Industrial Revolution, the average British family experienced material progress in the form of much wider availability of household items as basic as clothing and shoes, but including also pottery, rugs, carpets, mirrors, clocks, and tools like cutlery, pots and pans, nails, razors, and scissors. This was possible thanks to the shift from expensive handicraft to mass production and the cheap availability of iron and steel. Average incomes in England, and the West more broadly, were increasing many times over, making it evident that living standards, powered by technological progress, were edging well above those of other regions of the world.[44]

Colossal enterprises, such as the construction of the Eiffel Tower in Paris, the Suez and Panama canals, and celebrations of modern industrial technology like the Crystal Palace exhibition in London, also inspired the collective imagination, making it seem that anything was possible. In *The Moral Consequences of Economic Growth*, Benja-min Friedman shows that, at different moments in the 1800s, Charles Dickens, Thomas Carlyle, John Stuart Mill, and Henry George all expressed a degree of longing for a world that was lost to modernity. It can be hard to disentangle genuine social critique, such as Dickens's description in *Hard Times* of the horrible working and living conditions for London's poor in the early years of industrialization, from a degree of nostalgia for an idealized past that perhaps never quite existed.[45] Having pieced together much historical evidence, Friedman argues that, since those days, optimism and pessimism about human progress have moved in cycles, largely aligned to periods of improvement and stagnation in society's living standards.

Clearly, two World Wars dealt a powerful blow to the concept of progress, together with the Great Depression, which led economist Alvin Hansen to resurrect Mill's "steady state" concept and conclude that the United States was a mature economy with little potential for further growth.[46] Later, faith in progress flourished anew, as postwar prosperity kicked off and yet a new colossal enterprise—John F. Kennedy's pledge to put a man on the moon—captured the public's imagination with a vision of what science, technology, and the American capitalist system could achieve.

The past few decades have been a period of slow growth in which large chunks of society have experienced little material progress in living standards. The emerging skepticism regarding economic progress should not be seen as surprising, but rather as confirming the past two centuries' historical pattern. This lays out in clear terms the philosophical battleground for degrowth's advocates and opponents. If trust in progress is to be retained, the benefits of economic growth must once again be evident and tangible for society at large.

This closes the circle between the critiques of GDP, the calls for more widely shared prosperity, and the challenge of climate change. Fighting global warming by accelerating scientific and technological progress will require that the benefits of the accompanying economic growth are shared across society. If they are not, people's confidence in progress will continue to drop, and nostalgia for some past golden era will spread. This will provide fertile ground for degrowth sentiments to flourish, and for the potential demise of capitalism.

Where does all this leave us? Economic technicalities and historical digressions aside, the key takeaway from this chapter is that degrowth isn't in the same ballpark as other policy suggestions being touted by progressive thinkers. Policies like the provision of a universal basic income, free college education, and higher taxes on the wealthy may be costly and administratively cumbersome, but can still be designed in ways that are compatible with a capitalist system. Degrowth, however, would tear down the entire system and require a complete restructuring of society. It would not be just a matter of "giving up a political obsession with GDP."

To say so is not to prove that degrowth is a bad idea. It does mean, however, that one should avoid adding it en passant to a long list of desiderata, as if it were no greater endeavor. No doubt some people—in particular, some climate scientists and eco-biologists—fall into this fallacy in all good faith. Yet they should realize that arguing we should suppress economic growth is akin to suggesting we should abolish private property, marriage, or meritocracy.[47] Any such fundamental change would have ramifications for the whole organization of our society. It is possible to imagine there could be ways of doing it, but essential that careful reflection go into what the new system should look like, and what the transition would entail.[48]

To pragmatic environmental activists, the systemic implications of degrowth can be rather off-putting, because a climate crisis calls for the quickest possible action. This leaves no time and energy for the task of organizing an economic revolution, with all the societal soul-searching and uncertainty that setting up a new system would entail.[49] To others, it is obvious that knocking out a pillar of today's system would be massively disruptive, and for them that is the point. They see a window of opportunity to bring down capitalism once and for all. A large part of the degrowth literature of the past few decades has been devoted precisely to that end, identifying the relevant policy levers to pull to bring economic growth to a halt, and redesigning how life should be lived in a greatly transformed, post-capitalist, post-growth world.

In principle, it is possible that the radical transformation some degrowth activists seek would produce a better society. And, after all, capitalism does have many shortcomings, doesn't it? This is the question we take up in Chapter 3.

3

POST-GROWTH DYSTOPIA

Money can't buy happiness, but it helps.
—ANONYMOUS

Degrowth scholar Jason Hickel proclaims in *Less Is More* that, once we get rid of the obsession with growth maximization (or "growth-ism") and dispense with capitalism, entirely new possibilities open up. The fifty years that have passed since publication of *The Limits to Growth* have given the movement plenty of time to reimagine a post-growth world and sketch out the elements of its vision. While there are nuances to the proposals of various thinkers, most policy recommendations align with a recurrent theme. Let us go through them briefly, as their proponents explain them, before reflecting on their feasibility.

To begin with, a policy lever is needed to trigger the slowdown in economic growth. Generally, the favored route involves a sharp reduction in work.[1] Given Okun's Law (encountered in Chapter 2), this creates the tricky problem of decoupling growth from jobs—otherwise, degrowth would usher in large-scale unemployment and widespread malaise.[2] This problem can be solved in two ways. First, jobs can consist of fewer working hours, recalling the 1970s French labor union slogan: "work less, work all." If employers hire people to put in twenty hours per week instead of forty, the argument goes, twice as many people can share the remaining work. Second, a universal basic income can be introduced, topping off the reduced earnings from fewer hours and guaranteeing a living wage even to those who do not work at all. This subsidization scheme may avoid social unrest, but it creates the need to find the money to finance it. Generally its advocates plan to overcome that challenge with a significant wealth tax— hence the movement's appeal to Thomas Piketty, who so forcefully put wealth inequality and taxation on the agenda for policy debates with his *Capital in the Twenty-First Century.*[3]

Looking forward to the day when these solutions are in place, degrowth zealots proclaim one of their most well-known slogans: "our degrowth is not their recession." Their point is that, when recessions take place under capitalism, they cause unemployment, sharp falls in happiness, depression, anxiety, loss of self-esteem, alcoholism—even spikes in suicide rates, as seen in Greece during the eurozone crisis.[4] Degrowth, by contrast, entails a managed and equitable slowdown of the economy. As part of the menu, a "maximum income" is also typically considered, as a natural complement to a minimum or "basic" income in a steady-state economy, extending popular calls for caps on CEO pay. Recall that the degrowth movement has not only an environmental agenda, but also a strong social agenda, both responding vigorously to what is seen as the complex *problematique* of capitalism—a system inherently rife with inequality, pollution, poverty, and nearly every other ill observed in society.[5] Indeed, many degrowth scholars flirt more or less explicitly with the need to surpass private ownership.[6]

Freed from unnecessary toil and useless hyper-consumerism, people would then be in the condition to live their lives on a principle of sufficiency, rather than ever-increasing frivolous wants, and to dedicate themselves more to education, arts, sports, and literature. Parks, concert halls, and sport facilities would become prevalent.[7] As degrowth advocates push this view, they frequently refer to predictions by John Maynard Keynes, revered as a founding father by most macroeconomists. In his 1931 essay "Economic Possibilities for Our Grandchildren," he predicted that within a century most people would decide to work only three hours per day, and would dedicate much more time to cultural endeavors. As Tim Jackson notes, these recreational activities are low in greenhouse gas emissions.[8] Even if they still have some material footprint (as any human activity does), the shift minimizes environmental impact and maximizes people's potential to flourish. Sectors associated with these activities would therefore grow and employ more people, while polluting ones would come to an end. For degrowth advocates, it is self-evident that the economy as a whole need not grow (as captured by aggregate GDP). Only the "sectors that matter" should.

These more worthy activities should be sustained somewhat, if not mainly, by heavy public investment. To say so is to echo another important economist: John Kenneth Galbraith. In *The Affluent Society* (1958) he concluded that private material needs were already largely fulfilled in postwar

America. It was time, he believed, to focus on public goods, including education, public spaces, health, and social policy.

One issue for the movement is labor productivity. In a world with constant or shrinking economic output, becoming more efficient in the use of labor can only imply one thing: less work is available. As a consequence, degrowth advocates believe, labor productivity growth must end. One interesting way to achieve this builds on the 1966 analysis by economist William Baumol of what he diagnosed as the "cost disease." The idea is that certain sectors—Baumol's example was performing arts, but many other service sectors are susceptible—offer quite limited scope for productivity improvements. A string quartet, for example, is no more efficient at performing a Mozart concert today than it was two hundred years ago.[9] Turning a vice into a virtue, degrowth enthusiasts look forward to an economy shifting to these sectors, labeled by Jackson "care, craft, and culture" (he also refers to them as the "Cinderella Economy"), which are inherently labor-intensive and have low productivity growth and low environmental impact.[10]

Linked to the boost in craft work is the idea that production should take place on a small scale, ideally at a local level, to avoid heavily polluting, global supply chains—a vision far from the current paradigm of multinational corporations. Ecovillages—the sustainability-minded communities briefly mentioned in Chapter 1—represent a model to be extended to society at large. In aspects of its vision, degrowth blends well with trends currently on the rise in Western societies, and in particular among the highly educated, liberal, urban cohort, such as urban farming, DIY, farmers' markets, minimalism, preferences for local and traditional products, and the "slow food" movement.

Degrowth enthusiasts are not necessarily anti-technology; most believe innovation should continue. The argument goes, however, that innovation efforts should not be squandered on useless novelties and labor-saving technologies, as they tend to be by private, profit-driven companies. If innovation were channeled by a strong public sector it could make progress on what really matters: improving the quality of essential services and their environmental sustainability.

Hickel complements all this with the need to underpin a degrowth vision with a new ontology, shelving Enlightenment thinkers like Descartes with their anthropocentrism and human exceptionalism, and edging toward the ecological principles embedded in animism. Once people see

animals and plants as part of their greater family, they will take from nature only what is necessary to live, along the example of the Achuar people
of the Amazon rainforest.[11]

Following Richard Smith's *Green Capitalism*, the model of this post-
growth society could be referred to as *ecosocialism*, a term we will use here
as shorthand to signify this ensemble of policy recommendations and the
system that would result from them.[12]

The Utopia that wasn't

Many elements of the degrowth vision may strike a chord with much of
the population. Taken alone, some of the individual recommendations are
not even that radical. Various economists, including most recently some
at the International Monetary Fund, advocate greater redistribution of
wealth as a way to correct for widening inequalities.[13] Provisions of universal basic incomes are being tested in various jurisdictions. Some companies are experimenting with workweek reductions.[14] In the aftermath of
the Covid-19 pandemic, a more interventionist government is seen as
natural by many, especially to guide innovation.[15] As described, many vi-
sions of a post-growth society admittedly sound appealing. Who wouldn't
want a world where we retain the good things of our current system, but
at the same time have plenty more leisure and live in a greener, healthier
society, free from poverty, the wastefulness of capitalism, and staggering
inequality?

What is striking is how reminiscent the post-growth vision is of an early
sixteenth-century book: Sir Thomas More's *Utopia*. His title, a term that
has since taken on the connotation of an idyllic, imaginary society, is from
the Greek *ou-* (not) and *topos* (place)—in other words, *nowhere*. In a mas-
terpiece of sociopolitical satire, this Renaissance humanist (also high chancellor to King Henry VIII) describes an island society where, having abolished private property, people produce for the common good and take from
the common pot only what they need. Free of what we would nowadays call
hyper-consumerism or luxurious production, everyone wears similar,
simple clothing. Likewise, communal wealth in the form of gold is used for
diminishing activities (for example, to make chains for criminals and toilet
pots) to produce a healthy distaste for it among the Utopians.[16] As in many

contemporary ecovillages, everyone in Utopia is involved in agriculture and other basic trades, men and women alike. The workday is minimized and unemployment is nonexistent.[17] All are encouraged to use their leisure for education and culture. The welfare state is strong, and well-equipped hospitals provide free treatment. The comparison may be stretched, but it is remarkable that, five centuries later, the concept of an ideal society is strikingly similar. The fact that *Utopia* was written in 1516, two centuries before Adam Smith's time and the beginning of the Industrial Revolution, suggests that the degrowth vision is not so much a repulsive reaction to the excesses of capitalism or today's hyper-consumeristic society as an evergreen human desideratum.[18]

Still, one could argue that something that was only a dream back then is within reach today, thanks to the prosperity and technology achieved. This, then, is the question we turn to next: How likely is it that, by enacting the degrowth policy menu, we would arrive at the society envisioned?

The pub economics of degrowth

At the London School of Economics, Nicholas Barr used to kick off his public economics lectures with cases of what he facetiously called "pub economics," defined as economic truths everyone knows to be true, except that they are not. Quite a few degrowth policy assertions would seem to belong to this category.

For example, there is no empirical evidence that reductions in working hours, which have occurred in the past in France, Germany, Chile, and Quebec to name a few, lead to increased employment.[19] Economists refer to this as the "lump of labor fallacy"—the flawed notion of work as a fixed amount, so that one person's reduction is another's gain.

Another concept that does not seem to hold water is the idea that, in a stable or shrinking economy, we can perpetually finance welfare spending—and more specifically, finance a universal basic income—by taxing wealth. Ultimately, a shrinking economy will have fewer resources. We can temporarily finance policies out of accumulated wealth, but eventually this will mean getting by with less, and the reduction applies inevitably to government services. Turning around one of the degrowth mottos, you can't have infinite wealth tax flows in a finite economy.[20]

Yet another display of pub economics is the idea that GDP growth—or aggregate economic expansion—must mean growth spread indiscriminately across all sectors, rather than "prioritizing the sectors we care about."[21] In truth, the shape of the economy is constantly changing, and this flexibility is one of the strengths of a decentralized capitalist system. Each year, hiding under an aggregate positive GDP growth figure, some sectors expand while others shrink and occasionally even disappear. Coal extraction is a sector on the decline in Europe and the United States, as year after year more coal plants are decommissioned. Travel agencies are another example, as consumers increasingly shift to online booking. Production items that have disappeared in recent years include VHS cassettes, DVDs, floppy disks, and photographic film, all supplanted by modern digital technologies. The fact that the composition of the economy changes, as old sectors and production technologies decline and resources shift to companies making cheaper, better products that are more valuable to consumers, is one of the main reasons we can expect well-functioning capitalist economies to expand over time. Capitalism does, therefore, grow the sectors "consumers care about." Whether these are the ones they *should* care about is a question we will take up in Chapter 5.

There is another belief that sounds commonsensical, perhaps extrapolating business dynamics to a country as a whole, but that does not hold up to closer scrutiny. This is the perception that public policy, led by orthodox economists, is oriented toward maximizing GDP. In reality, no serious economist would base policy advice on such a simplistic criterion.[22] For that matter, in several types of situation, economics experts typically advise governments to take measures known to weigh on short-term growth. For example, we know that running large public deficits—an expansionary fiscal policy—temporarily boosts GDP growth. If its maximization were the sole objective, no orthodox economist would ever advise a conservative fiscal policy—the much-hated austerity. Likewise, economists have often recommended interest rate hikes to cool off an "overheating" economy, or to avert the financial market bubbles or inflation that can be associated with fast GDP growth. Many other examples could be cited, having to do with specific regulations or taxation. The point is that macroeconomists look at a wide variety of variables accompanying GDP growth—inflation, bank borrowing, employment rates, public deficit and debt, savings and investment levels, productivity growth, the diversification of exports and trade partners, and many others—as they think about what is needed for sustainable economic growth. Admittedly, until recently, their definition of *sustainable* has not done enough to factor in environmental sustainability.

In the realm of pub economics, additional points could be made, but they might strike readers sympathetic to the degrowth movement as quibbles—and in a way rightly so. We should recognize that the degrowth vision demands a complete rethink of society, and that is the level at which debate should be pitched. So let us move to a more abstract level of discussion.

Allocating resources in a post-growth society

Let's start with one specific consideration of leisure, which can then be generalized to a more fundamental critique of ecosocialism. This is the prediction that, once granted more free time, people will dedicate it to low-carbon, high-culture activities such as reading Proust and writing poems while sitting on the porch, or engaging in activities that rebuild social capital.[23] This reverie is possibly symptomatic of a degree of detachment from society.[24] Recently, a group of renowned economists undertook to revisit Keynes's "Economic Possibilities for Our Grandchildren," and recalled that Keynes was an avid member of the Bloomsbury Group—a vibrant literary circle also including Virginia Woolf. It was more than possible, they agreed, that in this famous essay he slipped into extrapolating his own experience to society as a whole.[25] If we look at the way people choose to spend their time off work now, it is generally quite carbon-intensive. Typically it involves flying or traveling, often to visit family and friends who tend to be much more geographically scattered than in the past. This applies particularly to Millennials and Gen Z, who are surely environmentally concerned, but at the same time eager to explore and live in other lands. These are the most diverse, curious, and open-to-the-world generations ever.[26]

This leads us to a wider point. What if, living in the ecosocialist world, even just a minority of people choose not to use their "regained freedom from work" in the ways envisioned by the master plan? Clearly, this could not be allowed to happen because, in a degrowth world—where human activities are inevitably coupled with greenhouse gas emissions and environmental degradation, and the price mechanism to coordinate demand and supply has been torn down—the only fair way to split the available carbon budget is by universally observed rules. Effectively, what the degrowth vision is missing is an allocation mechanism: by what process do we, as a society, decide how to spread our limited economic resources across competing activities, all of which inevitably have an environmental impact? All the abstract talk about the need to make these choices as a society fails to acknowledge that these decisions are pervasive in daily life and need to be made in the moment.

Imagine just a simple decision related to personal transportation. Under degrowth, we all know that cars are irremediably polluting and they destroy the environment, whether they have combustion, electric, or hybrid engines. We have agreed to cap emissions, meaning that there are only so many kilometers per year for our entire society to use. Now suppose that grandma is not feeling well. Should you drive her to the hospital? Is that a legitimate use of some of the societal kilometers available? Does it make a difference whether she is seventy-five or ninety-five? Who decides? What if instead it is your cat that is ailing? Or what if your mother lives alone and would like a visit? Under a capitalist system, prices provide incentives and signals so that these continuous allocation problems are solved in a decentralized way. In principle, you decide knowing that you will pay a price, and the price reflects the degree of scarcity of the resource.

Under ecosocialism, everything instead becomes a centralized problem of allocation and must be solved through common decision-making, or a rulebook for all possible situations.[27] This dramatically challenges the concept of personal freedom, casting it not as a general principle with some limitations, but as occasional latitude within omnipresent, constraining limits.[28] Liberal democracy would hardly have the required legitimacy to make these intrusive decisions by majority, and this is perhaps why degrowth advocates favor direct democracy, which in some ecovillages takes the form of governing by consensus. While this might solve the legitimacy problem, it hardly provides an efficient decision-making solution for the sheer volume of allocation problems constantly needing answers in large, complex societies.

At the most abstract political level, the degrowth vision might be internally and logically coherent, but it has not figured out a way to deal with opposing views regarding its societal blueprint. This inconsistency is somewhat hidden by magniloquent statements such as *"everybody* should be shifting preferences" and *"we* should free ourselves from the iron cage of consumerism and capitalism." Sounds good, but what if, as is likely, not everybody does? Incidentally, we are touching here on the essential problem that brought down the Soviet communist system. Once the "coarse" price incentives people use to reconcile their individual preferences with societal interests are removed, only full ideological alignment to the common cause can keep the system running. If that fails, the only way left to organize every aspect of society is to resort to convoluted, pervasive rules and coercion.

They all lived happily ever after

As it is framed by its advocates, ecosocialism sounds like a world in which human beings have finally reconciled themselves with the planet, and freed themselves from the tensions fueled (or artificially created) by capitalism. There are strong reasons for skepticism, even beyond the ecosocialist system's lack of a mechanism for dealing with dissent.[29] Let's turn to the problem of redistribution.

Economic growth expands the resources available to society in total. To use the common metaphor, it makes the pie bigger. Redistribution is the process of sharing this pie in a meaningful way. It turns out that it is much easier to distribute new resources, made available by the growth machine, than to seize and redistribute existing resources. (We will investigate the psychological underpinnings for this in Chapter 5.)[30] This is perhaps why most countries in the world have income taxes, while wealth taxes are less prevalent and, where they do exist, tend to be loathed. Examples include local property taxes in the United States and Italy, and inheritance taxes in the United Kingdom.

In a world in which total resources are fixed, or even shrinking, one person's gain is inevitably another's loss, bringing taxation dangerously close to a zero-sum game. Personal success comes to be associated with squeezing resources out of someone else rather than, perhaps, innovating and creating new value for society. With this attitudinal shift, the social and economic pact that binds our modern societies breaks down. When this is the case, we can imagine society devolving into homogeneous subgroups—based on income, race, class, geography, or even age—clashing for resources and political power. In stark contrast to the peaceful image painted by post-growth enthusiasts, we can envision social conflicts far beyond current levels.[31] Incidentally, this is the scenario described in Thomas Hobbes's *Leviathan*. Writing in 1651, Hobbes explained that, given the range and variability of human desires and the scarcity of resources to fulfill them all, life in the state of nature could only be "war of all against all." The preferred outcome of achieving peace in society was the justification for agreeing to the all-powerful (and illiberal) government he called the Leviathan.[32]

The conclusion we must reach is essentially the same—that ecosocialism is incompatible with personal freedom as liberal democratic societies have always defined it, and a strong, intrusive, paternalistic, and possibly illiberal government would be required to make it operational. Now, the counterargument could very well be that, even if that is the case, the Leviathan can be democratically elected (as Hobbes himself recognized), or can rely on direct democracy in the way that ecovillages do. But this raises another issue.

Ecovillages are not a proof of concept

From Jackson to Hickel, when degrowth zealots are questioned about the feasibility of their blueprint for societal reform, they at some point bring up ecovillages as a proof of concept. These are small communities, mainly in Europe but scattered across all continents, typically composed of 50 to 250 individuals who reject capitalist principles, embrace the desire to go back to the land, and strive to minimize their environmental impact. They do this by engaging in self-sufficiency, regenerative agriculture, and communal practices. These communities have their historical foundations in the "back to the land" movement that gained momentum in the 1960s and 1970s, but their ideological roots go back at least to late-1700s utopian anarchist thinker Charles Fourier and the *Phalanstère*—his model for self-contained intentional communities.

Within the degrowth literature some voices can be found calling ecovillages a sign of the movement's failure to persuade citizens broadly to embrace an ecosocialist vision; these communities are no more than retrenchments from the capitalist system to organize in an alternative way.[33] In any case, the fact that ecovillages exist and thrive across the world represents no proof of concept, if the concept is a scaled-up version of their organization to society at large.

First, by definition, these villages bring together ideologically homogeneous people with a shared vision of the world and common environmental values. This reduces the scope of potential societal tensions, while hardly showing the system to be able to cope with diverging views within it. Meanwhile, the very fact that a liberal democracy and capitalist society hosts ecovillages shows that the current system is able to incorporate dissenting views and diverse lifestyles within it—something ecosocialism would not.

Second, note that while there are ecovillages there are no eco-cities. In a small setting, complexity is minimized and governance can take the form of direct democracy, or even consensus decision-making. The same could never be true for a heterogeneous city—say New York City, with its eight million people—where energy supplies, food logistics, garbage collection, plumbing, policing, hospital care, firefighting, pest control, and many more services must be organized on a daily basis just to keep citizens alive. Making allocation decisions in such a setting in centralized, recursive ways, and by direct democracy no less, could hardly work. Either those decisions would be constantly behind the curve, or there would have to be a loose set of general principles, granting strong discretionary powers in their granular application to agents of the state. Here again we should recall the experience of the Soviet Union, whose centralized, five-year plans could not deal with the granularity of innovation, production, and consumption choices in a society that was complex and only growing more so.[34] Its initial vision of a just society organized bottom-up by the votes of proletarians through workers' councils (or *soviets*) led to large-scale inefficiency and then, to counter that, a rapid spiral into ever more centralized decision-making—all undermining the original desire for participation and justice.

Third, and most important, ecovillages might be superficially fenced off from capitalist dynamics, but ultimately they depend on them. Rather than descend to primitive conditions, ecovillagers make use of technologies developed and produced elsewhere. They benefit not only from the production of electricity, or of solar panels for that matter, but also from heating technology, repair and maintenance supplies, roads, and the telecommunications systems, smart devices, and internet they need to connect to other members of the Global Ecovillage Network. If this invites the counterargument that many essentials can be obtained secondhand or even for free, that only shows the reliance on capitalism to be stronger: it means the village model depends on capitalism's capacity for overproduction.

En passant, this last argument also relates to the current social tendency to celebrate individuals who embrace minimalistic lifestyles or "downshift" in significant ways. The story typically goes as follows: a young ambitious person climbs through the ranks of the corporate world, only to realize this lifestyle does not bring fulfillment, and quits. Then often follows a combination of abandoning the big city life, moving to a simple dwelling, perhaps in the countryside, living with only few items, and enjoying regained freedom in contact with nature. There is no contesting that such a change can bring joy to the individual, assuming the financials can be squared. For reasons detailed above, however, we must admit that the entire viability of this model depends precisely on the rest of society's not emulating it. The lifestyle is possible because it leans on a capitalist system continuing to provide the healthcare, security, education, energy, pest control, physical and digital infrastructure, heating, cooling, housing, and whatever other goods and services the minimalist keeps using.[35] An individual's downshifting experience is therefore no more useful a blueprint for an alternative frugal society than a wilderness camping trip is for the organization of a nomadic community.

A related point can be made about resilience, or the capacity of ecovillages to respond to unexpected shocks. Ecovillages are a fair-weather social construct, and function insofar as the sea is not rough. They count on the fact that the wider (capitalist) society is there to serve as safety net should things go south. What if a new pathogen spreads and agricultural crops are lost? What if an epidemic erupts? How about if antibiotic resistance becomes widespread? Or a natural disaster hits?[36] Or extreme weather events become prevalent?[37] Would ecosocialism be able to respond to these shocks effectively—or at least, would it be able to adapt and react as well as today's advanced economies do? In the context of Covid-19, would a centralized degrowth economy have been able to mobilize resources quickly and scale up the infrastructure to deal with a surge in hospitalizations, or the knowledge facility to rapidly develop and distribute a vaccine? These are all open questions, and the answers cannot be found in the experience of ecovillages, which have relied on the wider capitalist system to fend off such threats.

"You're not thinking fourth dimensionally!"

In a sequel to the film *Back to the Future,* brainiac time-machine inventor Dr. Emmett Brown chastises his young acolyte, Marty McFly, for "not thinking fourth dimensionally," by which he means not taking *time* into account. Some degrowth recommendations are tainted by the same lapse. Several policy proposals, designed to solve blatant problems produced by the current economic system, seem to have skipped the work of thinking dynamically about implications over time.

One symptom of this harks back to Jackson's hope for a "Cinderella Economy." Rich economies have managed to achieve certain successes—say, advanced healthcare and widespread education—that serve humanity and the planet well. The idea is that we can simply keep these while ditching the rest. Lately, proponents of this view have pointed to the useful start made during the Covid-19 crisis, when strict lockdown measures made it obvious that some jobs or activities were essential and others were not.[38] But the economic system as a whole is highly intertwined and what capitalism does, for good or bad, is to organize this complexity through the price mechanism. Is it really possible we can neatly disentangle the sectors or activities based on "care, craft, and culture" from the rest of the economy?

Let us start with an obvious example. Care in the form of nursing takes place in a hospital, which however is built of cement and steel—meaning that we need access to such production materials even just for maintenance work.[39] The hospital's advanced testing and screening machines, without which modern doctors would be severely limited, are in constant need of replacement pieces. Moreover, patients need medication, requiring a pharmaceutical industry to develop and produce them, and then update them, as is the case for antibiotics. These must also be transported, typically at cold temperatures, so care providers need access to truck manufacturing, logistics centers, refrigerator producers, and well-maintained roads. The list could go on, but the point is clear that all sectors are deeply intertwined in a complex web. The notion that we could easily identify the sectors we like, keep those, and be rid of the rest is ill-conceived.[40]

In a fascinating lecture delivered in 2015 at the World Economic Forum, growth expert Ricardo Hausmann illustrated this point by showing that the challenge in accelerating long-term growth in an underdeveloped country is not, for example, to bring advanced agricultural machines—say, modern combine harvesters—to its poor farmers.[41] That is easy and, if that were the secret of development, we would have solved world poverty a long time ago. The hard part is building the whole economic network that allows the combines to operate: fuel supply, maintenance, logistics, training for the farmers, and so on. Without all this, any marvel of modern agricultural technology will either be useless or rapidly fall into disuse. An analogy Hausmann frequently draws is to the game of Scrabble: rather than think that a single letter will score points, you must think about combinations. Lack some other crucial letters and you cannot form the valuable word. The challenge is to understand how productive networks are built and organized (usually self-organized), and then gradually try to create them in poorer countries. This is why development has been so hard to foster.

For the historically minded, another interesting parallel can be found in the fall of the Roman Empire. At its peak, this civilization reached highly advanced road systems and food logistics, a sophisticated legal system, and plumbing sufficient to sustain the mighty city of Rome as it grew to over one million people in the second century. After the system collapsed, the rulers that came next inherited all the aqueducts and other infrastructure. These, however, rapidly fell into disuse through lack of organized maintenance. Rome's population fell to as low as thirty thousand people. It took over sixteen centuries for a Western city, London, to reach one million inhabitants, a mark it handily surpassed after the Industrial Revolution. In *Collapse,* economic geographer Jared Diamond travels across centuries and continents to look at many other similar cases where societal capacities to sustain economic and social complexity were lost, and populations dropped precipitously.

The level of well-being achieved by today's society rests on the complexity organized by capitalism in a decentralized way. Again, the idea that we could take the parts we like—advanced healthcare, high-level education, technological innovation—because "they have been achieved," and simply ditch the rest, is misplaced. Likewise, the idea that we could manage a complex system by top-down rules, or else have it hold together because everybody embraced the principle of sufficiency, is a recipe for Diamond-style collapse. The challenge, instead, is to scale down production of polluting items without toppling, in a domino effect, many other things that uphold society and well-being.

The international dimension

Another dynamic element that receives too little attention in the degrowth literature is the international dimension of the policy agenda.[42] What happens if one country decides to walk down the ecosocialism path, but others do not? In that case, commitment to degrowth leads, in an act of self-preservation and contrary to the idyllic narrative, to an insular worldview and closure from the outside world.[43]

This is true, first, because the economic scale-down envisioned by eco-socialism reins in international trade. Production needs to be carried out on a smaller scale and as locally as possible to avoid environment-damaging global supply chains, so exchanges between countries will shrink. Along the same logic, international travel will need to be reduced, given the inevitable environmental repercussions of airplanes. As society moves full-throttle toward eco-autarky, we can expect less exposure to outside cultures, and possibly more xenophobia. So argues Benjamin Friedman in *The Moral Consequences of Economic Growth,* based on patterns of nationalism across US history. Indeed, this realization dates back to Enlightenment thinkers, who saw commerce as breaking through religious groups, political affiliations, and prejudices in its pursuit of economic gains. It inspired the well-known observation by 1800s economist Frédéric Bastiat that "when goods don't cross borders, soldiers will."[44]

Second, the degrowth society will become one of scarcity, however voluntary. When this happens, those who are not interested in such a project will flee. Who are these likely to be? Well, to begin with, probably the rich, along with many highly skilled and entrepreneurial individuals who see possibilities of greater rewards, self-expression, and self-fulfillment under democratic capitalism.[45] But of course, the ecosocialist society cannot allow this to happen—not only because the financing of its social agenda relies on heavy redistribution, but also because migration would belie the model's claim of superiority. This would weaken the ideological glue that is fundamental to keeping such a society together.

Once again, the Soviet Union comes to mind, and in particular the history of the Berlin Wall. After the Second World War, Berlin was divided in two, with one sector occupied by the Soviets and the other by Allied forces—but movement around the city was possible. As living standards started to diverge between the two sectors, people started voting with their feet: over three million East Germans fled to the West. For the reasons noted above, the Soviets could not allow this, and in 1961 they erected a barbed-wire fence overnight, which subsequently turned into a highly guarded wall. Thus, a Communist movement that first styled itself "The International" and professed the abolition of passports transformed into a Soviet system with an effective ban on foreign travel.

There are reasons to suspect that any degrowth society would need to erect substantial barriers to the outside world to keep its system operational. Once this is realized, it is hard to imagine this appealing to its potential constituency of environmentally savvy and open-minded international-ists—in particular, the urban, cosmopolitan, and educated members of the Millennial and Gen Z generations.

Innovation, freedom, and resources

If there is one point of confluence for all the issues discussed so far, it is the question of how innovation, science, and technology will continue to advance. In Chapter 2, we saw that innovation and growth are deeply intertwined at an abstract, philosophical level. Now that we have looked at the degrowth policy agenda, we can see this is also true in more practical terms.

In February 2020 in the *Financial Times,* economics columnist Martin Wolf ran the numbers of the degrowth agenda to show that halting climate change by taming growth alone is infeasible.[46] If global GDP growth were to go to zero permanently, and decarbonization were to continue at the current rate, then by 2050 CO_2 emissions would fall by 40 percent—far short of the UN scientists' recommendation of reaching climate neutrality. If we therefore chose to reach the emissions target by *shrinking* the economy, world output would need to drop by roughly 90 percent.[47] Taking into account population growth, this implies that global GDP per capita would need to go back to levels not seen since the 1870s—too much, probably, even for the most frugal champions of simple living.

Clearly, to succeed, the degrowth agenda also requires an acceleration in the development and adoption of green technology at scale. And degrowth advocates seem to be aware of this, which is why innovation is not ruled out in their vision. What is, however, worth inspecting in greater detail is the assumption (generally made implicitly) that, under ecosocialism, innovation will continue to churn out the striking advances in green technology we currently see under capitalism: hydrogen electrolysis, fast-improving electric battery storage capacity, reduced costs of photovoltaic panels and wind rotors, and so on. The validity of this assumption is far from established. To test it, let us connect what we know about innovation processes with the dynamics that would be unleashed by the degrowth agenda.

We saw above how our current economic system should be thought of as a network, the interlinkages of which make it impossible to just save a few favored pieces and dispense with the rest without very careful thought. Research and development are no exception. The capacity we see in advanced economies to generate innovation at an unprecedented pace—as evidenced by the development at breakneck speed of several Covid-19 vaccines, in contrast to what happened with the Spanish flu a century ago—builds on complexity and access to resources. These include state-of-the-art research centers, advanced microscopes and DNA sequencing machines, fast global information exchange through the internet, endless computing power, cooling capacity, chemicals and reagents, and much more. All this apparatus costs billions to nurture and maintain, and years of hard work are often required to make breakthroughs.

Long gone are the days when a lone genius such as Leonardo da Vinci or Benjamin Franklin could make significant technological advances in a workshop. Dividends of economic growth constantly get reinjected into this knowledge-accumulating process. Indeed, there is a clear positive relationship between level of development and innovation capacity, with rich countries significantly leading the pack (Figure 3.1). With shrinking resources available to societies and governments, and without an efficient allocation mechanism to govern complexity, ecosocialism would be incapable of sustaining such knowledge machinery. As Friedman notes, "laws and regulations are typically less effective when the desired behavior requires *taking* initiative or action, as opposed to *refraining* from unwanted action."[48] We can't, in short, foster creativity and innovation by law.

Moreover, studies show that cities are innovation hubs, mostly because they allow for a constant exchange of ideas and interactions, bringing together different types of backgrounds and skill sets.[49] Yet cities are most affected by the previously mentioned limitation of ecosocialism—that its allocation mechanisms and decision-making structure are unequal to the complexity of large human communities.

Finally, we know that innovation thrives when it manages to attract talent from multiple pools; a great research team, for example, is more often than not a wonderful potpourri of nationalities.[50] The capacity to attract and retain talent is therefore fundamental. Here again, ecosocialism is likely to be in a tough spot. Simplifying down to bare bones, a few things seem to matter most to scientists and innovators: freedom of inquiry and expression, access to networks of peers, and resources to carry out their endeavors. Such talent

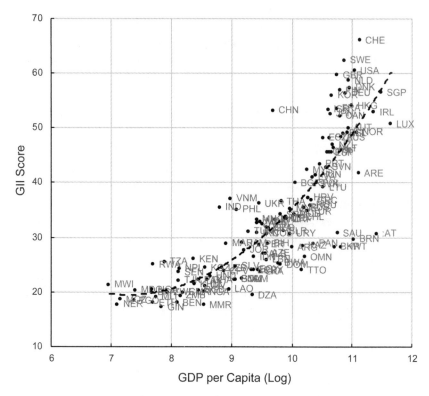

FIGURE 3.1: Innovation and GDP per capita, 2020

Data sources: WIPO (2020) and World Bank.

Notes: The Global Innovation Index combines eighty indicators to give a broad vision of innovation, including political environment, education, infrastructure, and business sophistication. It is published by Cornell University, INSEAD, and the World Intellectual Property Organization (WIPO), a specialized agency of the United Nations. GDP per capita presented in terms of purchasing power parity (in constant 2017 international dollars).

would not be attracted to a closed, eco-autarkic society where personal freedoms are routinely curtailed, personal success is vilified, the concept of progress is dethroned, and resource availability is trending downward.[51]

Mariana Mazzucato's work is worth mentioning here, given how frequently it gets dragged into this type of debate, generally to argue that it is a myth that innovation happens only, or chiefly, in the private sector. As she eloquently argues, everything that makes our smartphones smart— touch screens, GPS, voice recognition, and the internet—was made possible by publicly funded research. Likewise, the publicly funded and directed Apollo program not only put a man on the moon but generated innovation breakthroughs that reverberated through society for decades. It is true that the state has an important role to play in funding, fostering, and orienting innovation, as will be evident in Chapter 8. At the same time, it is no accident that, after government programs enabled the basic research, private companies were the ones to make new technologies applicable and useful to consumers. Driven by the quest for profit, they invested in the development and deployment of products. What makes the smartphone smart might be government-funded technology, but what makes the smartphone the fastest-spreading technology in history is profit-seeking private initiative.

The same thing could be said about space exploration. The Apollo program paved the way, but its cost was so high—at today's equivalent of $156 billion—that it could only be justified under a military headline during the Cold War, and its technological solutions were so ad hoc that NASA struggles sixty years later to meet its own deadline to send a woman to the moon by 2024. In parallel, private companies are striving to bring mission costs down to a level where space tourism could be a profitable business, also laying foundations for further exploration and colonization at scale. It is a great stretch, then, to argue based on Mazzucato's work that centralized technological innovation would thrive under ecosocialism, a system that would impair the mechanics of spreading and adapting green technologies across the economy. Chapter 6 will have more to say on the crucial interaction between innovation and capitalism.

Growth, well-being, and the political growth imperative

A key link in our system that is often vehemently contested is the one between economic growth and the production of what "really matters" and contributes to well-being. Among the common degrowth arguments is the claim that, while growth might be important early in the economic development process, that stage does not last. Once a basic level of income is reached and people have sufficient food, shelter, education, sanitation, and health, growth only undermines our serenity and yields useless stuff. To prove this point, Tim Jackson and others use charts illustrating, for example, that life expectancy increases with GDP up to a certain point and then only very slowly, and that there is a sweet spot (occupied by countries such as Cuba and Costa Rica), where life expectancy is higher than in the United States, at much lower levels of prosperity (Figure 3.2). Similar plots

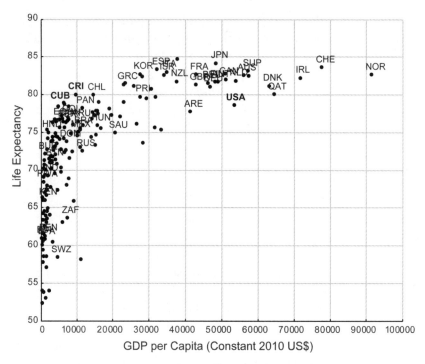

FIGURE 3.2: Life expectancy at birth and GDP per capita, 2017

Data source: World Bank.

Note: Cuba, Costa Rica, and United States are in bold.

of growth against levels of education, child mortality, and other variables lead Jackson in particular to conclude that beyond a certain, low level of GDP per capita, there are diminishing returns to growth.

In light of such data, a few observations are in order. First, two data points do not make a trend, but at least one big trend *is* evident in Figure 3.2: citizens in richer countries tend on average to live longer. As experts in global public health policy know, the positive relationship between income and life expectancy is not linear, but nonetheless is strong.[52] This J-shape reflects a reality of innovation: it is much easier to catch up on a trail others have cleared before you than to be in the group advancing the technological frontier. For this reason, a 3 percent GDP growth rate is seen as solid performance for a rich country like France, but a poor showing for an emerging-market economy like Turkey. The same dynamic is in play when life expectancy increases across the board at all income levels, mostly as scientific and technological improvements spread to poor countries via international health programs and development aid.[53] Antibiotics, penicillin, vaccines, and similar game-changers in modern medicine were developed and are often produced in advanced economies, but are now in use worldwide. As a result, poorer countries have seen sharp improvements in infant mortality rates, life expectancies, and other health measures, often in excess of what their level of affluence would have suggested.[54] This hardly negates the strong role of income and growth: it is the past innovation-led growth of advanced economies spilling over and positively affecting people in low-income countries.

Health is just one dimension of well-being. Across the board, for practically any variable that affects well-being, rich countries have a strong tendency to perform better than poor countries.[55] This includes both personal and public dimensions, market and non-market goods.[56] Higher levels of income, as measured by GDP per capita, are associated with fewer hours spent working and commuting.[57] Higher income also maps to better ratings in many other areas: capacity for remote work, ease of buying a house, clean air, access to and quality of education, spending on culture, number of theaters and museums, low incidence of murders, low levels of inequality, social mobility, interpersonal and institutional trust, strength of social ties (measured by the average number of people citizens say they can rely on), share of nature and marine reserves, overall amount of public goods and welfare (as proxied by government expenditure as a share of GDP), and even political participation, which the OECD includes in its eleven dimensions of well-being[58] (see Table 3.1). The correlations, moreover, do not only show up at the level of national averages. As Dani Rodrik always teaches in his first class on economic development at Harvard, whether the focus is on income or a wider set of indicators of well-being, it is much better to be poor in a rich country than rich in a poor country.[59]

Global surveys routinely find that the countries displaying the highest levels of self-reported happiness and life satisfaction are all rich countries.[60] There is nothing esoteric in this; it is the result of the intersection of capitalism, an effective machinery to push the possibility boundaries, and liberal democracy, orienting resources toward voters' desires. Once again, we encounter a seeming paradox given today's popular discourse. As argued by Iversen and Soskice in *Democracy and Prosperity*, well-functioning capitalism and democracy are not antithetical but, to the contrary, are symbiotic and mutually reinforcing.[61] At an abstract level, the essential role

Table 3.1 Country rankings on a set of well-being measures

Rank	Working hours per week	Healthy life expectancy	Homicides per capita	Green cities	Indoor cinemas per capita	Clean air	Government spending as % of GDP	Traffic commute time	Property-price-to-income ratio	Political participation
1	**Netherlands**	**Singapore**	**Bahrain**	**Austria** (Vienna)	**Estonia**	Bahamas	**Finland**	**Iceland**	**Saudi Arabia**	**Norway**
2	**Australia**	**Japan**	**Japan**	**Germany** (Munich)	**Norway**	**Iceland**	**Belgium**	**Cyprus**	**United States**	**New Zealand**
3	**New Zealand**	**Spain**	**Singapore**	**Germany** (Berlin)	**France**	**Finland**	**Denmark**	**Austria**	South Africa	**Iceland**
4	Rwanda	**Switzerland**	**Qatar**	**Spain** (Madrid)	**Switzerland**	**Estonia**	**Italy**	**Estonia**	**United Arab Emirates**	**Finland**
5	**Denmark**	**France**	**Austria**	Brazil (São Paulo)	**Finland**	**Sweden**	**Sweden**	**Lithuania**	**Qatar**	**United Kingdom**
6	**Norway**	**Cyprus**	**Luxembourg**	**United Kingdom** (Manchester)	**Greece**	**Norway**	**Norway**	**Slovenia**	**Iceland**	**Israel**
7	Belarus	**Canada**	**Norway**	**Portugal** (Lisbon)	**Denmark**	New Zealand	Hungary	Bosnia Herzegovina	**Belgium**	**Ireland**
8	Georgia	**Italy**	**Oman**	**Singapore**	**Czechia**	Canada	**Germany**	**Norway**	**Ireland**	**Sweden**
9	**Germany**	**Australia**	**Switzerland**	**Netherlands** (Amsterdam)	**Slovenia**	Australia	**Portugal**	**Slovakia**	**Netherlands**	**Germany**
10	**Austria**	**Iceland**	**Czechia**	**United States** (Washington)	**New Zealand**	Ecuador	**South Korea**	North Macedonia	**Denmark**	**Denmark**

Data sources: Working hours from the Labour Force Survey of the International Labour Organization (2017 or latest); healthy life expectancy at birth from the World Health Organization (2016); Green City Index from Resonance Consultancy; indoor cinemas from UNESCO Institute for Statistics (2017); clean air data by country from IQAir (2019); government spending from International Monetary Fund WEO (October 2019); traffic commute time index and property-price-to-income ratio from Numbeo (2019 midyear); political participation from the Economist Intelligence Unit Democracy Index (2019).

Notes: High-income countries (based on World Bank classification) shown in bold.

of the good policymaker is to maintain this balanced relationship, keeping impassioned forces of democracy (populism) from overturning capitalism, and perverse effects of capitalism (economic inequality, lobbying, and big money) from undermining democracy.

Taking the long-term perspective, economic growth is about reducing the need for tradeoffs by making more resources available to achieve more goals at the same time.[62] It is no surprise, then, that politicians—whose task in society is to allocate and redistribute limited (shared) resources—have such a penchant for growth. This is particularly the case in a liberal democracy, where they have powerful incentives to secure reelection and therefore to satisfy as many citizens as possible, despite the heterogeneity of needs. This deep interplay between politics (the art of the possible) and economics (the science of managing scarce resources needed to make things possible) is the real origin of the political growth imperative—not narrow-minded economists whispering in the ears of the powerful.

To clarify the concept, we can make a useful analogy to something we all experience: time, and specifically the twenty-four-hour day. We constantly have to choose how to "divide ourselves" over this time: some hours will have to be spent at work to earn an income, put food on the table, and accumulate savings. Others will be devoted to nurturing relationships with family members, to social life, to sleeping and personal care, to hobbies. If it turns out that not all these things can possibly be squeezed into a day, we have to make compromises, or tradeoffs. Economic growth is the equivalent of adding an extra hour to the day. The usual heart-wrenching choices still have to be made, but that much less painfully. As the growth process continues, at some point our days might last a hundred hours, much longer than our growthless neighbors, meaning we might have more time for all these pursuits, even as we continue making compromises among an expanding list of possibilities.

As we are unlikely to ever have enough time in life, the growth process becomes addictive. This is the ultimate growth imperative. Degrowth advocates are right to claim that for each single dimension there is a point of satiation: there can be too much even of a good thing, be it the size of one's house or time spent with one's kids. But economic growth presents the possibilities associated with having more of everything at the same time. And human desire, just like curiosity, seems to know no bounds.[63] If a lower-income country manages to excel on one dimension—say, its low working hours—this can mean it has made a societal choice to invest heavily on that

GDP AND BEYOND

Part of the "beyond GDP" agenda is the argument that, because economic growth is just a means to an end, rather than compulsively monitoring it with GDP, we should instead devise a more holistic measure that captures the various dimensions of well-being. This might take the form of a multifaceted scoreboard, of which Kate Raworth's "doughnut" approach is an example, or a single composite indicator, such as the OECD's Better Life Index.

In fact, such alternative well-being indicators are not new; almost half a century ago, in *Social Limits to Growth,* Fred Hirsch noted many already available. William Nordhaus and James Tobin produced calculations of economic welfare based on national accounts in the early 1970s (and, to the question posed by the title of their 1973 paper "Is Growth Obsolete?" they answered, "We think not.") The World Bank has compiled its Human Development Index since 1990, and the European Commission publishes all sorts of social, environmental, and even geopolitical scoreboards.

There is no reason to oppose more multifaceted indicators, which can help citizens and decision-makers identify policy failures and discuss trade-offs. Regularly published statistics based on standardized evidence-gathering can draw attention to gaps and issues more quickly than the democratic process usually does, and give much-needed ammunition to those pushing for improvements, whether in healthcare access, housing availability, greenhouse gas emissions, or environmental degradation. By greasing the liberal-democratic capitalist machinery, especially on its political side, measures beyond GDP improve its efficiency.

To be sure, GDP can be improved to become a better barometer of economic growth. As Diane Coyle argues in *GDP* (2014), it should do more to capture the value created by producers innovating, adding services, and expanding the variety of products; the value of services performed by the environment (such as pollination and other forms of "natural capital"); and the value of the vast services consumers now enjoy at zero cost in the digital economy. No doubt if GDP did account better for these, we would see that actual economic growth is higher than current calculations indicate. This is a point made by Aghion and Antonin (2018) and other endogenous growth economists to rebut Robert Gordon's claim that newer inventions are not generating the level of productivity growth that past ones did.

Debates are welcome over still other forms of value creation that could be incorporated in GDP. Coyle among others advocates the inclusion of work performed in homes and family settings which is unpaid but surely vital. Not everyone agrees (including myself), for reasons Coyle herself highlights. When the first attempts to measure national income began in the seventeenth century, the impetus was to improve taxation in preparation for war. To this day, the point remains to produce a rough metric of potential government appropriation. GDP answers a basic question: What is the hypothetical upper limit of resources that could in principle be mobilized toward a specific societal goal? From this standpoint, it makes sense to have specific budget lines (such as for R&D spending) and liabilities (public debt) expressed as percentages of GDP. After all, "statistics are of the state" (Coyle, 2019). But is it possible to tax unpaid activities carried out at home (or force individuals to switch from them to market activities that can be taxed)? In a liberal democracy, probably not. The point remains that any performance measure should continue to be challenged and improved.

In no way, however, should we expect any well-being dashboard to supplant the focus on GDP altogether, much less dissolve the growth imperative. GDP will remain the king of economic measures because, ultimately, it is economic growth that makes more resources available to society—and therefore makes it possible to achieve higher performance on all dimensions of any scoreboard of well-being at once. Rather than moving "beyond GDP," we should be calling for measurement of "GDP and beyond."

---

front. Given limited resources, however, it is an advantage that comes at the cost of other dimensions. That is fair enough, but surely less appealing than also having more on other well-being dimensions at the same time. For that to happen requires economic growth.

The American conundrum: So rich and yet so poor

One point about the United States must be made, as it is often shown that the richest country in the world performs poorly on many well-being dimensions, from access to healthcare and (average) quality of education to life expectancy and work-life balance. As a personal anecdote, when I was a Fulbright scholar in the United States, my group was invited to Philadelphia for a program reunion, bringing together scholars from all over the world. The day was organized around a set of important themes in American politics, from gun control to in-city severe poverty to lack of access to healthcare. I remember vividly the astounded faces of my Afghani counterparts, coming from a war-torn country, as one of them posed a question to the presenter: "But you are the richest country in the world—how can this happen here?"

Anecdotes aside, it is true that even in a rough summary of important components of well-being like Table 3.1, the United States makes only sporadic appearances. Recalling the symbiotic construct of liberal democracy and capitalism, we might look for a fault line more in the former than in the latter. In other words, structural problems are likely to have their origins in the country's political machinery, and potentially the value system underpinning it, rather than in the growth process per se, in capital, or in advanced capitalism.[64]

Given how central the topic of working hours is to the degrowth movement, a focus on them could be illustrative. A 2005 study by Alberto Alesina, Edward Glaeser, and Bruce Sacerdote shows that working hours in the 1990s were the same in Europe and the United States.[65] But then, Europeans chose to take a portion of their growth dividend in the form of more vacation days and shorter working hours.[66] Framed in this way, it sounds like a simple, individualistic choice, but this was not the case. The move resulted from a social struggle—a combination of labor union activism, street demonstrations, and savvy politicians spotting a nascent trend in the electorate and adjusting their campaign platforms. Of course, competing interest groups threw up roadblocks along the way, but these merely slowed down a process of societal choice.

This brings us to the United States today, where people are becoming increasingly unhappy with how the nation's growth dividend is distributed, not only among individuals but also across priorities; this is evident in the polarization of the political spectrum. If citizens' desires are not being transformed into policy, it means a wrench is being thrown into the works of "preference transmission" by something in the political system. This could relate to the strong influence of big money in US political campaigns, the two-party system entrenched in the winner-take-all electoral vote system in most states, the allocation of electoral votes (versus popular votes), or gerrymandering and voter suppression, to mention just a few factors. Net of specific policy actions, the structural challenge for the coming years is to improve the democratic transmission mechanism, and restore its balance with capitalism.[67] A failure in this respect will lead to a loss of faith in either the democratic process or capitalism, or both, with dire consequences.

This chapter began by considering what a post-growth society would look like according to its advocates, and moved on to question several of its assumptions and policy planks. This long journey into political economy proved full of twists and turns, from Sir Thomas More's *Utopia* to the future of GDP, and along the way the Apollo mission, ecovillages, and the political origins of the growth imperative. Where does all this leave us?

Degrowth, by definition, will reduce available resources and create some scarcity, albeit, according to its advocates, only for goods or services that are unnecessary or superfluous. The implicit assumption, or hazardous bet, is that this would be *voluntary* scarcity—that everyone, once free from the iron cage of consumerism, would gladly embrace it as part of a superior mode of simple living.[68] In this chapter we saw that, if this were not to happen, and the "infinite desire for more" (as Daniel Cohen calls it) were to remain prevalent, degrowth would likely turn into the proverbial road to hell, paved with good intentions.

Ecosocialism does not have an allocation mechanism, or a way to assuage tensions within society. The problems this chapter has described would rapidly cause the entire degrowth utopian vision to unravel, leaving citizens with some combination of perceived misery, social tension, autarky, emigration, and welfare reductions. An intricate web of rules would be spun up to regulate most aspects of life—all presided over by a strong central government at high risk of turning illiberal in its efforts to shore up its ideological raison d'être. The distance between ecosocialism and ecoauthoritarianism is shorter than one might imagine.

If the success of green capitalism hinges on a bet that technology will come to the rescue once more and help stave off climate change and environmental degradation, then degrowth hinges on a socio-psychological bet that, across society, people's attitudes and preferences will suddenly flip to a "less is more" worldview. Given the centrality of this assumption in determining the success or failure of ecosocialism versus green capitalism, Chapter 5 is devoted to inspecting the odds of this complete turnaround in desires, and why it is highly unlikely. But first, it is worth leaving the world of thought experiments to observe how some of the socioeconomic dynamics described in this chapter actually played out in the real world. That these are not just abstract ruminations is particularly easy to see in one rich country that has effectively seen no growth in the past thirty years: Italy.

4

THE ITALIAN CANARY
IN THE GROWTHLESS COALMINE

Your Italy and our Italia are not the same thing. Italy is
a soft drug peddled in predictable packages, such as hills in the
sunset, olive groves, lemon trees, white wine, and raven-haired girls.
Italia, on the other hand, is a maze. It's alluring, but complicated.
It's the kind of place that can have you fuming and then purring
in the space of a hundred meters, or in the course of ten minutes.
Italy is the only workshop in the world that can turn out
both Botticellis and Berlusconis.
—BEPPE SEVERGNINI, 2005

As an affluent country experiencing three consecutive decades of stagnant GDP growth, Italy should in principle be a positive case study for degrowth enthusiasts—and especially useful for those who see capitalism as inherently compatible with their vision. With a GDP of over $2 trillion and a population of about sixty million, it is an advanced economy, belonging to the OECD, the G7, and the G20. At least before the Covid-19 pandemic hit, it enjoyed an extremely high life expectancy: 83.4 years, one of the longest in the world. The country has a strong public-welfare system, including universal healthcare, and a government that is very active in the economy. It is strong on the cultural dimension, and as for employees working very long hours, they make up only a small fraction of the workforce: 4 percent, versus the average of 11 percent among its rich OECD peers. More broadly, Italy is number one in the OECD for time devoted to leisure and personal care: 16.5 hours per day on average, including sleeping, eating, and socializing with friends and family. The strength of the country's social capital is evidenced by survey data on access to people in times of need, and its robust civic participation shows up in indicators such as above-average voter turnout.[1]

And yet, despite its strong performance on many dimensions that eco-socialists would like policymakers to prioritize, Italy scores poorly in indices of well-being. On the OECD's Better Life Index, Italy ranks only twenty-fourth out of forty countries. More important, and independent of such top-down, technocratic scoreboards with their arbitrary weightings, Italians rank substantially below their OECD peers in their self-reported satisfaction with life. This poses a conundrum: how can Italians perform so well on "the things that truly matter in life" and yet be unhappy with how things are going?

In what follows, we will analyze the origins and nature of the Italian malaise between the early 1990s and the inception of the Covid-19 pandemic. We will then see that when economic growth dries up, this leads to a perception of relative poverty—even in places with high absolute levels of income by world standards—as well as greater emigration, less innovation, more social fractionalization and tension, and an inward turn. This is problematic for a degrowth approach to fighting climate catastrophe, which depends on the acceleration of innovation and peaceful sharing of reduced resources, both within and between countries.

A country of contradictions

Italy is a country of constant contradictions, as noted by journalist Beppe Severgnini in the epigraph to this chapter. Its history, monuments, architecture, art, and culinary tradition make for an incredibly rich culture.[2] In terms of natural splendor, Italy is blessed not only with its beloved landscapes, but also vast biodiversity by European standards.[3] These attributes, combined with an appealing lifestyle, a strong entertainment industry, and a vast array of luxury brands, make it a leader, year after year, in cultural influence and soft power.[4] It is one of the most visited countries in the world, with more than sixty-four million tourists in 2019—a number in excess of its own total population.

Net of these evident assets, Italy is well known for something else, particularly among economic commentators: its dismal economic performance. Since the late 1990s, it has increasingly become the poster child of poor macroeconomic outcomes among rich countries. It's much like the

country's performance in rugby. Since 2000, the Italian team has partici-
pated in the Six Nations Championship, an annual tournament among some
of Europe's best rugby teams. It has never won, and has often earned the
"Wooden Spoon" for coming in last—making it essentially the worst of the
best. In the macroeconomic league of advanced economies, Italy deserves
the same prize. When I talk to non-Italians trying to make sense of this situation, they constantly ask some version of the obvious question: How can a
country that has so much, and many prominent economists to boot, fail to
get its act together and manage to perform so poorly year after year?

It is important to realize that it was not always like this. Italy was "the
China of Europe" after World War II, achieving annual growth rates north
of 7 percent, only to then fall far behind.[5] This change in trend is depicted
in Figure 4.1. In the 1970s and early 1980s it grew above the OECD average,
and then roughly in line with the OECD average. From the 1990s onward,
it grew less than the rich-country average in good times, and contracted
more in bad times, showing poor resilience to shocks. Since the turn of the
millennium, and before the pandemic rebound, Italy has not achieved even
a 2 percent real GDP growth rate. Between 1999 and 2019, real GDP per
capita increased by 28 percent in Germany, 23 percent in Spain, and
19 percent in France. Over the same time period, Italy's was up 2 percent—a
steady-state economy for all practical purposes, yet implying a stagnating
living standard and falling behind by international comparisons.[6]

The Italian economy enjoyed extensive growth during the postwar recon-
struction phase, but once it reached the technological frontier, it struggled to
advance, except in a handful of sectors. It has arrived at the point of being a
manufacturing powerhouse and export leader—in Europe, second only to
Germany—while experiencing some of the lowest GDP growth rates in the
EU. Further contradictions show up on the geographical dimension. Some
of Italy's northern regions are among the richest in the whole of Europe, and
therefore the world. Yet some of its southern regions are among the poorest,
constantly in need of support from EU and national development funds.

Why did this happen? The prevalent narrative is that Italy missed the information and communications technology revolution in the 1990s due to
poor managerial skills, cronyism in local bank lending, and familism in
managing the very microenterprises that had been the backbone of the
Italian postwar economic boom. Others blame tight regulation of product
markets for preventing competition and the process of creative destruction,
or rigid labor markets for suppressing labor productivity.[7]

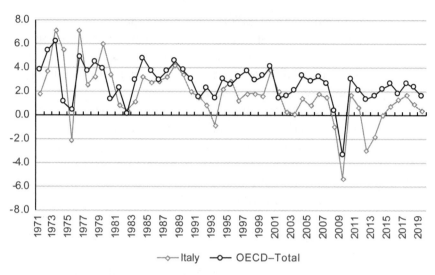

FIGURE 4.1: Real GDP growth, Italy and OECD average
Data source: OECD.

The competing explanations are of great intellectual interest, but not the crucial point here. Our goal in this chapter is to learn how the mechanics of a capitalist system play out in practice. As we view the Italian experience through the lens of the growth imperative, we will see the negative dynamics unleashed by flagging economic growth. While the rest of this chapter will look in turn at the big problem areas—public debt, low productivity growth, high emigration, and more—an essential reality to keep in mind is that these macroeconomic forces are highly intertwined. As much as possible, we will focus on their mutually reinforcing effects.

Also, before proceeding, we should recognize that Italy, as a member of the European Union and of the euro, has been shielded from some negative dynamics that would have been unleashed in a fully sovereign country. Normally, for example, a protracted period of sluggish economic performance would weaken a national currency substantially, making a country relatively poorer by international standards. Likewise, the fact that climate policy is defined largely at the EU level makes it hard to draw direct lessons on this front from the Italian experience. As we shall see, however, once we understand the socioeconomic dynamics at play, we can leverage the knowledge accumulated in the past few chapters and extrapolate some lessons for the climate agenda.

The great fiscal lever

We start our analysis by focusing on the most notorious symptom of the Italian malaise: public debt. In 2019, Italy was saddled with one of the highest debt-to-GDP ratios in the world, just after Japan and Greece in the OECD. The economic fallout of the Covid-19 pandemic has therefore only exacerbated a preexisting condition.

Italy's debt ratio was not always so high. At the beginning of the 1980s, public indebtedness stood below 60 percent—a frugal, even Germanic, fiscal position for an advanced economy. As its economic growth machine started to falter, however, two effects kicked in. First, the ratio of public debt to GDP expanded mechanically, due to the shrinking growth rate of the denominator. Second, policymakers responded to the growth rate slowdown by stepping on the throttle in terms of public spending. Seemingly, after thirty years enjoying the "Italian economic miracle," they were unable to believe this would fail to rekindle the fire of growth. Since then, belief in public investment has been an Italian economic obsession. The great Greek mathematician Archimedes proclaimed, "Give me a lever long enough and a fulcrum on which to place it, and I shall move the world." In the Italian macroeconomic context, policymakers of all stripes have labored under the conviction that a large enough fiscal lever could put any economy back on track.[8] When the lever was perceived as not long enough, either due to financial market pressure or European fiscal rules, the blame fell on external powers allegedly starving the country.[9] This misreading of the problem prevented policymakers from seeing the evident truth. The quality of public spending and investment was low, meaning that—whether due to incompetence or corruption—projects were poorly selected. While their impact on growth faded after a few years, their costs were left to linger on public debt balance sheets.[10] To return to the automotive metaphor, pressing down on the fiscal gas pedal is counterproductive if the driver lacks a sense of direction. It may only take the car in circles and leave it, at the end of the process, no closer to a desired destination and with far less fuel in the tank.

The process has continued for so long that, before ultra-low interest rates were ensured by the European Central Bank's response to the Covid-19 pandemic, in some years interest payments were so large that they exceeded public investment.[11] Take a moment to let that simple fact sink in: it is the ultimate sign that an economic policy is geared toward the past, or at least toward stabilizing the present rather than preparing for a brighter future.

It is, we might say, a negation of the concept of progress.[12] For Italy to refinance its debt and keep it from exploding out of control, Italian policymakers need to keep taxes high, but without providing commensurate public goods or services to citizens. Year after year, just to put enough aside to cover hefty interest payments, Italy has had to run large primary surpluses—meaning that its public revenues and taxes must significantly exceed its expenditures.[13] As one can imagine, this situation is not particularly popular with the electorate. This is why, as Barry Eichengreen and Ugo Panizza show, large and sustained surpluses are extremely rare in history: most subside before long, under the pressure of political discontent.[14]

In Italy, given that national income (or GDP) is broadly stagnant or shrinking, but the wealth that was built in the postwar decades of fast growth and accumulation remains high, a fixture in political debates is the topic of a wealth tax.[15] At the same time, wealth taxes are loathed by wide strata of the population. This is to be expected. At a point when people's faith in their country's future is faltering and they lack confidence in policymakers, a wealth tax can feel like selling the family silver. It does nothing to fix the imbalance of inflows and outflows that forced the measure, and therefore the same dire situation is likely to be faced a few years down the road, except with even less resources to respond.[16]

The lesson from a political economy perspective is that, when economic prospects sour and people no longer expect progress, uncertainty and concerns about the future grow. Discussions turn away from what would benefit society as a whole as each group becomes more focused on defending its own interests. It becomes much harder to reach decisions on allocation, redistribution, or even reforms. Recalling Hobbes's view from Chapter 3, we can see this as a soft preamble to the hard clash of all against all, when resources become limited and increasingly scarce.

The low-skill, bad-job trap

Aside from stagnant GDP, the most worrying concern from a growth economics standpoint is the flatlining of productivity growth. To degrowth advocates, this is part of the policy menu. But in a market economy, over the long run, wages and productivity move hand in hand. So it was that, as productivity growth stumbled in Italy, wage growth became subdued. This presented a dilemma: hiking wages faster would leave Italy less competitive in product markets vis-à-vis other countries. But allowing wage growth to remain low would spur more emigration, and also leave Italians with less purchasing power.

The process continued over the years, to the point that wages in Italy are now significantly lower than in other European countries, at practically all levels of expertise. Back in 1991, average wages in Italy and Germany were not substantially different, and both of these well exceeded the average in France. In 2019, the average wage corrected for differences in purchasing power in Italy stood at $39,200 per annum, compared to $46,500 in France, $53,600 in Germany, and $56,500 in the Netherlands. An Italian primary school teacher with fifteen years of experience has an estimated gross salary of $37,700, far lower than European peers in Germany ($77,600) and the Netherlands ($64,900).[17] The average salary for a mechanical engineer is €32,000 in Italy, €44,200 in France, and €49,500 in Germany.[18] Attracted by higher financial rewards abroad, many young Italians chose to take off.

One would expect that, under these conditions, foreign direct investment should flow massively to Italy, attracted by the availability of cheaper talent—and do so particularly from elsewhere in the European Union, since capital can move freely across countries. And yet, such investment is extremely low in Italy, at 22 percent of GDP—one of the lowest rates among European OECD peers. Surveys indicate that foreign investors are mostly dissuaded by Italy's mix of political uncertainty, high taxes, corruption and organized crime, and lengthy judicial proceedings.[19]

More broadly, a low-productivity, low-wage growth model is a dangerous bet for an advanced economy because pursuing it means venturing onto turf where emerging markets clearly have an edge. Italy has found it challenging, for example, to compete in car components and factories vis-à-vis Eastern Europe, and in textiles vis-à-vis China and Southeast Asia. China's explosive trade growth since its 2001 acceptance into the World Trade Organization has dealt a particularly significant blow to the Italian economy.

The dilemma referred to above belongs to a wider problem known as the "low-skill, bad-job trap." In countries where a large proportion of the workforce is not highly skilled, firms have little incentive to provide good jobs (that is, jobs that require high skills and provide high wages)—but if few good jobs are available, workers in turn have little incentive to acquire skills.[20] Indeed, Italy has one of the lowest rates of university graduates in Europe, just ahead of Romania. Only 17 percent of the working-age population holds a post-secondary education degree, against 33 percent in France and 25 percent in Germany. This remains true when the focus is just on a younger generation, such as the cohort aged thirty to thirty-four.

As discussed above, the country retains a special charm in the eyes of foreigners, and also high private-wealth levels. These positive attributes can turn into a mixed blessing, however, when income levels are stagnant. Consider the prices of houses, for example, which have not seen much correction to match the low wage growth; between an influx of international buyers and Italy's traditionally high level of home ownership, they remain high.[21] Younger generations find it harder and harder to buy homes and, for many, the only hope is to inherit one. As young adults struggle to strike out on their own, the age at which they leave their family homes has reached an average of 30.1—four years later than the European average and a whopping twelve years later than their Swedish counterparts.[22] Recall that access to housing is a classic indicator feeding into well-being scoreboards such as the OECD's Better Life Index.

By this point, the alert reader will have perceived the interplay among some of these dynamics. For example, low wages imply that it is harder and harder to retain or attract talent in an international labor market, dragging down the country's overall innovation capacity. A similar effect would take place in an ecosocialist setting, and Italy's experience suggests that a broadly defined high quality of life or access to welfare would hardly arrest the hemorrhage of talent.[23] This effect is also visible domestically, specifically in the case of Milan. This city at the industrial heart of Italy was, at least before the Covid-19 pandemic, on the upswing in terms of its national and international attractiveness and innovation. This meant, however, that it was a magnet for resources from the rest of Italy, in the form of young talent but also in the form of savings and capital.[24] Without debating whether this was a good or a bad development, what matters here is the transferable lesson: any degrowth plan would have to be considered in an international context and avoid the risks of setting off spiraling rebalancing dynamics.

No country for young people

Emigration deserves its own discussion, given the sheer size of this phenomenon. Young people in Italy, caught in the low-income, bad-jobs trap, saw wages being compressed and employment in innovative sectors languishing, and started taking off in large numbers. The figure is striking:

150,000 Italians move abroad every year, a threefold increase since the turn of the century. And this number is probably a low estimate since it accounts for only those who register their moves with consulates. Statistics show that those leaving tend to have higher educational attainment than the national average. As these numbers pile up, the latest figures indicate some 5.5 million Italians living abroad. Of these, the majority, almost three million, are in other European countries, where their right to study and work is safeguarded by EU treaties.

Technological innovation, wherever it may originate, has repercussions within Italy's borders, also due to the digital revolution. Amazon was not invented in Italy, but it is nonetheless reshaping the retail sector, and the pandemic only accelerated the process. Clearly, having a poor productivity growth record at home does not mean a country is free from technological unemployment.[25] Here again is a relevant, cautionary tale for the ecosocialist agenda and its plan to force innovation away from labor-shedding technology. As the old saying goes, if you are not sitting at the table, you are likely to be on the menu, and this applies to innovation, too. The alternative? Try to fence off your economy completely to the outside world, in the pursuit of autarky.

As jobs in Italy were either automated or shifted abroad, unemployment rose, reaching high rates by advanced economy standards. In 2019, before the pandemic hit, the country's unemployment rate stood at 9.7 percent, exceeded in the EU only by Greece and Spain. When jobs become scarce, the workers most affected tend to be the usual underrepresented groups: women, the poor, and the young. For those aged twenty to thirty-four, the unemployment rate in Italy was 28.9 percent, more than twice the EU average. Among the group aged fifteen to twenty-nine, those not in education, employment, or training (called NEETs) made up a staggering 27.8 percent, the worst showing in the OECD. Roughly a third of these were looking for jobs, while two-thirds were completely inactive.

Those remaining in the country and trying to secure jobs often found themselves applying for and accepting positions that did not fit their qualifications. OECD surveys show that Italy's level of field-of-study mismatch is among the highest in advanced economies, with 37 percent of workers stating they are employed in a field they did not prepare for as students, and that the country has high rates of over-qualification. Even for those who are not unemployed, this represents a poor outcome in terms of self- determination and realization—another important component of well-being.

In early 2019, in an effort to curb these perverse dynamics, the governing coalition introduced a new "citizens income." With the introduction of this element, the Italian case study becomes even more salient, given that some form of universal basic income is a central element of the ecosocialist vision.[26] While the jury is still out on the Italian version's impact, and scholarly assessments will take time, the results so far seem mixed at best. The program has reduced poverty somewhat, but not done much to budge employment, emigration, or for that matter, happiness and well-being. Of course, in the design and implementation of a policy measure, the devil is in the details, but this case casts at least some doubt on the idea that a universal basic income would alone ensure a stable ecosocialist macroeconomy.

Clashing over scarce resources

Perhaps it is not surprising that birthrates have also plummeted in Italy, in part because so many young people have left the country, and in part because those who remain face such uncertain futures. Italy has one of the lowest fertility rates in the world: 1.27 children per woman. Contrast this with 1.7 in the United States, for example, or 1.9 in France. The average age for first-time mothers in Italy is thirty-two, the highest in Europe.[27] As a consequence, the country's total population is projected to shrink from the current sixty million to thirty-nine million in 2100—the equivalent of setting Italy's population clock back to 1920. Moreover, the country is aging fast. With more than one in five people now sixty-five or older, it has the second-oldest population in the world, trailing only Japan. As recently as 1980, people of this age made up only one in eight. By 2050, they will be one in three.

From a narrow environmentalist perspective, this might all be good news. After all, as *The Limits to Growth* showed, population growth is a substantial contributor to climate change and environmental degradation, and particularly so in affluent countries. In the socioeconomic sphere, however, decreasing population has two effects. First, mechanically, a country with fewer workers, high public debt, no growth, and an aging population must increase the share of national income it devotes to pension payments and healthcare. This leaves less public resources available for other uses. Second, a political-economy dynamic arises as the composition of the electorate increasingly tilts toward older cohorts: politicians are unable to resist expanding the budget lines that matter most to the elderly. Already in early 2019, the same government that ushered in the citizen income opted to effectively reduce the retirement age—in a country with one of the oldest populations in the world. This is a tendency that goes far beyond this particular government; Italy has steadily grown its pension payments to the point that they are among the world's highest. By now, these add up to 15.6 percent of GDP, a higher percentage than any other OECD nation, representing almost a third of total government expenditure. Meanwhile, programs aimed at youth, such as baby bonuses and education programs, are comparatively underfunded. The young and the old are clearly intertwined by family ties. However, this simple fact does not imply a complete turnaround in the twisted political economy dynamics of a steady-state economy. The older and typically richer cohorts appropriate through the voting system a large share of national resources. Then, as they see fit, in their personal function as parents or grandparents, they use their higher incomes, pensions, and savings to help their offspring make ends meet. The affection or benevolence that one generation has toward the next, however, does not provide for larger resource allocations for the future. This is important to keep in mind for the climate agenda.

The broader lesson should be evident: when resources become scarce, whether due to lack of growth or greater spending on social needs, groups in society will fight over what remains—and, in a democracy, the largest group will win. In a non-liberal democracy, the strongest group will win. This competition for priority status leaves less resources available for other purposes—and typically, proposals to spend more on research and innovation, or other forward-looking investments, end up losing out. This is because, by definition, over the short term they benefit only a small set of people; their large, long-term societal payoffs accrue to non-voting younger or future generations. Older citizens' affection toward their children and grandchildren does not change this political arithmetic. This point is highly relevant to us not only because climate mitigation measures are of this nature, but also because, as we saw, the degrowth agenda still rests on an assumption of accelerated technological innovation.

Our case study is also specifically instructive on the topic of innovation. Among the world's great contributors to cutting-edge research are many Italians, but Italy as a country is not a leader.[28] The fact that international

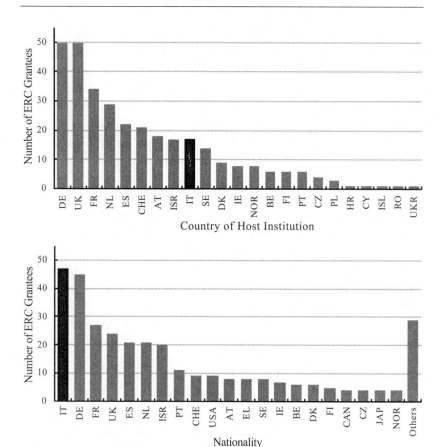

FIGURE 4.2: European Research Council research grants in 2020

Data source: European Research Council.
Note: ERC Consolidator grants shown by country of host institution (*top panel*) and nationality of applicant (*bottom panel*).

labs and top university departments and research centers are well stocked with Italian talent would be a point of pride, except that, in more of a brain drain than a healthy brain circulation, the talent too rarely returns.[29] While much of the evidence of this is anecdotal, one statistical sign of it shows up in the prestigious grants handed out to scholars by the European Research Council each year (Figure 4.2). When the 2020 recipients are sorted by the countries of their host institutions, Italy ranks ninth, with half as many grants as similarly-sized France. Breaking down the recipients by their nationalities, however, tells a different story: Italians (and in particular Italian women) take first place.

The economics of "us versus them"

As my friend Valerio Riavez, a strategist for political campaigns, likes to say, Italy has always been a fascinating laboratory for political dynamics. Often, trends first noticed in its discourse and decision-making go on to take hold in other advanced countries.[30] Examples include the country's election of a controversial tycoon as prime minister in 1994 and again in the early 2000s, its regional separatist movements beginning in 1991, its rise in anti-immigrant rhetoric and anti-globalism (in the form of anti-EU sentiments), and its recent anti-establishment populism—the latter propelling the Five Star Movement to win its plurality in the 2013 general elections.

Much could be written on these experiences, but perhaps the common thread running through them is an eagerness to find an external culprit to blame when economic and other prospects sour. For Prime Minister Silvio Berlusconi, still attached to an iron curtain worldview, the scapegoat was "the communists," a label broadly applied to any left-wing adversary. For the regional independence-seeking Lega Nord (Northern League), it was poorer southern Italians. For that same group rebranded recently as Lega (deemphasizing its northern focus), it was migrants from Africa and Eastern Europe. Then came the Euroskeptics, with their resentment of the EU, and particularly the Germans and French. Their views spread to a broad-based national feeling, deepened by many people's perceptions that Italy was abandoned during the 2015 migration crisis and in the early phases of the Covid-19 pandemic. And finally, for the Five Star Movement, the culprit has been the vague but despised "establishment," the ultimate grouping together of everyone who disagrees with that party's worldview.

This habitual process of looking for scapegoats has left the country more fractionalized, riven by multiple fault lines: young versus old, north versus south, poor versus rich, more educated versus less, natives versus migrants.[31] Once again, we see how a situation of scarcity leads to heightened sectarian tensions.[32] A heavy dose of wealth redistribution, along the lines advocated by degrowth zealots, would hardly salve these societal wounds. More likely it would stoke even further social tensions and intergroup conflict.

There is another political consequence of the Italian economic malaise:
A country that was in the past keenly interested to play a role in shaping
the global economy, as evidenced by its seat at the table of the G7 and G20,
has turned inward-looking. By now, Italians are completely self-absorbed,
wholly focused on domestic problems. Foreign policy becomes of relevance
only when it directly touches on national dynamics. In a fascinating survey
conducted in April 2019 by Eurobarometer, the European Commission's independent polling agency, one question asked citizens in each EU member
state which among various problems facing the world was the single most
serious. Based on how majorities of their citizens answered, countries were
clustered into three groups. The first, made up of the rich and Nordics (such
as Sweden, Denmark, Finland, Germany, and the Netherlands) answered
"climate change." The second, a small cluster, brought together conservative Eastern European countries (such as Bulgaria, Poland, and the Czech
Republic), which answered "international terrorism." The third, clustering
nearly all the rest, answered "poverty, hunger, and access to water." At a time
when global growth was strong, only one country stood apart from these
clusters, clearly looking at its own predicament: Italy, which answered "the
economic situation."[33]

As well as being inward-looking, Italian policymaking has for long been
marked by striking short-termism, with leaders seemingly incapable of
devising long-term strategies.[34] As its economic future dimmed, the country
became completely absorbed in navigating the rough present. When no time
and resources can be spared, however, to anticipate potential future shocks,
the result is a failure to create buffers or otherwise prepare.[35] The country is
thrown into disorder even by setbacks that could have been envisioned and
explored as "known unknowns," to use the parlance of foreign policy. This is
the origin of the lack of resilience noted above. All countries are hit by their
share of crises, but Italy seems always to be more affected by them.[36] Once
more, this finding is relevant to our inquiry, because climate change pre-
paredness will require strengths in long-term planning and risk management; action must be taken now to shore up against catastrophes that may be
of low probability but would have high impact.

The Five Star Movement should be of interest to degrowth advocates
globally, as that party has sympathized with their worldview for years.[37] In
line with the overall degrowth master plan, Five Star has strongly advocated
that the Italian parliament be abolished and replaced by direct democracy;

the online platform it proposed to use for voting was aptly named Rousseau. Within the party it used this mechanism to channel some bottom-up proposals, and occasionally for an "online referendum" on a party issue, such as whether to form a coalition with other political parties. More often, it was used to test top-down policy proposals, essentially serving as a large-scale focus group. Some critics therefore saw Rousseau as a deception, used not for direct democracy but instead in ways heavily managed by the party leaders, who put only selected few issues up for voting, and even then, made sly use of framing techniques to elicit the feedback they wanted. Once again, the jury is still out, as it will take years to evaluate the effectiveness and impact of this proto-direct-democracy experiment.[38] One thing that is clear, however, is that an online tool deployed at national level hardly became a game-changer. Far more than this is required to arrive at legitimate decisions about resource allocations in a society.

All in all, we can see the Italian case as vindicating the central argument that Benjamin Friedman made in *The Moral Consequences of Economic Growth,* based on his analysis of United States history. When growth wanes, even countries with deep democratic and liberal traditions take a nasty turn toward nativism, racism, parochialism, and a general closedness toward the outside world.

Adding insult to injury, economic stagnation and flat population growth have not particularly helped Italy to reduce its climate impact. Between 1990—about the time when the economy settled into its steady state—and 2019, Italy managed to reduce its territorial CO_2 emissions at roughly the same speed as its EU partners' average. If anything, it underperformed, as its emissions came down by 23 percent, in contrast to the 24.8 percent drop achieved by the EU excluding Italy. This suggests that climate targets and policies, generally set at the EU level, are much more important to determine the overall path of decarbonization than pushing an economy into steady state.

From Rome to Tokyo, and back

It would be natural if, at this point, at least some readers questioned the entire exercise of looking into the Italian predicament for broader lessons about degrowth. It could be argued that all these negative dynamics grew out of very specific ways that Italian policymakers mismanaged the situation. This is a non-falsifiable statement. Taking the objection even further, this line of argument might point to Japan as a country that is instead thriving, despite its own subdued GDP growth over a protracted period of time.[39]

On the surface, Japan displays some similarities with the Italian case. GDP growth used to be fast, to the point that in the 1980s it looked like its economy could soon overtake the United States. After the Asian crisis of the early 1990s, the fiscal policy lever was likewise pulled repeatedly in efforts to rekindle growth and, crucially, to exit deflation. Japan's public indebtedness is now the highest in the world, with a debt-to-GDP ratio of over 230 percent. On the demographic front, like Italy, the country has extremely high life expectancy and a low fertility rate. Unlike Italy, Japan has no comparable surge in emigration, its unemployment is extremely low at just over 2 percent, its government policy is efficient, and its society retains its legendary degree of cohesiveness.[40]

Once we dig beyond the surface, however, we realize that Japan has hardly embraced a degrowth path. While its GDP might be growing slowly by international standards, this is more due to a rapidly aging and shrinking population than anything else. Table 4.1 shows growth rates of GDP, corrected for inflation, comparing the performance of Japan, Italy, and the OECD average. We can see that Italy grew faster than Japan in the aftermath of the Asian crisis in the early 1990s. After that, the two countries displayed similar performance, except during the eurozone crisis, when Italy fared poorly. Over the entire period, both made a poor showing against the average of advanced economies. When we look at GDP per hours worked, however, which corrects for the pace of aging and shrinking population, we see a completely different picture. Japan grew much faster than Italy, and performed better than even the OECD average. We can attribute this to savvy public and private investments, which yielded innovation and exceptional labor productivity growth. This implies that many of the negative dynamics set off by stagnation in Italy did not materialize in Japan.[41] Still, the Japanese situation is not to be envied, even as it is self-assessed by the Japanese. In the OECD Better Life Index, the country ranks even lower than Italy in subjective life satisfaction: thirty-second out of forty countries.

Japan, it turns out, is both a bad parallel with Italy and a poor benchmark for degrowth advocates. It might be useful to someone willing to argue that population is what should be shrinking at a steady clip in rich countries, rather than their economies—but going down this path would create

Table 4.1 Growth rates of GDP and GDP per hour worked: Italy, Japan, and
OECD average

	Real GDP				
	1995–1999	2000–2004	2005–2009	2010–2014	2015–2019
Italy	1.3	0.8	−0.6	−0.8	0.9
Japan	0.6	0.9	−0.7	0.8	0.7
OECD	2.8	1.8	0.6	1.5	1.7
	Real GDP per hour worked				
	1995–1999	2000–2004	2005–2009	2010–2014	2015–2019
Italy	1.8	1.5	−0.1	0.7	0.7
Japan	5.9	4.1	4.2	1.8	2.8
OECD	. . .	1.6	0.7	0.7	0.8

Data source: OECD.
Notes: All figures computed in constant prices and converted to US dollars using 2015
purchasing power parity. Statistics on real GDP per hour worked are available for the
OECD average only from 2000 onward.

many other challenges from a policy standpoint. The advocate would have
to explain how population control could be made compatible with basic individual freedoms in a liberal democracy, and how it could be achieved quickly
enough to mitigate climate change, especially in countries like France and
the United States where population is still expanding.

Before moving on to a new chapter, it is worth pausing to reflect on what
we learned from the Italian experience. Readers from other rich countries
have no doubt noticed similarities between forces at play in Italy and their
own national socioeconomic dynamics, albeit to lesser degrees.

We started with a set of paradoxes, and we will conclude with one last
one. Several of the current grievances against capitalism and the growth
imperative identified by ecosocialists and others are actually problems that
result from a slowdown (or, in Italy's case, complete lack) of growth. Degrowth advocates got a first taste of this from one of their own prophets.
Thomas Piketty's argument for why wealth inequality is increasing and the
rate of return on capital (r) is higher than economic growth (g) is actually

that there has been a slowdown in the latter. As we learned from the Italian case, this same logic can be extended to a variety of dimensions. If you think your current society is already fractionalized, litigious, and gridlocked, having less or no economic growth will make the problem worse. In this respect, note that what matters is scarcity vis-à-vis your neighboring countries or your own past trend, not some absolute monetary (or moral) threshold. By absolute measures, Italy remains one of the largest economies in the world, and if we use the entire global distribution of income as a benchmark, its citizens are rich. But this simple fact does not protect the country from the toxic political and economic dynamics outlined above. The reasons for this will become more clear in Chapter 5.

This book's Introduction laid out other grievances against the current economic order, including the plights of so many young people: they do not have enough opportunities, they are underemployed, and home ownership is nothing but an impossible dream to them. They are forced to seek help from their parents, and the government does not allocate enough resources to help them out. These are also problems that an absence of economic growth will make even worse. More broadly, if you think there should be more redistribution, then you should hope for the economic growth that makes that politically easier to achieve. When the economy enters into a steady state and faith in progress wanes, every group in society becomes highly risk averse, and assigns top priority to the defense of its present entitlements. It might very well be, as Galbraith remarked, that economic growth emerged as a powerful substitute to redistribution, as it eases social tensions. But at the same time, the relationship is not symmetric: redistribution is not a substitute for growth. Redistribution alone, in an environment of no growth, cannot serve the function of easing social tensions. It is instead bound to exacerbate them.

If you think tackling climate change and environmental degradation will inevitably require an acceleration in innovation and new cutting-edge technologies, then you must see that a lack of economic growth will make that harder to achieve. Investment is choked off in times of scarcity, and when the most inventive and high-skilled members of a workforce look around and see poor reward and no opportunities, they progressively move away. Finally, you may think that climate change and environmental protection is a shared problem of humanity, requiring international collaboration, long-term planning, and a willingness to make some short-term sacrifices for benefits that will accrue to future generations. If so, you should also know that without growth all of this is less likely.

Whatever your primary concern—social, economic, political, or environmental—lack of economic growth can hardly be the answer. Conversely, in light of the toxic dynamics it prevents, economic growth should be seen as a force for good.

II

The Green Plan to Avoid Disaster

5

THE TRUE ORIGINS OF THE
GROWTH IMPERATIVE

Human nature will be the last part of Nature to surrender to Man.
—C. S. LEWIS, 1943

It's an evergreen joke among Italians that, whenever in doubt on whatever the issue, one should always blame the government. Even for bad weather, the response applies: *Piove? Governo ladro!* But there is another option. It is just as easy to blame capitalism.

As Francesco Boldizzoni has shown, the charges leveled against capitalism are as old as capitalism itself. Whether the problem is poverty or excessive affluence, imperialism or delocalization, globalization, war, human rights infringement, inner-city poverty, lack of affordable housing, climate change, environmental degradation, loss of morality, long working hours, misogyny, white supremacy, or some other wrong, fingers inevitably point to the usual culprit. The push for "unnecessary economic growth" is no exception, as degrowth sympathizers see behind it the evil juggernaut of capitalism. Blaming something on an abstract concept, or on its vaguely identified supporters ("capitalists") can surely feel liberating, in the way that attacking a generic "establishment" does. It allows the blame to shift outward, releasing individuals from personal responsibility. In the words of degrowth guru Jason Hickel, "People are victims of the system."[1]

At the same time, such an approach is highly unhelpful, because it prevents engagement in deeper inquiry. We saw when we delved into the mechanics of capitalism in Chapter 2 that, while there are some inbuilt features such as competition among firms which feed the growth imperative, our economic system could in principle be just as compatible with zero growth if this were what consumers desired. Along similar lines, we saw in

Chapter 3 how ecosocialism could in principle be made to work if a "less is more" ideology were embraced at once by the entire population of a country (or, ideally, all of humanity).

Typically, however, such lines of thinking are not developed very far before the critics of capitalism throw up their hands in helpless gestures. Going back to square one, capitalism is portrayed as the puppet master forcing human beings to act in ways they would not choose, for example by generating "artificial scarcity" to drive levels of consumption they do not even enjoy. Consumers are presented as behaving contrary to the wishes of humans, despite these being the same population.[2] Even John Kenneth Galbraith, fine economist that he was, fell back on this defenseless posture, arguing that people kept consuming more only because they were victims of advertising. And capitalism has more arrows than advertising in its quiver. From spurring shoppers with sales promotions to conjuring up new gift-giving traditions, it has untold ways to generate unnecessary purchases. Worst of all, perhaps, is that prime suspect: planned obsolescence.[3]

To the contrary, I see pervasive advertising and planned obsolescence—but also marketing and branding more generally—as practices that firms engage in to *escape* the mechanics of capitalism and the tough logic of innovation-based competition.[4] They are attempts to win some reprieve from the sentence to "innovate or die," and while important, their effects are circumscribed.[5] Rather than the evil tools of capitalism, they are, at worst, by-products of a system in which firms discover that product and process innovation are hard to accomplish. In any case, they are in no way central to capitalism's survival. Indeed, if we look at degrowth advocates' favorite examples of limiting these practices, such as the Norwegian ban on advertising to children under twelve, or Paris's ban on advertising near schools, it is notable that they all come from advanced capitalist countries.

This leaves us wondering: If these practices aren't the origin of the infinite desire for more, then where does it come from? As we shall see, we've had the answer under our noses all along.

Before we start, it is important to stress that reaching useful conclusions will not require us to render a verdict on human nature—whether it is good or bad, and what it should be—as Rutger Bregman does in *Humankind*.[6] We can leave that perpetual philosophical debate to others. Rather, our focus here is on some basic anthropological, cultural, and evolutionary traits observable in humans and how these relate to the growth imperative. If we can understand these we can avoid the risk of creating a dysfunctional society with policies that make assumptions about human morality and capacity for moral improvement—always dangerous to rely on, even in service of such a noble cause as avoiding climate catastrophe. In Chapter 6, we will build on the understanding we gain to explore how human traits and impulses could be deliberately leveraged (rather than combatted) to achieve an environmentally sustainable society.

Needs versus wants: A never-ending story

To start our analysis of the quest for more at a fairly abstract level, we might observe that everything begins with a fundamental dichotomy: *needs* and *wants*. This is one of those divisions of the world into categories that make intuitive sense, until we are asked to define them more formally.[7] What constitutes a capricious want versus an absolute need becomes an exercise in "you know it when you see it." Suspicion, therefore, should immediately arise because we know that any particular proponent of serving needs as opposed to wants has a somewhat different idea of what belongs on the list. Usually, it would feature a combination of health, food, and shelter provisions.[8] Most would specify clothing and some would add education, extend the health category from physical to mental care, or acknowledge some need for leisure. Such lists can seem quite arbitrary, and abstract, too. What, for example, does *shelter* imply? Does a roof over one's head fulfill this need? In that case, can we look at Dharavi—the slum of Mumbai where people endure overcrowding, faulty sanitation, and highly toxic fumes from cookstoves—and say their needs for shelter are met?

The best definition available, invoked by John Maynard Keynes in "Economic Possibilities for Our Grandchildren," is based on the insight that consumption brings joy through two channels: inherent and external. The inherent (or functional) component is the utility you gain from a specific object, irrespective of whether others are doing the same or not. Think of the toaster grilling the bread for your morning breakfast, making it warm and crunchy. It doesn't really matter to you whether your neighbors have one, too. The external (or nonfunctional) component is the utility gained from other properties of the good. For example, the joy of wearing a $10,000 watch does not derive solely from its inherent ability to tell the time. It comes largely with its external utility, as it signals wealth and social status to others, displays one's particular taste, serves as evidence of belonging to a specific group (the upper class), and perhaps even improves mating odds.

Writing in 1899, economist and sociologist Thorstein Veblen explored the extent and origins of the nonfunctional component, which he labeled *conspicuous consumption*.[9] In *The Theory of the Leisure Class,* he describes with repugnance how the new bourgeoisie produced by the Second Industrial Revolution engage in consuming certain goods only to display social power and prestige. Note that Veblen is describing what would later be derided as *consumerism*—fulfilling non-material needs with material goods—but doing so many decades before the advent of the so-called consumerist society after World War II.

Conventional wisdom, along with Veblen, sees items responding to non-functional needs as indulging capricious wants, even lowly psychological quirks. There is a logic to this. Conspicuous consumption derives its utility in relationship to other people, most notably in opposition to them. If I signal status for myself, I implicitly signal to others that they are less worthy.[10] This clearly reflects immoral sentiments such as greed and jealousy. To engage in consumerism is to play a zero-sum game, where my gain results in someone else's loss. If everybody had expensive watches, their signaling function would be lost. This brings us to the logical corollary: if, somehow, all luxury watches were abolished, people would be in the exact same condition as if everybody had one.[11] This is a powerful principle, which does indeed suggest that goods having only, or predominantly, external nonfunctional utility should be diminished—and suggests it even more if we factor in their environmental cost. As with all goods, extra CO_2 is surely emitted as they are manufactured and shipped around the world.

So, why not ban luxury and conspicuous consumption altogether, as provocatively advocated by Thomas More in *Utopia?*

The innovation treadmill

It turns out that jamming this mechanism completely is not only infeasible but undesirable.[12] It is infeasible because of the blurriness of the line between inherent and external utility.[13] Take the luxury watch mentioned above. Wearing it could be a consumerist act, entailing a status claim, but it could also be an act of love and respect, making use of an inheritance from a much-missed grandfather. In that case, it carries value that is inherent, producing satisfaction quite independent of other people. The difficulty lies in the fact that value is in the eye of the beholder (and, in aggregate, in the eyes of society), not inherent in the object. Drawing a line between inherent and external utility would force us to divine the purpose behind individuals' purchases at each moment in time, and lead to very cumbersome, if not arbitrary, bans.

Banning conspicuous consumption would, however, also be undesirable because it is an important part of how new and better products spread through society, with the process of innovation driving the diffusion.[14] Typically, upper-class buyers favor expensive and scarce goods. But another class of products also serves their desire to signal status, and that is new and better solutions. The first automobiles, the first electric lightbulbs, the first phones, the first television sets, and the first smart devices were initially expensive to make and priced for the rich. As manufacturers attract enough early adopters among the wealthy—and the wider group attempting to emulate them—they can gain the scale advantages to make products faster and cheaper. This can set off a cycle by which production costs continue to drop, the innovative product continues to spread, and, progressively, it loses its nonfunctional, status-signaling value. At the end of the process, a new or better product is available to the wider society at an affordable cost. Meanwhile, the upper class has moved on to yet newer and better products, getting other cycles started.[15] It is this "innovation treadmill" that powers the rapid spread of technology across capitalist economies. The key is the corrective mechanism: over the long run, democratization of consumption erases the external utility of the shiny, new object, leaving only its inherent utility and function.

The power of this mechanism is that it harnesses and channels some deep human tendencies—starting with the desire to stand out, but also to fit in. Both relate to the process of identity-building, and while they are psychological rather than material needs they are very real ones.[16] Regarding status-signaling, neurological studies of both animal and human subjects show that it produces serotonin, neuron firing, and blood oxygenation in the brain. The emulation of success and prestige is likewise deeply wired within our species. In *The Secret of Our Success*, evolutionary biologist and anthropologist Joseph Henrich combines a host of multidisciplinary evidence, from lab experiments to ethnographic studies across the world, to argue that imitation based on certain cues is a cultural evolutionary trait. Emulating the behavior of those most successful within the community accelerated the accumulation of human collective knowledge. This, in his view, is the secret behind human success vis-à-vis other species on the planet.

The effect of the innovation treadmill is that yesterday's crazy consumerism chasing "unnecessary things" yields today's basic needs.[17] This can be seen in the poverty indicators typically used by rich countries. In the European Union, for example, a household is classified as severely materially deprived if it lacks a color television, telephone, washing machine, or car, or if the family cannot afford a week-long holiday away from home. Seventy years ago, all of these would have been out of reach to all but the very upper class.

More broadly, needs come and go: a television set to my grandparents was not a need but a want. It became a need to my parents' generation—only to become, within two generations, something no longer important in an era of Internet access and large monitors. This logic could be extended to all sorts of goods. Is having a toilet in a house a need? What about running water? What about warm water? Or a shower or bathtub? Or an oven? For the majority of human history these were not available except to the extremely wealthy, but in any advanced economy today it would be impossible if not illegal to rent out an apartment without all these features.

The skeptical reader might object that the items above are still desiderata, given that they are not matters of life or death. The innovation treadmill applies to medical treatment, too. In 1836, Nathan Mayer Rothschild, arguably the richest man in the world at the time, died at fifty-eight for want of an antibiotic to treat an infection. In 1928, Alexander Fleming discovered penicillin, which made it to the market following medical trials in 1942. In the entire United States, there was enough to treat fewer than a hundred patients. By September 1943, there was enough to satisfy the demand of the Allied Armed Forces during the Second World War, at today's equivalent of $300 per dose. By 1950, production had increased to a point that it was available to the American population at four cents a dose: one-sixteenth the price of a gallon of milk.[18] By now, lacking access to antibiotics would constitute severe material deprivation, violating basic needs.

Note that the mechanics of the innovation treadmill extend beyond the gizmos of modern technology. An interesting botanical example is the lemon, an iconic symbol of the Mediterranean landscape and its bounty. Recent research carried out at the University of Tel Aviv, based on a detailed analysis of ancient texts, art, artifacts, and archaeobotanical remains, shows that lemon fruits originated in Southeast Asia.[19] Even during the Roman era, they were considered a luxury product due to their supposed healing qualities, symbolic use, pleasant odor, and rarity. But they were of scant culinary use due to their thick rind and dry, tasteless flesh. It was only after their spread was powered by their initial representation as a high-status, elite product of limited functional use that they started to be cultivated through ancient technology and know-how. Thanks to hybridization, varieties displaying fewer seeds, thinner skin, and more juice were produced. Now, a pound of them can be bought for less than two dollars in some US markets.

Over two centuries ago, Adam Smith was already describing the dynamics of the process by which technologies and products democratize, desires of the upper classes become basic needs of society, and lack of access to them comes to imply poverty. In his words:

> A linen shirt, for example, is, strictly speaking, not a necessary of life. The Greeks and Romans lived, I suppose, very comfortably though they had no linen. But in the present times, through the greater part of Europe, a creditable day-laborer would be ashamed to appear in public without a linen shirt, the want of which would be supposed to denote that disgraceful degree of poverty which, it is presumed, nobody can well fall into without extreme bad conduct. Custom, in the same manner, had rendered leather shoes a necessary of life in England. The poorest creditable person of either sex would be ashamed to appear in public without them.[20]

Smith was much more sophisticated in his thinking than many know, given his frequent association with simplistic "wealth maximization." He understood well the social nature of the relationships among commodities, affluence, and the achievement of basic living conditions.

This view aligns with more recent work by Nobel laureate Amartya Sen, who in the 1980s revolutionized the field of economic development. In *Development as Freedom,* Sen argued for moving away from standard emphases on earnings, opulence, utility, and ownership of specific commodities, to focus instead on *capabilities* to do or be what you'd like (the concept of self-realization).[21] This perspective turns the fight against poverty into a fight for (positive) freedoms, including freedom from hunger and short life, freedom to take part in the life of the community, and, building on Smith's observation above, freedom from shame or disgrace. While this might sound very philosophical, note what it implies. The same capabilities require more expensive goods and services in a society that is richer, and in which most people have means of transport, affluent clothing, radios, and televisions. Sen notes that richer societies progressively revise upward the capabilities accepted as "minimum."[22]

It becomes clear that there is no set of needs that can be called absolute or universal across time and geography—and thus there can be no quantitative threshold established beyond which further income or economic growth is no longer needed because it would only be funneled into unnecessary wants. Once we get rid of the artificial dichotomy between needs and wants, we see that need is, and always has been, relative and that it evolves as the benefits of growth and innovation spread through a society.[23] It is true that citizens in rich countries have much that others lack, but to the extent that these represent new solutions to old problems they open space for new needs.[24] We may live much longer and healthier lives than our ancestors, but we continue to have problems. Cancer and cardiovascular diseases are now the top causes of death in most advanced economies.[25] The fact that we live longer means that more people are afflicted by dementia. There is plenty of suffering still, from eyesight that cannot be restored to injuries that put victims in wheelchairs for life. What is possible determines what is needed.

The fact that needs are relative applies not only to health but to virtually every dimension of life, including leisure. Today, expressing a wish to visit another continent would not strike anyone as an eccentric or unreasonable desire. Five hundred years ago, it would have been received in the same way as someone today wishing to go to the moon. And even that is losing its capacity to surprise, given that 2021 saw the first "space tourists" in human history. It is not unimaginable that, across a few decades, space tourism will move from being impossible, to being unaffordable to nearly all, to becoming available to many.

Radical relativism

Is it possible that the argument here has turned too relativistic? After all, some things must be absolute needs, and this radical relativism might justify too many eccentric and unnecessary behaviors. For example, good health is absolute and everybody needs it, whereas extravagant entertainment involving, say, $2,000 bottles of vodka at a nightclub, is a waste of money—right? Well, let us unpack this statement. To begin with, what does *good health* mean? Or what does *long life* mean? Long with respect to what? In the prehistoric age, someone who lived thirty years had a long life. Making it to the age of sixty in Ancient Rome was considered a very good achievement. When that happens now in rich countries, we mourn someone who "left us too soon."

Likewise, different types of consumption, leisure, and entertainment are more or less extravagant in different eras. To enjoy tropical fruits such as pineapples and bananas in Europe was considered a great luxury as late as seventy years ago, reserved for special celebrations like Christmas. In today's world, my partner has a banana every morning with her breakfast. The same could be said of sugar, which is currently a staple, but was precious in the sixteenth century—even sculpted into table ornaments for aristocratic banquets to display status and wealth.[26] Ice could be obtained in the Persian Empire in 400 BCE, but the logistics involved in moving and storing it (in Yakhchāl ice houses) were so challenging that it served as an extreme display of wealth and technological prowess. Nowadays, of course, no bartender would hesitate to add it to your drink free of charge.

What about the rampant spending done only to show off, with no inherent value? As my mother has scolded, "your generation does things only so they can post them on Instagram!" Her example is a specific perversion of our era, but nonfunctional consumption has always been a habit of our species, responding to innate desires for status, prestige, sense of identity, and belonging. Even Veblen's *The Theory of the Leisure Class,* in the course of critiquing conspicuous consumption, traces it all the way back to tribal times, the birth of agriculture, and the early division of labor. Long before others took up the topic, Veblen recognized the evolutionary origins of cultural traits such as emulation and predation. Today's vodka in the club springs from the same impulse as yesterday's sugar statue at the banquet, or the conspicuous display of lemons in Ancient Rome.

Of course, one could argue that while the impulse has always been there, capitalism, through consumerism, has created a new situation in which goods are updated and replaced at an unreasonably fast pace, thanks to perverse inventions like fashion and branding. Clearly, this would be particularly problematic in a world suffering from climate change and environmental degradation. Here the picture is blurrier, but at least we can be sure there is nothing new about assigning external significance to objects.[27]

Anthropologists have identified symbolic, nonfunctional items in tombs of all cultures: rings and other jewelry, scepters of command, amulets, crowns, military decorations, special garments, and even things of specific colors.[28] Signaling status, prestige, and success is an important function, and it has always been done with goods. In Europe, in Asia, and in Africa, branding of products goes back thousands of years, originally devised as a way to guarantee the quality of the product.[29] For example, excavations in the ancient cities of Pompeii and Herculaneum have turned up a wide variety of goods bearing commercial inscriptions, or *tituli picti,* including wine jars, glassware, pots, ceramics, and even a type of fish sauce produced by *Umbricius Scaurus,* evidently much prized by Mediterranean diners of that day.

Fashion, grooming, and clothing respond to similar logics, powered by desires to participate in social collectivity, to communicate visually, to engage in self-expression, and to indulge in self-adornment. It is far from clear that capitalism has either created these needs or fostered a culture of high churn in a context that would otherwise see little change in clothing traditions. For example, historian Fernand Braudel documents that in Europe in the fifteenth century, grooming and clothing changed at such a fast pace that current art historians can detect the age of a painting within a five-year margin of error.[30] Romans are known to have changed fashion in clothing and hairstyle several times, in line with transitions of power. It does not seem, either, that this is a Western attribute. China during the Ming dynasty (1368–1644) is known to have experienced rapid changes in clothing fashion.

Veblen's concepts of emulation and predation also show up in architecture. An example comes from the Italian city of Bologna. In medieval times, families tried to build towers as high as possible to signal their power and wealth, as well as to serve some defensive and offensive purposes. It is estimated that, between 1200 and 1300, the city had perhaps a hundred towers, which must have made it look like a medieval version of Manhattan. By the thirteenth century, some had collapsed, while others were torn down or lowered in height for structural safety purposes. Yet those that remain are today of great artistic interest and serve as destinations for many tourists. More generally, most of the landmarks we venerate from the past, from the pyramids of Giza and the Colosseum in Rome to the Palace of Versailles and the Taj Mahal, go far beyond strict functional use. They were the conspicuous consumption of the past, aimed at displaying wealth, status, and power.

By now, two things should be evident. First, capitalism does not create needs we do not have, and its supposed instruments of darkness—advertising, fashion, and branding—predate it by millennia. Second, these techniques tap into deep-seated desires, rather than simply generating desires we don't have for stuff we don't need. Those who blame capitalism for such wrongs are barking up the wrong tree.

Freed from simplistic explanations that place the burden of all planetary evil on capitalism, we can start to understand why people behave the way they do. Faced with the crucial question of how to tackle climate change, trying to suppress cultural instincts that date back to our evolutionary beginnings is probably not the best course of action. It would be wiser to acknowledge these traits and then figure out how to leverage them to encourage behavior that is compatible with, or even instrumental in, protecting the planet.

The great moderation

A relative needs approach does not imply that anything goes. Nor should it be mistaken for a Panglossian belief that we already live in the best of all possible worlds. When even a staunch supporter like Milanovic, proclaiming capitalism as the system that rules the world, recognizes that hyper-consumerism has gone too far, we must acknowledge that something is fishy. Resentment today against the rich and their lives of excessive luxury is strong and widespread. How can we rationalize this?

Let us start with some historical perspective, recalling the constantly shifting borderline between needs and wants. Across time, people have objected at various points to excess and extravagance, and judged that consumption was spiraling out of control. Moderation, after all, is a key principle of Ancient Greek philosophy, as exemplified by Stoicism and Epicureanism, passed on to Roman moral philosophy, as evident in the writings of Seneca and Marcus Aurelius.[31] Likewise, it is a central tenet for various religious creeds, including Judaism, Christianity, Islam, Buddhism, Hinduism, Chinese Taoism, and Confucianism. As noted by philosopher Emrys Westacott in *The Wisdom of Frugality*, "for well over two thousand years frugality and simple living have been recommended and praised by people with a reputation for wisdom. Philosophers, prophets, saints, poets, culture critics, and just about anyone with a claim to the title of 'sage' seem generally to agree about this."[32]

At various turns in history, the feeling that this principle of moderation was being breached was especially acute. Many of the Macedonian generals led by Alexander the Great worried that the king was falling for the despicable excesses of the Persian tradition. Plato, Euripides, Rousseau, and Marx all came out against rapidly changing clothing fashions. Cato the Elder—a censor in Ancient Rome known for his frugality and moral integrity—inveighed against the excesses of his times. In most of these cases, the moralizers denounced behaviors which, to modern eyes, seem rather innocuous. For example, Cato—a man described by Plutarch as content with a cold breakfast, frugal dinners, and the simplest clothing—adamantly opposed "degenerate" Hellenism (meaning Greek philosophy, art, culture, and cuisine) as a threat to the rugged simplicity of the Roman tradition. He introduced luxury taxes on personal adornments and dresses, especially for women, and set limits on the number of guests at an entertainment event. Clearly, he failed, as Hellenism became prevalent—and made Ancient Rome the refined civilization we revere to this day.[33]

If conspicuous consumption has recurred throughout history, and reliably been chastised by people of those times, is there a pattern to be discerned across those episodes? A clue can be found in the workings of the innovation treadmill, where a glitch can slow down the democratization of consumption. If an expensive novelty does not spread to a wider population, or until it does, it continues to be perceived as mere luxury and showy display of entrenched wealth and inequality.[34] The period we live in marks no exception. As we saw in the Introduction, socioeconomic inequality is on the rise, particularly in some advanced economies, and it is not only a matter of perceptions. Studies show that, when inequality goes up, the upper class is more eager to engage in pure conspicuous consumption to signal their belonging to the "winners crowd."[35]

When the gaps between the rich, the middle class, and the poor expand, essentially separating them into compartmentalized chambers, the democratization of consumption and the spread of new products are greatly hampered.[36] In that case, it cannot be assumed that the fancy consumption of the upper class today will turn into the widely available products of tomorrow. Instead, this kind of process may only happen within the silos, so that the passage of years will simply cause the rich to be more and more detached in their habits from the rest of society.

If the objective is to reduce conspicuous consumption without hampering the process of innovation, the policy recipe calls for the same ingredients that would go into any serious effort to reduce inequality; it would ensure that the benefits of economic growth were spread broadly through society and didn't accrue only to the top.[37] As the rich, by virtue of their disproportionate consumption, also carry the heaviest environmental burden, any success in reducing today's striking inequalities would also take some pressure off the environment. Indeed, Lucas Chancel, drawing on a vast set of microeconomic studies, has been able to establish that environmental policy and inequality are deeply intertwined.[38]

To make a long story short, while calls for abandoning growth are misguided, it is of paramount urgency and importance to ensure that the benefits of growth are more widely spread. If policymakers could accomplish this, they would simultaneously relieve many other negative side effects of the current system, including its high conspicuous consumption and, at least to some extent, the large environmental footprint of the upper class.

Morality under capitalism

Luxury and excess can be curtailed top-down by government policy measures, especially when they reduce sharp inequalities—but is there a role for morality, too? In mainstream economics, we tend to shy away from this topic, just as we avoid other normative judgments of consumer preferences. And yet, it is interesting to note that even the founding father of classical economics and proponent of the invisible hand, Adam Smith, stressed the importance of morality in *The Theory of Moral Sentiments*. To Max Weber, Protestantism was central to capitalism's success because, among other things, it imposed strong moral restraints on the upper class. Keynes believed that the stability of the social pact hinged on the moral acceptance by the rich that they would reinvest most of their gained income in society, rather than spending it on mere consumption, as feudal lords did in the past. Reflections on the interplay between capitalism and morality also show up in the work of Schumpeter and Hayek. More recently, Milanovic recognizes that amorality is becoming an increasing source of tension in the capitalist system.[39]

At a minimum, in my view, upholding economic morality under capitalism boils down to having a broad perspective on the community, and recognizing the implicit covenant that keeps the societal fabric from fraying.[40] These considerations are always tightly interwoven with the overall condition of society at a specific moment in time, and therefore with prevalent socioeconomic inequalities.

Building on the work of Charles Darwin, evolutionary biologist David Sloan Wilson argues that what we commonly refer to as *morality* is strikingly similar across societies: it always includes honesty, bravery, unselfishness, loyalty, and kindness, to mention just a few qualities. He concludes in *This View of Life* that our moral psychology is yet another product of cultural evolution, which favored collaboration by early human beings, paving the way for our success as a species. Consider the *aurea mediocritas,* or golden mean—the moral principle that one should avoid extremes and tread a middle way, as between excess and deficiency. It is not hard to imagine that this guiding rule came about as an evolutionary cultural trait that made sociable living viable on a large scale. The fact that similar moral principles emphasizing virtues of moderation arose in far-flung civilizations hints that this may be so.[41]

At least some moderation can be found in the behavior of a subset of extremely wealthy individuals, who realize society gave much to them and therefore choose to give back abundantly. Some also refrain from certain offensively conspicuous forms of consumption, even though they have the means to indulge in them.[42] Clearly, this voluntary restraint is no substitute for government-led redistribution.[43] To the degree that economic morality is prevalent among the upper class, however, some lesser amount of government-led corrective intervention may be needed. Conversely, any lack of morality makes greater and more sustained redistribution inevitable, with all the social conflict this might entail.

Relative income theory

By this point, we have cleared away the "needs versus wants" smokescreen, and moved to a relative needs theory. Unsurprisingly, this spills over into considerations of income, which in the same way becomes relative. Relative income theory posits that the satisfaction a person gains from a certain income is not absolute, but depends on a comparator. Is $2,000 a month a little or a lot? As with most questions in economics, the answer is *it depends*. This opens up an entire debate on what the comparator is. The most promising empirical explanations combine two broadly stated insights. First, people tend to look to the past for comparison: are they earning more than they used to?[44] Second, they look to a contemporaneous "reference group," usually made up of family, friends, neighbors, coworkers, and people of the same age cohort, to make a social comparison.[45] The latter speaks to the idea of, in popular parlance, "keeping up with the Joneses."

Suppose you were asked whether you would prefer to live in state A, where the average salary in society is $25,000 but your yearly income would be $50,000, or in state B, where others on average earn $200,000 but your income would be $100,000. Goods, you are assured, cost the same in both places. What would you answer? Those championing radical egalitarianism might be appalled, but again and again surveys and experiments by both neurologists and behavioral economists have found that people would prefer to live in a society where they earned more than others, even if this meant earning less in absolute terms.[46] The skeptical reader might accuse capitalism and commodification of having created this relativity monster but, once again, we should recognize it as more of a human feature, predating capitalism and industrialization by far and probably as old as human history.[47]

Some evidence of this can be found in the warnings that religions and life philosophies have issued for millennia against placing high priority on comparisons to others. After all, envy is a universally disdained sin. Catholicism came the closest to condemning an exact formulation of relative income theory. In the Gospel of Matthew, Jesus admonishes against interpersonal comparisons, teaching a parable about a group of day laborers who were angry to see others toiling less yet being handed the same pay. Specifically: "And on receiving it they grumbled against the landowner, saying, 'These last ones worked only one hour, and you have made them equal to us, who bore the day's burden and the heat.' He said to one of them in reply, 'My friend, I am not cheating you. Did you not agree with me for the usual daily wage? Take what is yours and go.'"[48]

Clearly, social comparisons were a tendency already two millennia ago. Small groups of people, self-isolating in temples and monasteries, might have managed to escape them, but while religious or philosophical admonitions might have done something to tame these human tendencies, they could not extinguish them altogether.

These relative preferences are not even limited to the human species. In 2012, a YouTube video featuring an experiment with capuchin monkeys went viral. Primatologist Frans de Waal filmed a monkey who was perfectly happy to receive pieces of cucumber until it observed a neighbor being given grapes. Seemingly offended by the injustice, the monkey tossed away the next cucumber offered as an inferior good.[49] Even monkey preferences are reference-dependent.

Once we accept relative income theory, we are able to make sense of several economic and political dynamics. For example, these comparisons to the past-self (inter-temporal) and to a reference group (social, or inter-personal) are why the precept is widely accepted that growth serves as a substitute for redistribution. When there is steady economic and income growth, people focus somewhat less on reference groups, including upper classes, and somewhat more on whether they are doing better than in the past.[50]

The implications of relative income theory can also serve as a framework of analysis for broad historical phenomena. Economic growth responds to the logic of intertemporal comparison, while redistribution responds to concerns stemming from interpersonal comparison. The two are clearly tightly intertwined, as sketched in Figure 5.1.

With a matrix defined by low and high levels of, on one axis, growth, and on the other axis, redistribution, the diagram is subdivided into four quadrants. In the top-left corner, we have a situation like early-stage capitalism. Economic growth was high but, with limited redistribution, interpersonal inequality was huge and accelerating.[51] This is the dismal reality described by economist David Ricardo in his *Principles of Political Economy and Taxation,* where wealth was increasingly concentrated in the hands of those owning the means of production.[52] We saw in Chapter 2 that this system would have collapsed on itself, as foretold by Karl Marx, had it not responded by successfully increasing redistribution. It did this by creating both regulations (antitrust laws) to restrict the excesses of accumulation and a welfare system. This pushed the system into the top-right corner, toward a more sustainable equilibrium, characterized by a combination of growth, redistribution, and shared prosperity, of the type seen in most advanced democratic capitalist economies, especially in the aftermath of World War II.

In the lower half of Figure 5.1, we have scenarios of limited or no growth. The bottom-right corner refers to the case where overall resources are fixed and redistribution is large, albeit likely to be nonconsensual. This situation implies oppression and value extraction by the most powerful group (for

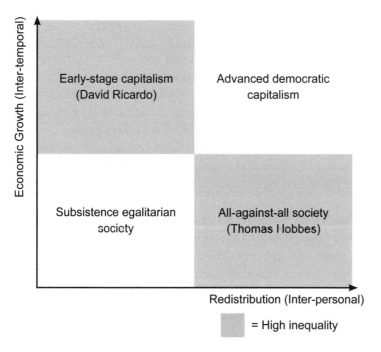

FIGURE 5.1: Four possible scenarios for societal equilibrium

Note: Focusing on two important variables—a society's level of economic growth and its level of redistribution—yields a matrix of possibilities. In advanced democratic capitalist systems, both are relatively high.

example, the aristocracy) over another (for example, commoners), and is likely to spiral into social unrest and a Hobbesian, all-against-all scenario of clashes among groups. When resources are limited, redistribution can be high, but not as we normally intend it—meaning peacefully from the rich to the poor—but rather, violently from those with less power to those with more. Also, constant war and pillage of neighboring societies to increase appropriation of an overall fixed set of resources could be framed as high (nonconsensual) redistribution. As such, the bottom-right corner can be seen as characterizing the unstable equilibrium of feudal systems or medieval Europe, and possibly most of preindustrial human history—defined by "continual fear, and danger of violent death" and where life, in Hobbes's words, is "solitary, poor, nasty, brutish, and short."[53] We saw in previous chapters that this is neither a desirable nor a stable situation. It leads to a constant struggle for power, illiberalism, and political tension within and among societies.

Finally, in the bottom-left corner, we have the society that ecosocialist advocates aspire to, with no economic growth but also limited need for re-distribution in a post-capitalist, egalitarian system. Note that all the various small subsistence communities showing the way to an environmentally sustainable society—like the Ju/'hoansi Bushmen community in Namibia that James Suzman held up as an example, and Jason Hickel's case of the Achuar in the Amazon Forest—would fall into this category.

This brings us to a question: Under what conditions is the bottom-left quadrant an equilibrium and therefore capable of being socially sustained? Recalling relative needs and income theory, three elements come to mind.

- *Isolation.* Due to relative income theory, interpersonal comparisons extend to neighboring countries or tribes. In the absence of limited to no contact with other groups, the mechanics of "keeping up with the Joneses" kick in, meaning that eventually there will be an incentive either to accelerate growth (moving upward on Figure 5.1) or to enter into competition, including war and plunder (moving to the right).

- *Limited size.* To act independently of the price mechanism, and to build instead on extended sharing and trust, requires reciprocal monitoring, which can take place only in relatively small, tight communities. Moreover, to retain the egalitarian character of the system, as noted by Veblen, no great specialization can be allowed; it would open the path to task diversification, formalized exchanges, and eventually, inequality. By extension, agriculture should be limited, perhaps even absent, as it leads to private property. In turn, only a small population can be realistically sustained under these unspecialized production arrangements.

- *Limited innovation and knowledge accumulation.* Even in a subsis-tence society, there are relative needs which can in principle be solved by technology—for example, to save children from infant mortality, or to respond to unexpected shocks like cold winters and bad harvests. Indeed, every community has its own environment-specific know-how, as extensively recorded by the anthropological literature—from the Inuit, who know how to survive in a cold and harsh hunting environment, to the indigenous populations of Australia.[54] But there must be barriers to sustained innovation, which would create a growing economy and a society of increasing complexity, clashing with the requirement noted above to remain small and unspecialized. On the demand side, one such barrier could be a pre-scientific value system, whereby events are seen as the result of whims of the gods or nature, not governed by deterministic laws which can be twisted by human action. On the supply side, a barrier could be a strictly oral or limited writing tradition, to limit the accumulation of know-how.[55]

Clearly, these social sustainability criteria can help us understand the case studies brought forward by ecosocialist advocates, but they also challenge whether these cases are of much relevance to those in advanced societies searching for environmentally sustainable models. This is because the starting point is completely different. Modern societies have citizens in the tens and hundreds of millions, living in close proximity with neighbors in a world highly interconnected by travel and communications technology. Most crucially, they have embraced a scientific-method mentality. For them, the genie is out of the bottle: they know that something could be done to solve a current (relative) need, and that they are not at the mercy of sheer bad luck or destiny, so they will not resist the pulsion to do something about the situation. The desire for human self-determination will keep igniting and propelling the engine of innovation, growth, and progress.

Working hours

In Chapter 1, we noted a part of the standard degrowth critique of society: its attack on the persistence of long working hours despite large increases in GDP. Recall that Keynes, who in 1930 almost perfectly predicted the increase in productivity over the century to come, wildly overestimated how this would yield reductions in working hours. This has led to abstract accusations that capitalism has forced people to work more than they would like.

Let us first get the facts straight. Working hours have been on a steady decline over the past two hundred years in industrialized economies, and the countries with the lowest working hours in the world are practically all high-income economies. But only by embracing relative needs and income theory can we make sense of Keynes's failed prediction. Only with reference to it can we answer the question: Why do people continue working as long as they do, in societies such as the United States where real GDP per capita has increased six times over since 1950?

As it turns out, a strong driving force behind longer-than-expected working hours in rich countries, despite their high aggregate income levels, is concern for status. Keynes acknowledged that there is functional and nonfunctional utility derived from consumption, but he fell prey to the usual fallacy of assuming that needs are absolute and satiable, rather than relative and shifting. As a consequence, he underestimated how far individuals would continue down the path of their quest for more. This is something that was crystal clear to Adam Smith, who asked in 1759, "To what purpose is all the toil and bustle of this world?" and quickly provided the answer: "It is our vanity which urges us on." En passant, this also explains the paradox presented by Milanovic that, in today's society, people from wealthy backgrounds, who in principle could sit back and enjoy their inheritance, nonetheless tend to work long hours in highly competitive sectors such as consulting, venture capital, big tech, and finance. The job is a carrier of status, too.

Working hours do decrease as resources available to society increase. People do take advantage of the extra resources made available by economic growth in terms of more leisure, at the individual and societal level. They do so, however, at a slower pace than we would expect. And this is due to pressures to earn status and keep up with their peers.

Capitalism and the quest for more

The main topic of this book is the relationship between capitalism and the growth imperative, and this is an essential relationship to understand as we consider a societal rethink to address the challenges posed by climate change and environmental degradation. The point that must be accepted is that economic growth is a result of human desires, rather than a top-down imposition by capitalism on otherwise indifferent individuals. The quest for more will endure as long as humans populate the planet, whether we tear down capitalism or not. It is best to acknowledge this and craft our sustainability strategies accordingly.

While we have looked at the origins of the growth imperative from a more individual, or microeconomic, perspective, it is also possible to do so at the country level. Skeptics will surely point to historical estimates of GDP,

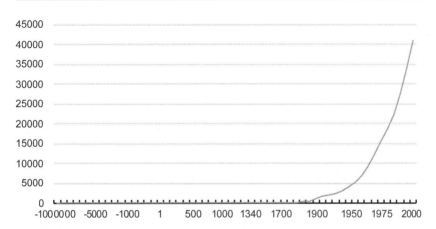

FIGURE 5.2: World real GDP, long-term estimates

Data source: De Long (1998).

Note: In billions of 1990 international dollars.

such as the ones displayed in Figure 5.2, to claim that capitalism and in-dustrialization brought a growth-obsession phase to an economic history previously characterized by simple stability. Such an interpretation of the data is understandable, but fallacious. Aggregate economic growth has ac-companied humankind throughout its history, but it accelerated enormously from the Industrial Revolution onward thanks to certain conditions: the Scientific Revolution, better protection of property rights, presence of capital, an extensive banking system, high wages in England, and more.

To understand a hockey-stick chart like Figure 5.2, one should appreciate that behind the astounding growth since the Industrial Revolution is a pre-vious economic history that has been obscured by it.[56] More important than the calculation of how much economic growth was realized before that point should be the recognition that the quest for more—the growth imperative—was in place. Aggregate growth could materialize only to a limited extent, as new technologies were developed and spread at a slower pace (and in some cases, were forgotten altogether when their inventor died or a civilization col-lapsed).[57] But that does not mean that early humans did not crave it.

In *The Measure of Civilization*, Ian Morris offers an index to gauge a so-ciety's capacity to harness energy. It combines measures of not only food (for humans and their livestock), but also fuel (for cooking, heating, cooling, firing kilns and furnaces, and powering machines, and including wind and

waterpower as well as wood, coal, oil, gas, and nuclear power) and raw materials (for construction, metalwork, pot-making, clothing, and any other
purposes). Using this comprehensive, standardized index, Morris shows
that energy-capture capacity (and the technology to enable it) rose steadily
from its baseline of four to five kilocalories per person per day in a typical
pre-agrarian, hunter-gatherer society, accounting for food, cooking, and
little else. By the peak of the Roman Empire, it had reached over thirty
kilocalories.[58]

Advancing energy capture and technologies could sustain an increasing
population. At the onset of the Agricultural Revolution, ten thousand years
ago, the human population of the planet was roughly four million. This was
about what you would expect for a medium-sized mammal like *homo sa-
piens*. It reached two hundred million, however, by the time of Christ. This
was made possible by great enabling inventions like domestication and
breeding, language, metallurgy, agriculture, irrigation, cities, writing, money,
and commerce. Each time, more available resources spurred population
growth, which eventually led to a crisis of overpopulation. Each time, demographic pressure against the limits of available resources spurred inven-
tiveness.[59] Through constant struggles, the result has been a slow but steady
rise in population over the centuries.[60] Thus, human history can be interpreted as a constant struggle against limits imposed by nature, then shattered
by tinkering and ingenuity. While climate change, and environmental degradation more broadly, is in a different league in terms of its complexity and
scale, it would seem to belong to the same category of problems that humans
have faced throughout history.

Famine was not the only disaster holding population and economic
growth in check. There were also plagues, like the Black Death that killed
perhaps 200 million people in Europe in the mid-fourteenth century. And
there were wars, too. If one society managed to start accumulating, it quickly
attracted the attention of its neighbors, which saw violence as a legitimate
way to increase their resource availability. Some civilizations failed to sustain the complexity they achieved, or to respond to changing climate or
environment, and either collapsed or were conquered.[61] Still, it would be a
mistaken interpretation to see a patchy track record of achieving and sustaining broad-based economic and population growth as an historical absence of the quest for more.[62]

As evidenced by invasions and wars, relative income theory applies at the
country level, too. In a context of sustained interactions with other societies, economic comparisons are not limited to the past, but also extend to

one's reference group—in this case, neighboring countries. The broader issue here is that, in a geopolitical context of confrontations among countries, economic growth is highly linked with political clout and military expenditure. This generates a geopolitical growth imperative akin to the desire to keep up with the Joneses, even just to deter foreign interventions. As noted in Chapter 3, an instructive example comes from the experience of the Soviet Union, which was not built on a capitalist system, and so was free from the price mechanism, Schumpeterian competition between firms, and other in-built growth imperatives. It had not even adopted GDP as a metric to guide economic policy. Nonetheless, it was subject to a geopolitical growth imperative due to its Cold War with the United States, and the desire to protect and project its value system in what was then called the third world.

Over the course of history, some countries actively chose to withdraw from the iron law of continued economic growth and progress, and to avoid comparisons to others, by embracing self-isolation. In all cases, domestically, this was associated with sentiments of national superiority, insularity, and less innovation. When the curtain came up, these nations were revealed to be relatively poorer, and at the commercial and military mercy of others. Two examples stand out: Ming-dynasty China and Tokugawa Japan (see Appendix B).

Harking back to the opening of this chapter, capitalism is blamed for many ills, but it is better understood as a great enabler. It is an efficient way to foster growth. It rewards ingenuity. It prods people to invent practical things and add to the store of useful knowledge. When firms compete for customers, they have an incentive to master the latest technology and learn how to produce it as cheaply as possible. This appeals to consumers, who display all the emulation and status-signaling behaviors we have discussed. In short, capitalism is efficient at delivering what people want, and people seem, due to cultural-evolutionary traits, to constantly want more. This did not start with the Industrial Revolution. The quest for more has characterized most of human history, both in periods when growth was achieved, and when it was not.

Climate change and value systems

Not everyone finds examples drawn from the past to be persuasive. Today, thanks to advanced science, people know that current methods of production and consumption of goods and services are harmful to nature.

THE HAPPINESS PARADOX EXPLAINED

Degrowth advocates often point out that economic growth does not map neatly to changes in happiness as measured by various surveys—a lack of correlation known as the Easterlin paradox. Their rushed conclusion is that, at least in rich countries, dialing back economic growth would have limited, or perhaps even positive, impact on happiness and well-being.

We can resolve this apparent paradox with the help of relative needs and income theory, which emphasizes the role of social comparison. If everyone's income increases in a country, they may be materially better off but average happiness levels don't rise by much. Whatever joy results from the intertemporal comparison fades quickly because of another psychological effect: habituation. As people have more, they rapidly get used to it. What used to be desires are recast as needs, and people move on to wanting more. This is known as the "hedonic treadmill."

Like the ones in health clubs, the hedonic treadmill works in only one direction. Habituation is not symmetric; losses are harder to live with than gains. People's unhappiness at being deprived of something—say, a dollar of income from an initial reference point, is much greater than their happiness on receiving a windfall. The power of this cognitive bias, known as "loss aversion," has been documented by Daniel Kahneman (2013) and Amos Tversky in a vast variety of experimental settings—a line of work acknowledged by the 2002 Nobel Memorial Prize in Economics.

Even as it seems clear, however, that any happiness gains from economic growth are not direct but rather intermediated by relative income theory, there are still some facts worth keeping in mind:
- In advanced as well as developing countries, richer people are on average happier than poorer people.
- Econometric studies at the level of individuals find strong correlations between boosts in income and positive changes in happiness.
- Populations grow unhappier in times of recession.
- Studies spanning multiple economic cycles of rise and fall in inflation rates and GDP growth find fluctuations in happiness tracking with these ups and downs.

These basic patterns of human sentiment lend no support to a degrowth agenda, or any transition to a "less is more" paradigm. Particularly hard to sell would be directives for rich countries to reduce their consumption to open up space for poor countries to expand. The idea of forcing such change might make perfect sense from a moral perspective, but from the standpoint of relative income theory it will trigger loss aversion and social comparisons across countries—a recipe for widespread unhappiness.

This is especially true given that the reductions needed in rich countries would be huge. Calculations by degrowth advocate Peter Victor show that, to respect planetary boundaries at the global level and at the same time open space for poor countries to catch up, an advanced country like Canada would need to shrink its income per capita to half its 2005 level by 2035—leaving their people earning just a quarter of what they would have in a business-as-usual scenario (Victor, 2012). The same simulations can be used to show that a shift to no-growth would fail to result in sufficient CO_2 reduction, bringing it down only to roughly 80 percent of 2005 levels by 2035—and that a steady-state economy as a middle-ground solution would fail to achieve the climate targets set by the UN's Intergovernmental Panel on Climate Change.

Between basic arithmetic and foreseeable unpopularity, it is not surprising that very few mainstream parties worldwide have so far embraced a full-blown degrowth agenda. "Make America Small Again" just doesn't work as a campaign slogan—and nor does shrinking hold appeal in other nations.

Therefore, as the message spreads, and global protests such as Fridays for Future are noticed, people will simply stop wanting more, and will resist the pressures of relative income theory. Rationality will surely prevail over vague psychological instincts, will it not?

While the trend is surely positive, and we see younger generations enthusiastically embracing the environmental cause in ways their older siblings and parents did not, there are reasons to believe that a complete turnaround in preferences remains unlikely.[63] For a long time, the presumption in climate change communication has been that, once scientific evidence was made available, people would quickly act upon it. But being aware and concerned about climate change does not necessarily jolt individuals into action.

Over the past twenty years, as a large gap between knowledge, concern, and action developed, so did a wide literature on the topic. A review of over two decades of research, ranging from behavioral and environmental economics to cognitive sciences, social psychology, health policy, and marketing, finds many scholars trying to understand the cognitive traits that hold people back.[64] With regard to climate change specifically, the problem is that shared gains, often erroneously perceived as far in the future, are constantly weighed against personal, immediate costs. Given that, in our current technological state, CO_2 emissions originate from basically every aspect of modern lifestyles, most of the actions required to bring them down clash with elements that have been, until recently, at the core of our values, identity formation, and belief system.

These are the same human traits we saw above as we explored the non-functional utility of consumption. For example, for over half a century, it has been a status symbol to have a luxury car, and jet-setting around the globe has been a clear marker of success—think of the iconic James Bond with his deluxe Aston Martin and secret missions involving travel to exotic locations. It has been the tradition in many countries for people's diet to center on meat, a very CO_2-intensive form of nourishment. Such norms do not evolve quickly. No matter the level of education or degree of climate awareness, when systems of beliefs clash, individuals experience cognitive dissonance and look for ways to avoid change. This conflict has not boded well for climate action, as people have relied on all sorts of justifications to evade a sense of guilt and stick to their carbon-intensive habits. Some feel compelled to question the science behind climate change, and stress the wide uncertainty surrounding estimates. Others fall back on fatalism ("we are all going to die in any case"), or the odd argument of being too busy to act. Yet others engage in wild discounting of the timeframe and geographic scope involved ("it's going to affect people far away from here, in a future far away"), or use blame-shifting techniques ("why should I do something when China / politicians / corporations are not doing anything?").

Fighting climate change will surely require changes in behavior by citizens, which will hinge on a complex set of incentives and regulations, as Chapter 8 will explore. We should not hold our breath, however, waiting for a spontaneous and complete turnaround of current value systems. Much less should we make this, as the degrowth agenda implicitly does, the central element of the fight against climate change and environmental degradation.

Resilience as societal self-determination

According to Amartya Sen, people also seek positive freedom from uncertainty, economic and beyond. As argued by Galbraith, this has been a concern for humans throughout history, and becomes increasingly so the more they have to protect.[65] Initially, people seek basic reductions of uncertainty, whether they are farmers worrying about drought or crop failure, or workers at risk of job and income loss, or anyone fearing sickness and accidents. Progressively, as societies have more resources at their disposal, they do more to reduce uncertainties, fending off price collapses that would ruin merchants, for example, and creating systems to ensure income in old age. Societies in aggregate also crave resilience—the ability to withstand shocks and recover quickly—and therefore devote some of the resources gained by economic growth to building capacity to respond to more or less known unknowns.

Rich countries find many ways to use resources to prevent or react to shocks. As we have seen, the most developed social welfare systems are in the world's richest countries, and typically include old-age pension, emergency healthcare, and stricter provisions for on-the-job safety. As for shocks to industries, rich countries can offer bailouts, as seen in the large packages put in place to soften the impact of the 2008 and 2020 financial crises. The tough standards applied to food, livestock, and produce in rich countries are all to prevent disease shocks. A major investment in science, and especially in basic research, can be seen as an insurance policy against the unknown; it certainly proved to be that after the Covid-19 crisis hit, when vaccines were able to be developed in record time. On an even larger scale, a planetary defense program at the US space agency NASA aims to avert the disaster of a large-scale asteroid or comet crashing into Earth. Many more examples could be provided, but the point should be clear. These are all measures to reduce the chances of shocks and increase resilience, and they all cost big money.

Through the centuries, economic growth, innovation, and technology have been geared not only to increasing the resources available, but also to protecting them from uncertainty. Protection from foreign invasion or interference belongs to the wider category of resilience to shocks. In this respect, economic growth contributes to a Sen-type capability or self-determination capacity on a societal level.[66] Economic and technological progress allow countries to protect, if not project, their value systems and lifestyle. It may be that a society living at a lower level of complexity and technology could sustain a happy life for its citizens, in line with the subsistence equilibrium of Figure 5.1 (and given the conditions of isolation, limited size, and limited innovation discussed above).[67] But it would still score poorly in terms of resilience in response to shocks beyond its control.

In many cases, the devastating shock to such societies has come in the form of imperialism and colonization, but disasters are not all man-made, as shown by the cases of the Easter Islanders and the Anasazi of southwestern North America. There can be a catastrophic change in the environment or climate, or an invasion of pests or disease.[68]

The advance of science and innovation has been powered through the centuries by a desire to put more and more of societal destiny into human hands. Seen in this light, the quest for more is just another side to the human desire for self-determination.

We often hear that economics and institutions are just social constructs, and therefore that the possibilities for them are limited only by lack of imagination.[69] This inspirational talk holds some logical truth, given that no laws of physics are involved in social constructs. Institutions need to be compatible with incentives, however, and incentives in turn channel impulses of human nature.

A useful metaphor comes from building riverbanks. One must first understand the direction of the flow of a river, and how it changes during wintertime freezes, during the summer, and so on. Then, interventions can be built to channel water to where it is needed, or otherwise make the flow optimal. These can be for good purposes—perhaps to lend energy to a renewable energy power plant—or worse ones, but the river cannot simply be dammed, or else it will spill over in all directions, wreaking havoc. If only a dam is built, then cursing the "evil" principles of hydrodynamics will not help.

If, as we imagine alternative institutions, we choose to ignore basic human cultural traits (perhaps blinded by false dichotomies such as whether people are inherently good or bad), then the mechanics of our society will rapidly turn perverse, as people find ways to go around those institutions. With this understanding, we can see why people would not suddenly embrace a narrative of sufficiency or a steady-state economy. We can appreciate that they would not simply ignore the fact that neighboring countries would begin to have more, thanks to their sustained economic growth and technological progress. We can recognize that impulses to emigrate would begin under ecosocialism. And we can see why redistribution alone, without growth, would lead to an unsustainable equilibrium. In sum, we can understand why capitalism has proved a successful way to organize society hitherto. It channels some basic human impulses.

Rather than fantasizing about changing human nature altogether, it is more promising to acknowledge these traits and try to align them with the environmental imperatives laid out by scientists.[70] The odds of achieving this should be stacked up against the odds of accelerating technological change—which, as we saw, is the chief mechanism humanity has used to solve problems throughout its history. It is to this topic that we turn in Chapter 6.

6

UNDERSTANDING THE WINDS OF CHANGE

Mater artium necessitas. [Necessity is the mother of invention.]
—ANCIENT ROMAN PROVERB

Each year, a high-level conference in the capital of the European Union brings together government leaders, finance ministers, central bankers, CEOs, and world-class academics to discuss the main economic themes of the moment. Belonging to none of these categories, in 2018, I was just a participant in the audience of the Brussels Economic Forum, watching a debate between economist Bob Gordon and futurist Jeremy Rifkin. Gordon was pushing his well-known view that productivity growth is largely over, because none of the modern gizmos can match the revolutionary power of past inventions. He sees forecasters of future technology as prone to exaggerating major developments and how near-term they are. To illustrate, Gordon referred to *Wired*'s 1997 predictions for the upcoming decade, which included a complete cure for all forms of cancer, and a fledgling human colony on Mars. Sitting on the other side of the fence, Rifkin did what Rifkin does best by presenting his inspiring vision of the Third Industrial Revolution now underway, powered by artificial intelligence (AI), robots, renewable energy, and autonomous transport.[1] Weighing their competing ideas felt very much like a challenge of faith, as if technology were a religion. One either believes in it or not. And those who do believe are eager to be prepared, because the day of reckoning is nigh.

An article of faith for a growing number of techno-enthusiasts today is that the "singularity is near," meaning that we are on the verge of seeing general AI that surpasses human cognition, after which an unprecedented explosion in technological growth will ensue.[2] The information technologies associated with the singularity are, like the telegraph, steam engine, and electric motor of the industrial age, general-purpose technologies. As such, we should expect them to unleash a host of complementary innovations that will yield dramatic productivity improvements. If we are not seeing the effects in GDP and productivity statistics, the argument goes, it is because the latter do not properly account for the digital economy.[3] Or, in any case, it is just a matter of time for the revolution to materialize.[4] The wonders promised by the prospect of unlocking quantum computing are further feeding this hype.

Peter H. Diamandis and Steven Kotler, in *Abundance,* bring this world-view closer to our matter at hand. Part of an enthusiastic trilogy, it tells the stories of modern technologies on the verge of solving human problems in many domains: water, food, energy, healthcare, education, and even freedom. Diamandis and Kotler showcase the pioneering work carried out by such revolutionary innovators as Elon Musk, Larry Page, and Craig Venter. Their conclusion: "the future is better than you think."[5]

But there is something else that is abundant, and that is cause for doubt in the face of such a sunny narrative, especially when it fails to explain why all this should happen precisely now. This is particularly true in the green sphere. In truth, renewable energy is as old as civilization: the ancient Romans were already using water and wind power in mills.[6] Solar panels are an old technology, too, dating back to 1884. Scientists have known about climate change at least since the 1970s, and in 2002 French president Jacques Chirac was already decrying at the UN's IV Earth Summit that "Notre maison brûle et nous regardons ailleurs" (our house is on fire and we look the other way). The degrowth school is therefore right to ask why a "green industrial revolution" should take place now, after years of unfulfilled promises and disappointments. Why should we expect greenhouse gas emissions to drop at a much faster rate than has been seen over the past decades—a rate that, projected into the future, will be far from sufficient to stave off catastrophic climate change? This will be the topic of this chapter, but to go beyond stories about great efforts by single innovators, we will need to venture into the mechanics of innovation, understanding the forces that feed it and shape it.

Perhaps inspired by caricatures of past inventors crying "Eureka!," there seems to be a widespread belief that innovation comes about because, periodically, someone sitting in a lab or research center is hit by an idea. Coming from a family of natural scientists, I am under the impression that even some of the people sitting in labs believe they and their research teams are the ones who alone define the relevant topics, based on their personal initiative and inquisitiveness, and are therefore solely responsible for scientific and technological breakthroughs. The same basic mistake leads to the misconception, articulated in Chapter 1, that green growth plans are betting the future of the planet on a mad gamble, assuming some stroke of luck will save us all. Innovation rests on a process that is much less fatalistic, and much more deterministic, than is often believed. Because climate change is the greatest of all environmental risks, this chapter will focus mostly on it, but the logic of innovation also extends to other environmental issues we face.

Relative needs also rule in innovation

Let us start with some basics on how innovation, and its accumulation into knowledge, comes about. Throughout history and across continents, as extensively documented by anthropologists, all societies have developed their own know-how in ways deeply shaped by their local environments and local challenges. Take the Inuit, for instance, who developed snow goggles made of wood or whale bone with a slit cut into them to reduce glare and protect the eyes. Cultures are generally quite advanced in solving local problems because solutions have developed through repeated trial and error, and accumulated over generations. The catalyst for all this is *hyper-emulation,* one of the behavioral traits behind the success of the human species. Humans copy the behavior and solutions of the most successful, prestigious, old, and skillful, and in the process, collectively learn to solve problems and build know-how. In this way, humans have developed the ability to survive in extremely diverse habitats, from the Arctic tundra, where they had to devise ways of hunting seals, to the African desert, where it was crucial to learn to build wells. Centuries of constant improvement led to sophisticated technologies such as the water systems that used to supply the desert city of Petra, in Jordan, designed to optimize flow, minimize turbulence, mitigate leakage, and prevent clogging.

In *The Secret of Our Success,* evolutionary biologist Joseph Henrich gives the example of what he calls the "lost European (or American) explorers' files." These stories, from Sir John Franklin's search for the fabled Northwest passage in the Arctic to the Burke and Wills expedition to the interior of Australia, follow a standard pattern. A small group of hapless explorers get lost, cut off from their civilization in seemingly inhospitable locations. They eventually run out of food, their shelter is insufficient, and disease often follows from encounters with unfamiliar plants and animals. The explorers' ability to travel deteriorates. At this point, two things can happen. In the bad scenario, things get extremely nasty, scarcity leads to violence (even episodes of cannibalism), and then the explorers all die. In the stories with happier endings for the explorers, they receive help from indigenous populations. These societies, having survived and thrived in these "inhospitable" environments for centuries, have accumulated locally relevant know-how and technology. The legend of the first American Thanksgiving hews to this pattern. The Pilgrims celebrated a good harvest after native people taught them how to cultivate corn—a huge relief after the first year's tough winter and scarcity of provisions had killed half the European settlers. Taking a broad perspective, all these stories align with a history of humans with relative needs, struggling with "natural" limits, and eventually surpassing them thanks to human ingenuity.

Clearly, some societies were more successful than others in developing and applying new ideas. Through interactions with others, however, useful ideas could spread, and be adopted, imitated, and modified for local contexts.[7] Think of gunpowder, thought to be discovered by a ninth-century Chinese alchemist trying to create an elixir of life. The Mongols deployed firearms during their invasion of Europe in the thirteenth century, and soon enough this technology was rapidly copied by various European armies. In the fourteenth century, production methods were improved and the European technology developed in its own way. As discussed in Appendix B, gunpowder and its dreadful weapons were then reimported into China and Southeast Asia in much more advanced form.[8] On a more peaceful note, the same could be said of paper, invented in China around 100 CE, spread to the Islamic world six centuries later, and then passed on to Medieval Europe in the thirteenth century, with significant adaptations along the way. The process by which ideas develop and spread bears a resemblance to natural evolution—Charles Darwin himself was the first to hint, in his *Descent of Man,* that culture, broadly defined as "inherited habits," might be shaped by similar dynamics.[9]

Often a solution to a problem of the day laid the seeds for new problems to arise.[10] For example, at the end of the nineteenth century, chemist Sir William Crookes sounded the alarm that a shortfall of nitrogen in the soil would soon lead to massive hunger. At the time, ammonia, rich in nitrogen, had to be recovered from tropical islands covered in guano to fertilize the fields. Not only was this very expensive, but soon, it was predicted, even those reserves would fail to satisfy demand. It became a pressing matter to research potential new sources of ammonia, as without it, millions would starve.[11] The irony was that 80 percent of the Earth's atmosphere is composed of nitrogen gas. Industrial scientists Fritz Haber and Carl Bosch were able to devise a method using energy (mainly natural gas) to convert atmospheric nitrogen into a compound, which could then provide nutrients to plants. The environmental boundary set by soil nitrogen availability was therefore overcome, paving the way for the greatest-ever expansion in human population. If we fast-forward to present day, however, the Haber-Bosch process accounts for a large part of the CO_2 emissions associated with agriculture, and to disruptions of the global nitrogen cycle.

The takeaway is that humanity solves problems as they come, employing science and technology as they exist at the time, in a constant process of experiential learning. Underscoring the point, there is no evidence that our brain capacity has changed significantly over the past hundred thousand years. The Early European modern humans (Cro-Magnons) painting the caves of Lascaux and people creating cave art in Sulawesi were us.[12] With the benefit of hindsight and hundreds of years of knowledge accumulation, solutions of the past might look basic or myopic.[13] From today's standpoint, the Haber-Bosch process looks energy-inefficient and heavily polluting. But we should not feel we have more capable brains than our ancestors. At most, we have a bit more knowledge.

It is also worth stressing that quests to innovate in response to challenges and needs, however they might be depicted in popular culture, do not always have happy endings. History is full of instances when this did not happen, or did not happen fast enough, causing human suffering and, in the worst cases, spurring mass migrations and even the collapse of civilizations. This is crucial for us to recognize because it tells us we cannot be complacent about climate change, even if we believe technology will be the main tool to address it. Nor should we adopt a patient stance of waiting for innovation to run its course. The speed at which innovation is developed and deployed will make a great difference to the human consequences of climate and environmental change. Complex systems like modern economies are held back by inbuilt inertia and new technologies take time to diffuse. The steam-powered machines of the original Industrial Revolution took at least fifty years to spread, even with their obvious prospects for enormous productivity improvements. Electrification of the United States following the second Industrial Revolution took roughly seventy years.[14] In the face of looming climate catastrophe, UN scientists tell us humanity has thirty years at best to complete the transition to a carbon-neutral economy. This sets the scene for the wide-ranging policy interventions needed to accelerate the green transition that will be outlined in Chapter 8.

Before delving into whether capitalism is malfunctioning or myopic in fostering innovation, we can say this: technology and the accumulation of know-how respond to needs that are specific to locations and moments in history. Jared Diamond's accounts in *Collapse* of past civilizations whose own decisions were their undoing (as with the Easter Islanders' deforestation of their land), show that myopic mistakes have been made throughout human history. To say that it is capitalism forcing people to focus only on the present—and that, without it, individuals and societies would be prudent and carefully take into account all future scenarios, including in their innovation strategies—is an untenable claim. Heavily discounting the uncertain future is a standard animal trait, and humans, while doing so less than other species, are not immune from the tendency.[15]

Innovation under capitalism

Innovation under capitalism follows precisely the same logic it has throughout human history. Capitalism simply oils the wheels of the knowledge machine with features like clear property rights, contract enforcement, competition, and decentralized decision-making.[16] In particular, capitalism fosters not only the development of new ideas, but more importantly their diffusion. The latter is the crucial part of the innovation process in which learning, imitation, and feedback effects arise and enhance the original innovation.

As odd as this may sound, very often it is not so much about having big ideas, but making it possible to operationalize them. Leonardo da Vinci in the fifteenth century had loads of ideas, envisioning the airplane, the diving suit, the parachute, the submarine, and the helicopter. Most of these were not credibly produced until three hundred years later. Methane was known in antiquity, and allegedly it contributed to the clairvoyant capacities of the Oracle of Delphi in Ancient Greece. Accounts of a Baghdad battery suggest that two thousand years ago the Persians knew electricity existed, although they had little understanding of the phenomenon, and few applications for it. Thus, it remained mainly an item of intellectual curiosity, perhaps applied in medical therapy. Likewise, in 600 BCE, the Ancient Greeks knew that rubbing cat fur on amber (fossilized tree resin) caused an attraction between the two, in what is known today to be static electricity. But it was not until 1830 that electricity became useful for practical uses, when Michael Faraday created an electric dynamo to generate current in a continuous way. Looking to more recent history, early experiments with self-driving cars were conducted in the 1920s, but only now is the technology available for a possibly viable product. As an engineer friend, CEO of a technology venture, tells me, the hardest part of a start-up is not having the good idea for it—there are plenty of those—but creating the business that can deliver the reliable product at scale.

Capitalism provides strong incentives, in the form of personal rewards, to shift from the realm of ideas and intellectual curiosity to that of practical production of solutions that respond to needs of the moment and serve large customer bases. As usual, drawing a contrast with the Soviet Union is instructive. The Soviet system featured strong investment in education, brilliant scientists, and well-equipped research labs. As a consequence, it generated major scientific breakthroughs—such as Sputnik, the first battery-powered satellite in 1957, which orbited around the Earth for a few months. In 1961, the Soviets managed to send the first man safely into space: the famous Yuri Gagarin. In the field of nuclear physics, they created the hydrogen bomb, just a few months after the United States. At least for circumscribed, military, or centralized innovation, the socialist Soviet model worked marvels. When it came to spreading innovation and technology through the economy, however, it fell short. Product and process innovation stalled, because there were no incentives for it beyond some fears of being chastised or punished by the party leadership.[17] Without diffusion, raw innovation has very limited social and economic impact.

The strength of the capitalist system on the innovation front is not that it creates more geniuses, but rather that it is an efficient organizing principle, providing strong incentives for people to develop ideas.[18] Even more crucially, it fosters the deployment of these ideas, matching them with sources of capital—firms on the hunt for novelty to maximize returns and eager to win in (Schumpeterian) market competition. In the process, capitalism also nudges acute minds toward what historian Joel Mokyr calls "useful knowledge," as opposed to, say, astrology, numerology, or other intellectual endeavors with no impact on well-being. Because capitalism bridges the gap between abstract scientific curiosity and practical applications, it is an efficient machinery to produce economic growth. Indeed, historians of the "Great Divergence" that separated industrializing Europe from the old world view some of capitalism's standard features, such as strong property rights protection and contract enforcement, as instrumental.[19] Strong economic growth in turn opens up space to use more resources for the future, encouraging investment in research.[20] Innovation generates economic growth, which in turn fuels innovation—a dynamic that has yielded sustained exponential growth since the Industrial Revolution. Recall the high correlation between innovation and GDP per capita (Figure 3.1). This shows up not only in terms of absolute spending by rich countries on research and development (R&D) but, more crucially, in terms of R&D as a share of their GDPs—a good example of how economic growth reduces the need for trade-offs (exemplified in Table 3.1).[21]

Innovation serves perceived societal needs, whether it is government funded, particularly through a democratic process and public accountability, or privately funded, seeking to address consumer relative needs and capture demand.[22] The process of innovation has an exploratory nature, and is by no means perfect, including under capitalism. We often hear of abandoned research programs which, in hindsight, would have produced real bang for the buck, and also see others turn out to have been a waste. These mistakes are bound to happen given that foresight is imperfect. All we can do is align incentives so that those investing in new technologies will try to collect as much information as possible and use it in making decisions. This happens when they can profit from successful R&D. The same applies to government spending in a healthy democracy, assuming that poor investment of taxpayer money by politicians does not go unpunished at the ballot box.[23]

Pulling all this together, we see that researchers and scientists are but the tip of the innovation iceberg, below which a societal organization supports their work. This might be news to some toiling away in the hard sciences, members of a profession proud of being insulated from politics and business thanks to institutions such as academic tenure. The truth is that the bright people sitting in labs are there, with access to research funds, expensive instruments, and scores of PhDs and post-docs, because society has decided their research topics are worth resources and attention, and expressed that through some combination of government, businesses, and philanthropy. A picture therefore emerges of innovation and inventions happening, in whatever domain, not by chance but rather as a response to societal priorities.

In capitalist systems, most resources are privately owned, and indeed, most R&D in rich countries is funded by the private sector.[24] The frequent argument, including from the degrowth movement, is that the private sector focuses on the wrong things, led by its desire to maximize private profits rather than solutions to societal needs. This supposedly results in "useless innovation," rather than innovation in areas that "truly matter."[25] To understand how frontier innovation really works in the private sector, we must look to the central element of the capitalist system—the price mechanism—and we can start with an anecdote.

In 2018, Lant Pritchett, known as a maverick in the field of development economics, was invited by Larry Summers to present a guest lecture to his *Globalization and Its Critics* class at Harvard. Pritchett's opening words— "Tesla is dumb!"—generated the expected murmur in a room full of digital natives, who immediately sized up Pritchett as a Luddite. The fact is that Tesla (like Google, for that matter) invests many millions of dollars in self-driving cars as it responds to outstanding incentives, including the relatively expensive price of labor in the United States. At a world level, however, human labor is very cheap, and there are millions of people who might take work as chauffeurs, charging a fraction of the price of a self-driving car. The same could be said for cashiers in supermarkets, whose work is being increasingly automated. What makes these investments locally sensible but globally illogical is clearly the existence of national borders. For cultural reasons that one can agree or disagree with, we have created borders which are rather closed, and this distorts national prices, in turn taking innovation in different directions.

Pritchett's provocative statement highlights that companies respond to price incentives but that, because of existing distortions, the result might not be the most efficient allocation of resources. From his development perspective, and relating to globalization, the worst part is that developing countries tend to import and imitate the technology solutions of rich countries. A museum in India, for example, installs automated gates that project an image that it is quite modern, but this is a country with countless workers willing to take tickets for a dollar per day. This speaks to an important insight. We know that, under capitalism, prices are fundamental in causing supply to meet demand, but that regulation defines the parameters. Innovation is no exception. It is possible for government to steer the direction of innovation by changing these parameters. To do that within a democratic process, however, policymakers need majorities of people on board.[26] One can blame capitalism because it focuses on "useless innovation," but it is more a reflection of people's revealed preferences than anything else. If borders were made more permeable, as Pritchett has suggested on multiple occasions, the focus on labor-shedding innovation would diminish. But politically, or culturally, people do not seem inclined toward such a solution. The green dimension marks no exception to the rule—a subject we turn to next.

Capitalism's planetary epic fail

If capitalism is such powerful innovation machinery, why have we not seen progress on the green agenda so far—or in any case, seen too little progress, given the foreseeable disaster of climate change and environmental degradation underway? In today's internet lingo, one might call this an epic fail—indeed, of planetary proportions.

Some, building on Marxist theories of value, erroneously conclude that economic growth is only due to the surplus value extracted from natural resources, and specifically petroleum products. Some declare that capitalism cannot get out of fossil fuel because the two are deeply intertwined in a doomed tango, dancing to the tune set by economic growth. The end of petroleum extraction would spell the end of economic growth and the collapse of capitalism.

From an economic standpoint, there is nothing special about petroleum.[27] The Industrial Revolution started with wood, then it moved on to charcoal, then to coal with the deforestation of England. Only in the 1900s did it move to oil and gas, following the invention of the internal combustion engine.[28] There is no question that, as Ian Morris has charted through history using his social development index, an economy of growing complexity and population requires more energy.[29] At the onset of the Industrial Revolution, people simply used the technology of the time to increase energy availability and meet the growing demand of the new machines. There is thus no support for the assertion that capitalism, and its inherent mechanics, are necessarily antithetical to the health of the environment. Indeed, when yoked with democracy, and therefore responding to citizens' concerns for the environment, capitalism has produced less damage than non-capitalist systems. Witness the Soviet Union, immune to capitalist logic, and the catastrophic environmental outcomes it has seen.

As obtuse as this may sound in an era of climate emergency, reducing greenhouse gas emissions simply was not perceived as a pressing need until quite recently.[30] Part of the explanation for this cavalier indifference is that, for a long time, climate science did not exist. Even as that science developed, starting in the 1970s, most voters and therefore politicians only paid lip service to the green cause and were unwilling to put their money where their mouths were. This included richer and more educated cohorts. The capitalist production system responded as always to price and demand incentives, including to the prevalent regulatory environmental standards.

It is appalling that some companies with vested interests have undertaken active efforts to undermine the science of climate change—in the case of oil-producing Exxon, even discoveries by its own scientists.[31] This invites a comment on lobbying more broadly, similar to one made already about advertising. That lobbying can heavily influence policies and delay action is known—and despicable, as we know that timing is crucial and delays can only mean more human suffering. The economic and social dynamics shaping the course of history, however, are much stronger than any company or interest group. When climate change started to show its dire consequences, and became a priority for voters, the moment arrived when energy companies could only realign their agendas with the green transition. As Keynes puts it in his *General Theory,* "Soon or late, it is ideas, not vested interests, which are dangerous for good or evil."[32]

Consumer buy-in is a fundamental component to bring about a shift to a new productive equilibrium. Without it, smart inventions will still be thought up, but they will be locked away in drawers, waiting to be pulled out in better times. Consider photovoltaic cells, invented by Charles Fritts in 1883 and first installed on a New York City rooftop in 1884. Past that point, significant investment would been have required to make them relevant for any implementation, given their limited energy-conversion efficiency (roughly 1 percent). This did not materialize because they were not addressing a consumer need of the moment, given the cheap availability of energy from fossil fuels in the early twentieth century. Interest in the technology picked up instead in the 1950s, as the need arose to power space satellites.

Another beautiful example comes from *The Knowledge,* by Lewis Dartnell. As incredible as it may sound in our era, as electric vehicles seem to be just entering the market, it was far from clear at the dawn of the twentieth century, when cars were in their earliest days, what type of engine should power them. Three technologies were effectively competing neck-and-neck: steam, gasoline, and electric. In Chicago, electric vehicles dominated the nascent market. Likewise, on the other side of the Atlantic, they did well in Berlin. In Manhattan, a taxi service was successfully run on electric cars in 1900, with rapid flat-for-full exchanges of batteries at service stations. In the end, however, gasoline won. The reason for this early defeat of electric cars has nothing to do with "big oil" lobbying, or capitalism being hopelessly intertwined with oil. Rather, gas-powered cars won the cost-benefit battle. Electric cars were quieter, mechanically simpler, and free of exhaust gases. But gasoline allowed greater range and autonomy—and exhaust was considered a minor annoyance rather than a toxic pollutant, especially as compared to the nauseating smell of horse manure and urine that had filled the air of cities for centuries.[33]

Eventually, thanks to economies of scale, gas-powered cars became cheaper, and experienced a boom with the famously affordable Ford Model T. This in turn meant that infrastructure, including a network of gas stations, was expanded only for this technology, leading to an early demise of electric cars. At the time, there was no knowledge of CO_2 emissions, or climate science, so environmental factors did not enter into the cost-benefit equation. Of course, with the benefit of hindsight, we can mourn that moment in which the course of technological history changed, and with it the climate's destiny, but it is important to understand why it happened.

Governments, once supported by their electorates, can intervene to use their taxing powers or public spending to generate demand or twist relative prices. Specifically on CO_2 emissions, a tax can adjust prices to recognize that using fossil-fuel technology generates a societal loss that companies would not otherwise take into account in their standard investment and production decisions. Once this happens, the entire capitalist system adjusts to take things in a new direction. In the short run, companies that rely on carbon-emitting technologies increase their prices. These increased prices force consumers to think harder about alternatives. Innovative start-ups and rival companies using less-polluting technologies spot opportunities to serve new levels and forms of demand. Likewise, companies respond by reallocating some of their capital investments to implement low-carbon production processes. Overall, because the capitalist system relies on the price mechanism, it automatically readjusts when prices shift. The beauty of this, as opposed to having, for examples, hard limits set by regulation, is that it does not require a regulator with deep knowledge of production and consumption processes.[34] This, in a nutshell, is why so many economists see carbon pricing as the gold standard to tackle climate change.

The green tipping point

Even if they accept the mechanics of the price mechanism, staunch skeptics could point out that economists have been saying the same old thing about carbon pricing for ages. As early as 1997, more than 2,700 economists, including nineteen Nobel laureates, issued a public statement ahead of the Kyoto Protocol international negotiations, arguing in favor of carbon pricing to address climate change. Over twenty years later, carbon is still largely underpriced and we are still grappling with the problem. On top of the changes in citizen concerns and voting preferences that we now see, something else makes the green tipping point credibly near.[35]

Year after year, slowly but surely, environmental regulations have been piling up, investment in green R&D has been accumulating, and government-supported pilot projects have proliferated. As a result of the actions and investments of visionary first movers, philanthropists, and dedicated early adopters, renewable technologies have been rolled out more and more. We saw in Chapter 5 what happens when this is the case: production prices fall, as firms benefit from learning curves. The learning curve for solar photovoltaic is particularly stunning, as production prices are down 99.6 percent since 1976, from $106 to $0.38 per watt (Figure 6.1). Even professional forecasters knowledgeable about the business have been taken aback by the speed at which this has happened. In 2014, the International Energy Agency predicted that average solar prices would reach $0.05 / kWh by 2050, thirty-six years later. In fact, this took six years.

As a result of such rapid progress, 2020 marked a turning point: in that year, new energy production became cheaper for renewables, specifically via solar and onshore wind power, than for the cheapest fossil-fuel alternative (Figure 6.2). This powerfully implies that, going forward, the most cost-effective way of producing electricity will also be climate-friendly. This shifts the green transition from a luxury consumption choice, reserved to rich countries and citizens, to a profitable investment opportunity—and, at the macroeconomic level, firmly into new economic growth model ter-

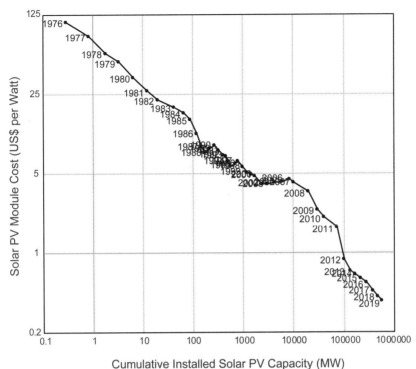

FIGURE 6.1: Price per watt of photovoltaic modules

Data sources: Our World in Data; Lafond et al. (2018).

Note: Prices adjusted for inflation and presented in 2019 US dollars. Both axes use a logarithmic scale.

ritory. It is of greatest importance that this has happened for electricity generation, because this is a fundamental building block. Most climate mitigation plans specify a two-step approach: first, electrify everything that can be electrified (which is why, for example, we see today's strong push toward electric cars, substituting for combustion engine vehicles); and second, ramp up the share of renewable energies in the electricity generation mix.[36]

We are on the verge of a world where capitalism's wind fills the sails of the green transition and we no longer have to rely on the goodwill of philanthropists and dedicated early adopters willing to take one for the team. Following profitable opportunities, financial market money will start flowing to clean technology producers at scale, getting out of older, polluting sectors. As these "brown" sectors decline, they will lose economies of scale,

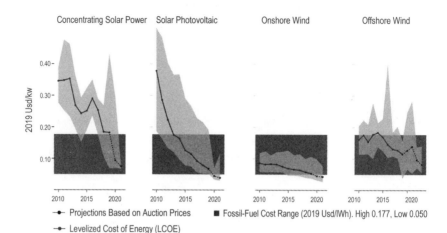

FIGURE 6.2: Cost of energy from newly commissioned power generation technologies, 2010–2023

Data source: IRENA (2020).

Notes: The levelized cost of energy (LCOE), a metric for comparing costs across energy sources, represents the minimum constant price at which electricity must be sold for the power plant to break even over its lifetime. Projections, displayed by a dotted line, are based on current auction prices.

effectively falling backward on their industrial learning curves. Accessing loans will become harder for them, and so will attracting talent. We will see a steady move from a world where it is necessary to decide, even at individual consumer level, whether to go for the cheaper option or the greener option, to a world where both motivations point to the same choice. From an economic perspective, the beauty of this is that we return to a decentralized logic, where we no longer have to rely on the hope that people are willing to make day-to-day sacrifices to benefit future generations and society at large—or be stuck in a world where poorer households need to decide between end of the month and end of the world. Personal and societal incentives will align as if led by a *green* invisible hand—without the need for the complete ideological conversion that ecosocialism demands.

From a political economy perspective, once green technologies also become the most profitable investment opportunities, it becomes politically much easier to introduce or ramp up carbon pricing. It will be important to do this to avoid the famous rebound effects and toxic dynamics unleashed by Jevons's paradox, as discussed in Chapter 1. Increased carbon prices will avoid a situation where, as society moves away from, say, coal, the price of coal then plunges and turns it back into an appealing energy source, thwarting the green progress that was being made.

Infinite innovation on a finite planet

Diehard degrowth advocates might now point out that such a technology-led plan to tackle climate change will require large material inputs to get everything electrified and get new renewable power plants built. And because "you can't produce stuff out of thin air," as their slogan goes, that is the end of innovation; we are simply kicking the can down the road because, on a finite planet, we are bound to run out of raw materials.

On this point, as President Obama liked to say, "let me be as clear as I can be." Innovation, and therefore economic growth, will not run out of materials, because the very definition of a production material depends on what is available. This is something we have observed throughout human history. Societies solve problems using what they have on hand. Think of a widespread basic technology like writing, developed by many societies independently to meet exigencies of commerce first, and then knowledge accumulation more broadly. Depending on location and existing technology, several writing materials have been used throughout history—stone tablets, clay tablets, bamboo slats, papyrus, wax tablets, vellum, parchment, paper, and copperplate—all depending on local availability. The same could be said of basic navigation, which has been with humanity since prehistoric time: early vessels were built using techniques and materials from bamboo and logs to reeds and leather, based on material availability and local needs.

In 1968, biologist Paul Ehrlich caused a sensation by predicting, in *The Population Bomb,* that humanity would run out of resources and people would be starving to death before the next decade was out. The only escape in his view was strict population control, of the type implemented by China with its "one-child policy" ten years later. It was the Malthusian story redux, and just as in Malthus's time, the power of innovation was underestimated. At the time, economist Julian Simon made a famous scientific wager. He challenged Ehrlich to pick any five commodities, saying that regardless of the choices he would bet on their prices being lower ten years later. Ehrlich picked ones that looked likely to run out: copper, chromium, nickel, tin, and tungsten. By the end of the decade, all were priced lower. It is important to understand that it is not that Simon got lucky. Rather, he understood the dynamics of capitalist markets: as the price of a commodity goes up, it activates a rationing mechanism whereby buyers shift away from that material—and this prods the innovation machine to find alternatives, or devise more efficient ways to use what is available.[37] The reason that the price of tin, for example, went down is that aluminum became a cheaper and better substitute for it.

Aluminum has a neat story of its own. It is the most abundant metal in the Earth's crust. And yet, up until the 1880s, when a method for melting and electrolyzing its ore was invented, it was prohibitively expensive, meaning it was not a production material, but treated more as a precious metal. Dartnell reports the curious episode of an imperial banquet organized by Napoleon III in the first half of the nineteenth century. In this setting, aluminum cutlery was displayed as an ostentatious show of power, reserved for the most distinguished dinner guests, while less favored guests used knives and forks of simple gold. The very definition of what is a commodity, and useful in production, varies across centuries, and depends on availability.

The concern that some input material will come to an end, and that this crisis will deal a fatal blow to civilization, is a recurrent theme in history. In Britain, the fear was expressed in the age of sail that when the country was depleted of oak, the fundamental wood used in shipbuilding and masts at the time, the country would become defenseless at sea. Clearly, this did not happen; eventually iron replaced the need for wood, and in the meantime, sufficient supplies came through trade with Scandinavia, the Baltics, and North America. For goods that cannot be replaced, prices generate strong incentives to become better at extracting and recycling them.[38] Think of copper, which has good properties as an electrical conductor. As electrification was in full swing in the early twentieth century in the United States, geologist and copper-mining expert Ira Joralemon warned that "the age of electricity and of copper will be short. At the intense rate of production that must come, the copper supply of the world will last hardly a score of years. . . . Our civilization based on electrical power will dwindle and die." One century later, we can see that this prediction did not age well.

It is not even a necessity that the bulk of our technological solutions must be based on geological materials. Think, for example, of recent research on fertilizers, based on the discovery that certain bacteria can capture atmospheric nitrogen, without the high pressures and temperatures of the polluting Haber-Bosch process described above. Likewise, the research frontier is shifting toward new materials, opening up the possibility of producing consumer goods with algae-based bioplastic. It would seem that if the twentieth century was characterized by the triumph of inorganic chemistry, the twenty-first might well be the century of organic materials, repurposed to meet human needs. This age is being ushered in by the advent of biotechnology which, also thanks to recent advances like Crispr / Cas9, will provide powerful tools to address current challenges, from reducing the use of pesticides and herbicides in agriculture to fostering natural carbon sequestration.

Innovation not only yields new technologies and products, but also allows more value to be extracted from what is already available. Under the heading of more efficient material use falls, for example, the idea of the circular economy, in which society is organized to reuse or recycle extensively, and rely less on new material extraction. The digital revolution has also caused the economy to be more and more "dematerialized" in production and consumption, as it becomes more based on intangible goods.[39] This is what economist Danny Quah refers to as the "weightless economy." Another push toward more efficient use of reduced material inputs is known as the "sharing economy," as seen recently in urban settings with cars, mopeds, home rentals, and coworking spaces. Under this model, people shift from individual ownership of an asset to an arrangement, usually employing digital technology, by which it can be shared by multiple users, collectively extracting more utility out of the same amount of material.

Andrew McAfee shows in *More from Less* the extent to which the American economy has dematerialized since the 1970s, including by slashing its consumption of virtually all the seventy-two material resources tracked by the US Geological Survey, including minerals, timber, and cement, whether domestically sourced or imported. In its agricultural sector, the United States managed to increase crop production by 35 percent between 1982 and 2015, while reducing the use of fertilizers and returning to nature an amount of land equivalent to the state of Washington (45 million acres). The same combination of productivity enhancements and dematerialization has allowed many advanced economies to see reforestation and economic growth progressing together over the past three decades.[40]

To sum up this discussion, an old quip might suffice: The Stone Age did not come to an end because we ran out of stones. Likewise, the age of oil and gas will not end because we will run out of those, and innovation will not come to an end because we run out of material inputs. While in theory, and strictly speaking of physical inputs, it is conceivable that a finite planet could be depleted of all materials at once, as a practical reality this is not a concern. Assuming that point were ever reached, it would be in a future so distant that, by then, our material sources would probably not be confined to this planet.

Is technological progress inevitable?

As we saw earlier, there is an evolutionary logic to innovation and knowledge accumulation, and as in natural evolutionary processes there is no inevitability of success. At first blush, this might seem like a pessimistic statement. After all, human-made innovation is a faster and more efficient mechanism than natural evolution, given that technical solutions, unlike mutations in Darwinian natural evolution, do not happen by chance.[41] Innovations are designed to solve the problems of the moment.[42] This implies that, while the concept of progress does not make sense in biological evolution, it could in social progress, if that is interpreted as sustained knowledge cumulation and increased societal complexity.[43]

So why is it not the case that humanity is on a safe stride toward neverending progress, powered by technology and knowledge accumulation? Is it that we are at risk of running out of ideas? In the end, if everything is bounded, shifts in prices and needs notwithstanding, can't science, innovation, and ultimately human ingenuity be bounded as well? For practical purposes, the answer can safely be *no*.

Nobel laureate economist Paul Romer has built productively on the concept of ideas as instructions for arranging atoms and for using the arrangements to meet human needs. For thousands of years, silicon dioxide provided utility mainly as sand on the beach, but now it delivers utility through the myriad goods that feature computer chips. Throughout history, humans have mixed elements together at different temperatures and pressures to see what comes out. The Bronze Age began with an idea to mix tin and copper. Later, mixing carbon and iron turned into another powerful metal: steel. Later yet, in the 1980s, mixing copper, yttrium, barium, and oxygen created a superconductor. Taking the periodic table, and applying combinatory calculus to just the first ten of its elements, Romer shows there are more possible combinations than there have been seconds since the Big Bang—and this does not even take into account that different proportions would generate different compounds. "There have been too few people on Earth," Romer concludes, "and too little time since we showed up, for us to have tried more than a minuscule fraction of all the possibilities."[44]

In a less mathematical vein, having reviewed thousands of years of human technology development, Dartnell shows that all new inventions build on previous ones. Developments in glass, for instance, made it possible to create new instruments, including telescopes and microscopes, which in turn opened new worlds for research.[45] The more that is discovered, the more new opportunities are revealed.[46] This is particularly the case where the scientific method is employed, and investigations are guided by hypotheses based on theories. Where there is no guiding theory, as noted by Mokyr, advances associated with new inventions tend to fizzle out, and fail to lay foundations for next steps.[47] Simple tinkering and learning-by-doing soon reach a point of diminishing returns, or require a serendipitous event to make another productive leap.[48] In the face of all this, evolutionary biologist Stephen Jay Gould bluntly concluded years ago that we are so far from the limits of knowledge that it is not even worth posing the question of when they will be reached.

Given that the more we discover the more we open up space for new discoveries, and that the upper bound of knowledge is so far off that it is not even worth discussing, one might assume that humanity is on a sure march toward betterment. Unfortunately, that is not quite so. Up until now, we have discussed only the feasibility side, which simply implies that unbound progress is possible, not that it is inevitable. We know that extremely fast technical progress brings with it social complications. On the economic side, for example, innovation generally entails some degree of technological unemployment, creating winners and losers. With regard to economic inequality, increasing R&D spending can make wealth gaps wider, as it helps skew income distribution in favor of high-skilled labor.[49] Innovation also creates cultural issues, as a changing world can undermine long-standing values at a speed that feels unsettling to some. As a consequence, feasible though it may be, technological development has not progressed linearly through history. In China, for example, innovation flourished during the Tang and Song dynasties, in cultural conditions that were conducive to it, but then stagnated in the Ming and Qing dynasties. Often across history fast innovation has been followed by a period of retrenchment or stagnation, as conflict and disorder occur.[50]

Calestous Juma shows in *Innovation and Its Enemies* that China is by no means an isolated instance. Analyzing the conditions for acceptance or resistance to innovation, across nearly six hundred years of technology history, Juma surveyed the introduction of the printing press, farm mechanization, electricity, mechanical refrigeration, recorded music, transgenic crops, transgenic animals, coffee, and even margarine. Innovations that threaten to alter cultural identities tend to generate intense social concern.[51] Fear of loss can lead individuals and groups to oppose innovation, even if that means forgoing important gains. Objections may focus on material losses, but they also include intellectual and psychological dimensions. Juma's message, that societies with great economic and political inequities are likely to experience heightened technological controversies, aligns with the idea of an "innovation treadmill" and the importance of social cohesion.

The period we are currently living in is not exempt from these dynamics. On the one hand, we need a speedy technological transition to fast-track green, sustainable solutions. On the other hand, the changes will shatter entrenched beliefs, disrupt jobs, and damage some industries. Juma's work reveals that, perhaps paradoxically, the feasibility of the green transition will rest more on the careful management of the socioeconomic dimension than on science and technology per se. The latter are the pinnacle of innovation, but a well-functioning economic system provides the foundations for it.

Aside from preserving economic growth and ensuring that its benefits are spread widely, the other socioeconomic priority to underpin the green innovation vision is maintaining the fundamental principles of the Enlightenment and capitalism that made the Great Divergence possible. These include fair competition, checks and balances within government, freedom of speech, tolerance, and liberal democracy. Those eager to do battle with strawman enemies like "capitalism" and "the establishment" might wish that a regulatory hard limit could simply be imposed on emissions tomorrow. But if the process of climate mitigation is incompatible with the social and economic dimension, popular uproar will keep it from happening.[52] Again, the Yellow Vest protest movement comes to mind, sparked by a 2018 tax on fuel in France—a carbon-reducing measure that then had to be reversed. This is why European Commissioner and climate czar Frans Timmermans frequently reminds people that there will either be a just green transition, or there will just be no transition, because people will oppose it.

The potential limit on technological innovation, and human progress, is therefore social—echoing Fred Hirsch's *Social Limits to Growth*—rather than technical. Its success will rely on the solidity of the socioeconomic infrastructure that sustains and nurtures it.

With great powers come great responsibilities

One last aspect of innovation should be noted which is subtle and perhaps abstract, but still crucially important: it is not only the economic system, but also the value system, that provides the foundations for successful and sustained innovation. Recall that the origin of the Great Divergence was a shift to a growth model based on sustained innovation—the seeds of which had been planted several centuries earlier by Bacon, Newton, Galileo, Kant, Descartes, and Spinoza. As argued by Mokyr in *A Culture of Growth,* this led to cultural change in Europe.[53] Effectively, it all started from the realization that nature operates through regular patterns, rather than the whims of capricious gods or inscrutable spiritual principles, and could therefore be understood by science and harnessed by technology. As the scientific method transformed natural philosophy into an empirical activity rooted in experimentation, modern science was detached from the rest of philosophy. A new set of principles, falling under the label of Positivism, laid the foundations for the Scientific Revolution and, in turn, the Great Divergence, providing the epistemological support to propel innovation.

It is important to stress that, before Positivism was a school of thought, technological development and deployment had always been about making deliberate and substantial changes to the natural physical environment. Throughout history, every human society has modified its surroundings to further its material interests, at times with dramatic ecological effects, predating industrialization by far. Henrich notes that, when ancient hunter-gatherers arrived in the Americas, 75 percent of megafauna went extinct, including horses, camels, mammoths, lions, saber-toothed cats, giant sloths, and dire wolves. Similar patterns emerged when humans arrived at different times in Madagascar, New Zealand, and the Caribbean. Hannah Ritchie states it plainly: "the romantic idea that our hunter-gatherer ancestors lived in harmony with nature is deeply flawed. Humans have never been 'in balance' with nature."[54] The introduction of agriculture ten thousand years ago similarly led to natural upheavals, as it involved slashing and burning forests, then farming the resulting soil to exhaustion, with no notion of regenerative crop rotation. In Roman times, hundreds of thousands of square kilometers of Mediterranean forest were eliminated to make way for cereal crops, such as wheat and barley. In Asia, rice paddies replaced the extensive forests of China, Indochina, and some of the largest Indonesian islands.

As happens today, ancient technological solutions at times generated by-products affecting the Earth's atmosphere, albeit not on a scale to change the world's climate. Starting around 100 BCE, methane concentration started to rise in the atmosphere. The ancient Romans kept domesticated livestock—cows, sheep, and goats—which excrete methane gas, a by-product of digestion. Around the same time, in China, the Han dynasty expanded its rice fields, which harbor methane-producing bacteria. Also, blacksmiths in both empires produced methane gas when they burned wood to fashion metal weapons. The combination of all such sources modified the composition of the atmosphere. Evidence trapped in Greenland's ice shows that a second methane rise in the atmosphere occurred in Medieval times, as Europe's economy emerged from the Dark Ages.

We therefore see a history of continuity. What has really changed since the Scientific and Industrial Revolution, followed by the Great Divergence, is that fast-accumulating technology and know-how massively increased human abilities, including that to shape the environment at a planetary level, eventually giving rise to what is now commonly referred to as the Anthropocene. Note that, for the longest time, and at least up until the 1970s, the very thought that human activity could affect the planetary atmosphere and ecosystem was considered preposterous.

While arguably it is taking much too long, the intermediate phase in which we are living is one in which the human species is still getting used to its "increased powers," in what has otherwise been two hundred thousand years of an uneven match with nature. This can be seen in a practice that humans have engaged in for at least forty thousand years: fishing. This activity started on riverbanks, but soon enough fishermen went to sea, then went progressively farther, hoping for good luck and catching what they could. Today, using advanced sonar and navigation technologies, humans have incredible advantages and are able to pull fish from the sea at a much faster rate than they can reproduce, putting us on a path to depletion.

Just because we *can* do things like this does not mean we *should,* but accepting this principle might require a change in mentality. As Albert

Einstein famously observed, "We cannot solve our problems with the same thinking we used when we created them." In 1793, the French National Convention proposed a norm that was later invoked in speeches by Winston Churchill, and eventually in the film *Spider-Man:* With great power comes great responsibility. This is what humanity will need to accept more and more going forward.

Rather than idealize nature as perfect, and human action as necessarily nefarious, and rather than call for dramatic retrenchments from the natural world, we should embrace the responsibility that comes with our dominant position on this planet. The very concept of the Anthropocene should evolve from our defining it as the era in which humans can affect Earth's geology, ecosystem, and climate. We should see it instead as the era in which humans acknowledge their power to do this and take full responsibility for the planet's future. A balanced view would take into account that there are other living forms and nonliving elements on Earth that should not be disregarded, but rather safeguarded precisely because we depend on them.[55] This perspective does not require that we attach the same value to human life and well-being as to all expressions of nature, and therefore retains the human-centric principles of Humanism and the Enlightenment.

The argument might be made that everything nature-related should be blindly preserved.[56] To the extent that we believe this, we open the door to a return of a pre-scientific mentality, which frames any manipulation of physical settings as unnatural and therefore to be rebuked.[57] Still, we must aim to understand and map the interconnections, to manage our complex impact on the planetary ecosystem. In this spirit, two practical examples are, in the private sphere, Microsoft's planetary computer, and in the public sphere, the European Union's Destination Earth initiative. Both aim to harness the power of current technologies, including high-resolution satellite imaging, big data, and cloud computing, to produce a *digital twin* for the planet—that is, a comprehensive model connecting data on the climate, sea levels, land use, biodiversity, water management, polar areas, and cryosphere, supporting more informed decision-making on human activities and their interaction with nature on various dimensions.

Mokyr makes an important point in his historical analysis of the constraints that held back innovation in the preindustrial era: "If the culture is heavily infused with respect and worship of ancient wisdom so that any intellectual innovation is considered deviant or blasphemous, technological creativity will be similarly constrained."[58] Applied to our modern times, the same could be said of the natural world, and the nostalgia toward an idealized "natural" past, when humans were in truth at the mercy of natural processes they hardly understood. Positivism—Baconian thinking—must be retained, or else our capacity to develop and spread far-reaching technological progress will come to an end. And with it, so will end the possibility of managing the challenges posed by climate change.

Embracing a green growth model, powered by technological innovation, is not about bracing for climate impact while hoping that some technology god intervenes to save the day. It is about understanding the process by which technology is developed in general, and particularly under capitalism. Eroding the capitalist system that makes resources available at scale—including for innovation and for powering its deployment and take-up at societal level—would undermine, rather than facilitate, the fight against climate change. If instead, the capitalist system can be harnessed to a green transition, because citizens are on board with it, progress toward a green economy can be rapid.

It is a lazy intellectual logic that declares, in our own era or any other, that "this time everything is different." Fundamental human needs, desires, hopes, fears, plays for power, and pains of inequality are better understood in the light of continuity.[59] This includes human relationships with both technology and nature. Historically, population expansion and the efforts to meet human needs have always entailed struggles against nature, the local environment, and the climate. Against these challenges, technology— accumulated into know-how through trial and error in less complex societies and through the scientific method since the Enlightenment—has always been humanity's preferred response. While the sheer complexity and potentially catastrophic implications of climate change seem unprecedented, our challenge has elements of continuity with the past. This is not to imply that we can sit back, relax, and let technology run its course; we must consider the key variable of time, and the tremendous inertia in the system. The longer it takes for new green technologies to spread through the productive system, the more other forces will come into play—as more catastrophic events follow from increases in average global temperatures, and more human suffering ensues.

There is much that policymakers, companies, and individuals can do to minimize both damage to the planet and harm to the human species.

Before moving into policy territory, however, we might take a moment to reflect on that term, *human species*. While climate change and many of the environmental problems we experience are global in nature, societies are organized along other lines. Policies are decided mainly by nation states— and climate change is expected to affect different regions of the world differently, carrying important political implications. In the next chapter, we will confront the geoeconomics of climate change.

7

THE GLOBAL FIGHT FOR PLANET EARTH

There's one issue that will define the contours of this century more
dramatically than any other, and that is the urgent threat
of a changing climate.
—BARACK OBAMA, 2014

I think we're lucky to be living when we are, because
things are going to get worse.
—SIR DAVID ATTENBOROUGH

In 1817, David Ricardo published his magnum opus *The Principles of
Political Economy and Taxation,* establishing the theoretical foundations,
together with Adam Smith, for what is commonly known as classical
economics. Fifty years later, Karl Marx published perhaps the most famous
book of the nineteenth century, *Das Kapital,* laying the theoretical foundations for communism. From a political standpoint, the two books could not
stand further apart on the spectrum. From an analytical point of view,
however, they have strong elements of commonality. Describing early-stage
capitalism, Ricardo had reached the conclusion that, left unchecked, the
inherent mechanisms of the system would have led to an impoverishment
of the masses, the enrichment of those possessing the means of production,
and conflict between wages and profits. Ricardo fell short, however, of
imagining an alternative. Marx took the same premises on capitalism and
turned them into a revolutionary call, making it his mission to identify
fault, place blame, urge change, and enlist disciplined belief.[1] Mutatis mu-
tandis, the current discussion on climate change and the economy carries a
striking resemblance.

It is now widely accepted in climate science and economics that economic growth, in its present form, based on fossil fuels and unfettered carbon emissions, cannot be sustained for much longer. The fact that the old ways inevitably need to change in turn opens the door to all sorts of imaginative scenarios on what could come next.[2] Indeed, books on climate change and the future economy have a very strong tendency to take a normative stance. Authors in this genre get easily carried away by idealistic future predictions, based on their own aspirations, rather than sticking to impersonal analyses of social facts—and typically adopt a moral angle, enrolling the climate crisis to advocate a broader, preexisting worldview or ideology. Many view the need to reshape capitalism to address climate change as an occasion to fix all past wrongdoings, such as global and local inequality, colonialism, patriarchy, racism, and even conspicuous consumption. This tendency is not confined to ecosocialist, post-growth visionaries. It is shared across the spectrum, all the way to mainstream green growth advocates. Take economist Jeffrey Sachs, for instance, who in *Common Wealth* casts climate action as an opportunity to fulfill the internationalist dream toward one-worldism.

This book offers a different perspective. First, it focuses more on what is likely to happen and to be feasible than on what would be ideal or desirable based on a particular moral compass. Climate change alone will be a colossal challenge, and adding extra layers of complexity will hardly make it easier to address—so conversations about the future of capitalism in light of climate change could benefit from some realpolitik. Second, this book has avoided using a generic "we" except when discussing quite basic elements of human culture and evolution, or the general process of economic growth. This is because *we* refers to a dimension—humanity—that makes sense from a moral standpoint, but doesn't as much from a sociopolitical, and therefore policy, perspective. This chapter turns, however, to that general concept of humanity, and then, indeed, explores what *we* can realistically expect in terms of cooperation on climate change and the environment going forward.

One planet, one people

In economics and political science, climate change falls under the general heading of *collective action failure*. This phenomenon was popularized in 1968 by an influential paper in *Science* titled "The Tragedy of the Commons."

In it, ecologist Garrett Hardin, building on an 1832 lecture by an obscure political economist, William Foster Lloyd, describes a hypothetical case of overuse of a shared resource. The resource in question is a patch of land (a "commons") used by many local people to graze their cows in the tradition of English villages. Each livestock owner adds more cows to the common, attracted by the greater personal gains that result from taking advantage of free pasture, and unconcerned about potential overexploitation. In the jargon of economics, this is a problem of *free riding*: members of a community failing to bear individual costs to maintain a shared resource. When eventually too many cows on the commons cause the complete depletion of the grass, everyone loses. Therein lies the tragedy. Hardin concluded that the only way to avoid it would be for the parties involved to collectively agree to some sort of coercion, in line with the Enlightenment concept of a social contract. Because the focus of his article was on human overpopulation on the planet, walking squarely in the footsteps of Malthus, Hardin effectively advocated for population control.[3]

So much of the "sustainable future economy" literature focuses on humanity, I believe, because of an understanding (which is true, albeit simplistic) of the atmosphere as a common pool resource, or commons, and therefore an understanding of climate change as a shared problem that will affect humanity as a whole. This challenge, the argument goes, which is also formidable in its magnitude and wide-reaching in its consequences, must be tackled by all countries agreeing on legally binding and coercive international treaties, in line with Hardin's conclusions.[4] Humans must therefore set aside old ideological fault lines and artificial national barriers and act in unison. In the words of Jeffrey Sachs, in light of climate change, "the very idea of competing nation-states that scramble for markets, power and resources will become passé."[5] This narrative has a certain inspiring appeal, especially for those who grew up in multicultural settings, traveling, studying, and living abroad. I suspect the future will look somewhat less idyllic.

In the past, expectations for international collaboration against climate change have fallen short. Reliably, the optimistic internationalists have read these repeated failures as simply part of dealing with a new challenge, and concluded that we just need to get used to cooperating in ways we have not cooperated before; members of the intelligentsia should therefore be even louder in expressing the need for deep cooperation and the benefits it would bring to all.[6] Of course, it is expected that these calls will include generic attacks on "dumb politicians" who are myopic and fail to see the obvious. A subgroup of those holding this worldview, the institutional (or global) governance reformists, see the main barrier impeding such global cooperation as the lack of a proper forum for it, criticizing the fora that do exist, such as the UN Security Council and the G20, for their flawed composition and operating rules.[7] In the words of globalization expert Ian Goldin, "global governance is *the* challenge of our time."[8]

In line with the Marxist tradition of placing blame, degrowth advocates see the roots of insufficient global cooperation on climate change in the greed of rich countries: blinded by "growth-ism," and at the mercy of multinational corporations, leaders in industrialized nations are not willing to shrink their economies and greenhouse gas emissions to open up space for poorer countries to develop. Such explanations are likely too simplistic to account for a complicated reality. More helpfully, historical precedents abound, and these can inform our expectations about the future.[9]

Climate change: From abstract to practice

The frequent initial logical fallacy is to imagine climate change as a single, homogeneous shock that affects all of humanity. The more complicated reality is that climate change will be manifested in increasing numbers of extreme weather events—droughts, floods, heat waves, hurricanes, and wildfires—and in rising sea levels and desertification. In terms of impact on their lives, people will experience decreased agricultural yields, failed crops, job losses, destruction of housing, forced relocations from coastal areas, shortages of clean water, frequent wildfires, and more. Uncertainty will rise to a permanently higher level, and relative scarcity will become worse.

Most climate experts agree that certain parts of the world, and specifically the Global South, will be hardest hit. In many cases, this is largely a matter of unfortunate geography, but there are also economic reasons. Climate adaptation, meaning the set of measures needed to respond to a changing climate, is expected to be expensive. Considerable resources will be needed in specific places to reinforce river dams or embankments to avoid flooding, relocate cities from coastal areas, rebuild infrastructure when it gets destroyed by tornadoes or hurricanes, prevent or contain wildfires, build new pipelines to secure access to water, and reinforce healthcare systems.[10]

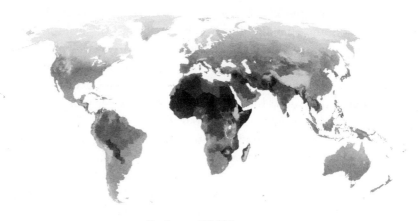

Deaths per 100,000

-1000 -900 -800 -700 -600 -500 -400 -300 -200 -100 0   100 200 300 400 500 600 700 800 900 1000

FIGURE 7.1: Mortality risk of future climate change due to changes in temperatures
Source: Carleton et al. (2020).

A granular, empirical model developed by environmental economist Tamma Carleton and colleagues allows for estimates of how rising temperatures due to climate change will affect death rates across 24,300 regions of the world. As shown in Figure 7.1, today's poor nations bear a disproportionately high share of the global mortality risks of climate change, due to geographical locations and insufficient resources to counter the threat. Current incomes, as well as current average temperatures, are strongly correlated with future climate change impacts. While the worst outcomes can be expected in poor, hot regions like sub-Saharan Africa, some cold (and rich) regions, such as Oslo, Norway, might even see life expectancies increase as a result of global warming.[11] Nonetheless, Carleton's analysis shows that, to keep mortality low, most high-income countries need to make large climate-adaptation expenditures.

Insufficient adaptation in the Global South is also likely due to the fact that governments in developing countries are more constrained in their borrowing on financial markets, further exacerbating a negative equilibrium, whereby those most in need of resources are those less likely to find them.

Climate change is therefore likely to cause, on one side, a rise in poverty, global inequality, famine, and epidemics. On the other side, rich countries will need to devote increasing resources to buffer the climate shock, possibly having to cut spending on other priorities and budget items or increase taxes at various levels of government.[12] Heterogeneity of outcomes and scarce resources, in turn, are likely to spark a surge in migration.[13] And this will likely destabilize internal politics and the relationship between wealthier countries and the Rest, just as it did in Europe following the outbreak of the Syrian civil conflict.

A point that does not get enough attention in the popular discourse, and yet is omnipresent in the technical literature, is that, whether we like it or not, this will to some extent happen.[14] Even if all polluting economic activity were to magically come to a halt tomorrow, some degree of climate change would take place due to the greenhouse gases already accumulated in the atmosphere. This should not come as a surprise, given that effects of global warming are already with us. In March 2019, Cyclone Idai took the lives of more than a thousand people across Zimbabwe, Malawi, and Mozambique; it was also the most expensive weather-related disaster in African history. In January 2020, Australia lived through its worst-ever bushfire season, following the hottest year on record, which had left soil and fuels exceptionally dry. In Venice, two of the worst floods in the past twelve centuries occurred in 2018 and 2019.[15] Five of California's six largest fires burned in 2020. In the same year, the Atlantic Ocean experienced its most active hurricane season on record, with thirty named storms—too many for the usual alphabetical sequence of names. In 2020, Kenya was afflicted by the worst infestation of desert locusts in seventy years, due to exceptionally wet weather in the eighteen months beforehand. In June 2021, Canada experienced what is known as a "heat dome," and recorded its highest temperature ever: 49.6°C (121°F). The list could go on, and the simultaneous occurrence of such exceptional events is not sheer coincidence.

This should not lead people and policymakers to fatalistic conclusions. Rather, climate mitigation has a strong risk management component, aimed at avoiding the worst of the worst, and even small changes in degrees of global warming can make a huge difference. At the same time, this is the global backdrop against which we should consider future prospects for cooperation.

Climate change will be experienced in the same way that nature and uncertainty have been in humanity's past. Global warming effectively sets the clock back to a time when humans were generally less in control of their environment. As in the past, people will aspire to develop and use new and more advanced technologies to push back against shocks in the local environment, or in the climate more generally. We are heading toward a world that is structurally more uncertain, where additional resources might be needed frequently and on a large scale to adapt to climate change and respond to disasters. All countries will be affected, but because poorer countries will be hit particularly hard, large-scale migration will occur.

Climate change as an abstract concept will affect all of humanity just as, in abstract, sickness afflicts all humanity. In practice, illness is more treatable in rich countries, because they have the resources required for healthcare systems, medical research and training, and state-of-the-art diagnostic instruments. Similarly uneven advantages will affect experiences as people cope with the practical consequences of climate change. The forces that underlie the quest for more—the growth imperative—will drive responses much as they have with past needs of humanity.

Against this background, what can be realistically expected in terms of global cooperation on climate and the environment? As in past chapters, let us start by taking a step back for a broader sense of context. In particular, evolutionary science can contribute useful perspective.

A primer on in-group bias

Nicholas Christakis's *Blueprint,* published in 2019, is the culmination of a titanic scientific and intellectual effort, pulling together evidence from the social sciences, evolutionary biology, genetics, neuroscience, and network science. In it, the physician-turned-sociologist sifts through decades of cross-cultural work to identify what he calls the "social suite"—a set of attributes at the core of all human societies. These include the capacity for tight friendship, social learning and teaching, and the preference for one's own group known as "in-group bias."

The work resonates with Joseph Henrich's research, noted in the last chapter, indicating that humanity's secret to success has been its capacity for cooperation. This gives rise to a kind of collective brain with no peer among other species on Earth. Tight cooperation, which opens the door to social learning, is facilitated by humans' predisposition to form groups and prefer those within their own groups. Humans create social dividing lines all the time, even on the most trivial bases. Randomly distribute red or blue sticky notes to students as a classroom experiment, and they will start generating in-group and out-group dynamics. A 2011 experiment found five-year-olds showing intergroup bias for other children who were randomly assigned the same T-shirt color. In this context, it is crucial to understand that we need an out-group to define the in-group.[16] Who we are *not* says much about who we *are*.

The origin of this in-group tight cooperation dates back all the way to the Pleistocene, the so-called Ice Age. At the time, the climate was more volatile, creating an environment in which competition for scarce resources favored groups that had brave and self-sacrificing members—in other words, in-group altruism in the service of out-group conflict. As noted by evolutionary biologist David Sloan Wilson in *This View of Life*, the stronger the outside pressure, the more humans tend to come together, displaying social behaviors in groups that are generally small compared to the total human population. This is a trait we share with other animals, including fish schools, bird flocks, and lion prides. Incidentally, building on Charles Darwin, Wilson shows that in-group bias allows scientists to make sense of the "paradox of goodness"—that is, the need to reconcile Darwin's "survival of the fittest" imperative with inherent human feelings of kindness, altruism, and self-sacrifice.[17] Natural selection for these traits has occurred at the group level, ensuring the survival of the fittest, most cooperative groups.

Interestingly, and preempting some objections that may arise in pages to come, studies from cross-cultural settings show that in-group bias, and the distinction between *us* and *them,* is strongest in collectivist societies such as communism.[18] Collectivist societal models depend on a strong ideology that elevates the importance of the group. Liberal capitalist societies allow for decentralized coordination and greater individual autonomy.

At a time of rampant populist and widespread xenophobia, it is worth stressing that in-group biases do not make "us versus them" rhetoric inevitable. As noted by Christakis, surveys and lab experiments suggest it can simply imply an "us *and* them." At a country level, this is the fine line between patriotism and nationalism. But, of course, the in-group bias primes humans for prejudice against "the other," and this leads us to another fundamental point. The fact that a trait is natural does not make it morally right. The fact that it is part of our "social suite," however, means we should be aware of it, as it will inform our understanding of human social history, and define the contours of future policy action.

Nation states as sophisticated in-groups

In 1983, political scientist and historian Benedict Anderson gained renown for his work exploring the origins of nationhood. In *Imagined Communities,* Anderson argues that nation states are built around the concept of *us* in antithesis to *them,* defining themselves through "differentiation" by what they are not.[19] He calls these communities "imagined" because "the members of even the smallest nation will never know most of their fellow-members, meet them, or even hear of them, yet in the minds of each lives the image of their communion."[20] By fostering such feelings of self-sacrifice and comradeship, nation states achieve in-group bias at scale, forging the cohesion needed to create and sustain larger and more complex societies.[21]

In her more recent book, *Symbols of Nations and Nationalism,* sociologist Gabriela Elgenius describes the sheer efforts that go into fostering cohesion and strengthening in-group sentiments within nation states. National symbols—including flags, monuments, anthems, and ceremonies on national holidays—heighten awareness of membership and illuminate the boundaries of non-membership by honoring and validating official founding myths and identity.[22] Celebrations make use of national colors, traditional clothing, and carefully staged choreography to transform the reality of diversity into the appearance of unity and similarity. Nationhood can be seen as a sort of secular religion.[23] Especially in moments of uncertainty, of which the Covid-19 lockdowns are just the latest conspicuous examples, groups see the greatest displays of flags and other symbols, encouraging them to be more certain of their collective identities. Traditions, heroes (typically leading revolutions and liberations), tales of sacrifice for community, and histories of sacred, golden-age pasts are often created, or conveniently magnified. Generating a collective narrative gives a sense of joint destiny, "chosenness," and membership in a moral community.

In 1971, John Lennon envisioned a different world in his classic song "Imagine," picturing a world where national borders would fall and humanity would come together in peace, sharing the planet.

In light of what we have learned about how an *us* requires a *them,* perhaps the song was well named. Humanity has not come together in this imaginary way. Rather, we observe what economists call a spatial discount factor: the farther apart people are geographically, the less they are willing to sacrifice for each other's benefit. This is true between families, towns, and states—and certainly, at the global level, between nations.

The sense of us

Understanding that all this has deep ramifications for public policy, economist Ricardo Hausmann has recently focused much of his interest, and the work of his Harvard Growth Lab, on "the sense of us" and its role in whether nations fail or thrive. How citizens define their in-group—all the people who make up *us*—determines the feasibility of all sorts of policies, from taxation and redistribution to immigration laws.

As a somewhat facetious example, Hausmann likes to tell the story of England's shock when the football team it sent to the 2016 Euro Cup lost to Iceland, a country of a mere 350,000 inhabitants, whose national team is a club of amateurs. Hausmann points out that Great Britain overall was represented by five separate teams: England, Northern Ireland, Scotland, Gibraltar, and Wales. Wouldn't it have fared better if it had competed as one team, just as Spain does and Germany does, despite having their distinct regions? This wasn't an option that Great Britain considered, Hausmann explains, because most of its citizens, even if such a combined team won, would not feel the thrill that *our* team has won. The same could be said of the European Union. If it competed at the Olympics as a united team, its share of medals would surely be on par with the United States and China, but it decides not to.

Turning to more serious realms of contention, most of the EU's internal policy battles can also be interpreted through this lens. It matters whether the *us* in the in-group is defined expansively as Europeans or narrowly as Germans (or French or Dutch or Italians). This profoundly affects combined efforts from creating common federal budgets to managing internal and external borders, and from dealing with public debt during the euro crisis to purchasing Covid-19 vaccine supplies at EU-level to avoid rivalry among Europeans.

What is feasible in terms of risk pooling or redistribution within a country is also based on how cohesive a nation is and, effectively, a people's perception of who counts as *us*.[24] In a fundamental contribution to the political economy literature, Alberto Alesina and Edward Glaeser analyzed why the United States does not have a European-style welfare state. In *Fighting Poverty in the US and Europe,* they work through a series of potential arguments to conclude that America's troubled race relations, and therefore ethnic fragmentation, is a major factor, because it reduces willingness to support redistribution and the enlargement of the welfare state.[25] In this, America is part of a broader pattern, as studies published since, looking at communities from small villages to cities to entire countries, suggest that ethnic fragmentation leads to smaller provisions of public goods in many places.[26]

What does all this imply for our matter at hand? As climate and environmental stresses escalate, we should expect increasing uncertainty and hardship. When threats of disaster and uncertainty reign, people tend to cling more tightly to their in-group, looking to overcome adversity together, including through greater risk pooling and sharing. In practice, this is likely to generate a turn of the tide in favor of a stronger role for government. For this to be feasible, however, there must be enough cohesion in the in-group which, given historical legacies and the sophisticated symbolism machinery of nation states, is likely to occur at national level. For countries that are not cohesive enough to step up risk-sharing arrangements and weather the climate storm together, low resilience over time might lead to fractionalization, social tension, and in the extreme case, civil war.[27] This in turn will reestablish an equilibrium characterized by smaller, more cohesive groups. For nation states that manage to stick together, efforts to reinforce cohesion will probably come at the expense of some out-group.[28] This will happen based on standard historical patterns of how the in-group bond gets fortified.[29] In plain English, we can expect nationalism, rather than internationalism, to increase as a result of climate change.

Reading through the evidence of how civilizations over millennia responded to environmental crises, and in some cases collapsed, makes evident that climate and environmental stress impact complex systems like human societies in indirect ways. Devastating as they may be, the direct impact of extreme events such as floods, fires, desertification, or hurri-

canes did not tip societies into chaos. Rather, it is mainly through the social unrest and tense political dynamics that these events, repeated in
time, contribute to generate or acquiesce that climate and the environment wreaked havoc.[30] As we concluded in Chapter 6, on the subject of
innovation, in the coming decades significant political attention will
therefore need to be devoted to safeguarding socioeconomic cohesion to
successfully navigate a changing climate.

Humanity versus climate change

This chapter has so far taken an extremely long-term perspective, drawing
conclusions from evolutionary human traits that were developed during the
Ice Age. The skeptical reader may wonder: what relevance can the Stone Age
possibly have for the twenty-first century? Thanks to immense human advances in scientific knowledge and cultural sophistication, combined with
a highly globalized economy and cosmopolitan leadership, surely humanity
will not fall prey of fractionalization, and rather face the common challenge
of climate change as one, right? Personally, I am a great fan of sci-fi movies,
especially of the alien-invasion and asteroid-collision genres. One element
I find fascinating is that there are some standard features in these Holly-
wood blockbusters, and one of them is that humanity quickly joins forces
to combat the challenge of the moment.[31] Could a common enemy, or an
external threat like climate change, unite all humanity? Beyond anecdotal
evidence, or cinematographic references, social psychology lends potential
support to this hope.

Professor Christakis reports of a psychological experiment conducted in
1954 in Oklahoma on twenty-two fifth-graders who did not know each
other. The kids were taken to camp in a state park, and the experiment took
place in three stages. First, they were divided in two isolated groups, where
basic in-group solidarity was fostered by the camp counselors through standard techniques like choosing a group name or designing team T-shirts.
Rapidly, each group adopted its own symbols and preferred songs, and created an internal hierarchy with leaders and followers. In the second stage,
the two groups were placed in conflict with each other, playing zero-sum
games, such as baseball, football, or treasure hunt. Over the course of
the competition, the groups came to dislike each other more and more,

displaying derogatory attitudes, and cementing a strong in-group feeling (which did not imply the complete absence of occasional in-group bickering). In the final stage, a common threat afflicted both groups, when scientists secretly sabotaged the camps' shared water tank. Rapidly, the kids started collaborating to get water flowing again. A few days later, the experiment was over, and quantitative measures of out-group negative stereotyping had declined while intergroup friendship had risen. The overall takeaway, or general principle, is that in-group boundaries can rapidly broaden, and particularly so by a shared agenda, which facilitates cooperation.

There are unfortunately plenty of reasons to suspect that climate change in the twenty-first century is rather unlikely to serve as a uniting foe. First, and most tellingly, because it is not the first time climate change interacts with humanity, even in more recent history. And when this happened, it increased intergroup conflict over resources, rather than leading to bury all hatchets.

In *Global Crisis,* historian Geoffrey Parker painstakingly connected the dots of what happened four hundred years ago, when economic and social upheaval sparked throughout the globe, from England to Japan, from the Russian Empire to sub-Saharan Africa, North and South America to boot. The seventeenth century saw a proliferation of wars, civil wars and rebellions, and more cases of state breakdown around the globe than any previous or subsequent age. Parker argues that the so-called "Little Ice Age," and its peak in the seventeenth century, was at the origins of the turbulence. All of this echoes the work of archaeologist Brian Fagan, who had reached similar conclusions already at the turn of the millennium.[32]

In 2017, economists Murat Iyigun, Nancy Qian, and Nathan Nunn constructed a dataset that records the date and location of each conflict in Europe, North Africa, and the Near East from 1400 to 2000.[33] Then they merged this with historical temperature data to show that, indeed, the period of cooling associated with the Little Ice Age led to an increase in conflict. They were even able to leverage their database to show that the phenomenon was symmetric: in periods of agricultural abundance, as for example when potatoes were introduced from the Americas, there were sharp and persistent reductions in conflict, mainly driven by declines in civil wars.

The evidence piles up beyond the Little Ice Age. Studies show that the introduction of drought-resistant sweet potatoes led to a persistent reduction of rebellions in China between 1470 and 1900.[34] On the other hand, weather shocks, in the form of reduced rainfall, have been shown to have contributed significantly to the onset of civil wars in sub-Saharan Africa, and the effect is confirmed also in richer African countries.[35] Similar patterns were found in Egypt, where Nile floods across centuries were associated with social unrest.[36] Recent interdisciplinary archaeological and paleoenvironmental research in the Arabian Peninsula confirm the pattern.[37]

There is also no reason to believe that this dynamic was at play only in the ancient past. In 2015, a group of California-based economists reviewed over fifty quantitative studies and found evidence in the modern period (1950 onward) that extreme temperatures and rainfall have led to extraordinary rises in intergroup conflict.[38] Their takeaway is that, across cultures and time, irrespective of the level of development, an increase in perceived scarcity creates the conditions for a scramble over available resources. Conversely, abundance paves the way for cooperation, between groups and in international relations.

Second, we saw what climate change will look like in practice. It is not a single-shot abrupt event, but likely to be the new state of the world for decades, if not centuries. It will not resemble, for instance, the 2009 financial crisis, which spread quickly, visibly, and tangibly. In that case, G20 countries came together to muster a strong coordinated fiscal response.[39] Against more slow-moving threats, the global governance forum has proved much less effectual.

Third, the starting point is one in which countries are highly unequal, in terms of not only wealth, but also population growth, aging profiles, and technological advancement. This translates to different needs and priorities. Moreover, it fuels resentments about unfairness, in the same way that such tensions arise among groups and individuals within societies, making collective efforts difficult. Climate change and extreme weather events affect different countries each time, and some disproportionately overall.[40]

The world's response to the Covid-19 pandemic was a foretaste of how this will all play out.[41] Incidentally, more frequent epidemics represent just one way in which climate change will materially affect lives. In principle, a pandemic is a homogeneous shock: nobody is immune, all countries are affected. But reality is much more multifaceted, with countries having different age profiles, different access to financial markets, different political situations, and different value systems—and these differences in turn determine which policies are feasible and which are not. In the face of "a global problem that would require a global solution," we saw that people's reaction was to stockpile personal protective equipment. Perhaps even more concerning, twelve months into the global pandemic, as soon as they were developed, vaccines were hoarded by rich countries. When vaccine supplies were scarce, rules for local producers were put in place under the banners of "Britain First," "India First," and "America First."[42] Of course, from a moral perspective, we can see it as despicable that a young, healthy Israeli could be vaccinated earlier than an older person working as a healthcare first responder in an African country. On the other hand, we can see it as predictable: in the face of scarcity, in-groups privilege their own. Note that this is the case even if, rationally, it cuts against self-interest. On vaccines specifically, it would have made sense to act in unison: if the virus is not eradicated worldwide, the risk remains that mutations will arise and push the light at the end of the tunnel further away for all.[43] Unfortunately, such rationality runs counter to the social contract on which in-groups like nation states are formed and held together: first, they take care of their own. To be perceived as violating this principle is to invite social upheaval. This is why it is crucial to generate abundance—in this specific case, enough vaccine supplies to ensure wide global availability. With abundant supplies, and international cooperation initiatives such as COVAX and donations of doses to poor countries by the G7, it becomes possible to beat Covid-19.

Geopolitics amid climate change

As climate activist Greta Thunberg likes to remind her audiences, when the house is on fire, you stop whatever you are doing and focus on extinguishing. The message is that in our climate emergency all other, petty concerns must be set aside. Applied to international relations, this aspirational thinking has some believing that climate change will imply the end of conflictual geopolitics.[44] There are practical reasons why this type of thinking might have gathered momentum. After all, much of the world's recent conflict history, including war after war in the Middle East in the 1990s and early 2000s, was driven in large part by issues related to petroleum. It might seem logical that once economies, as part of their climate mitigation efforts, switch to renewables like solar and wind, conflictual international relations will become a thing of the past. Unfortunately, that change is very unlikely to represent the end of geopolitics.[45]

First, the overall premise is wrong, as oil and gas will not disappear any time soon. Rather, we will observe a gradual shift in the energy mix from traditional energy sources toward renewables.[46] An effort to fast-track the demise of highly polluting coal, while renewable capacity is still building up, might initially lead to a counterintuitive increase in the use of methane—which is, at least for Europe, another source of conflictual geopolitical relations, in its case with Russia.

Second, as discussed in *The New Map* by energy expert and Pulitzer Prize winner Daniel Yergin, we are by no means close to seeing the end of geopolitics and power politics in the world. Specifically in the realm of energy, new materials such as rare-earth metals will become focal points of contention. Because these minerals are key inputs to the construction of wind turbines and electric cars, and will become fundamental to our technology, varying levels of access to them will reshape the world order and create new imperatives to ensure the safety of supplies.[47] Military strategies and foreign policy priorities will be reoriented toward that goal, just as attention has been focused over the past decades on securing oil supply bottlenecks like the Strait of Hormuz and the Strait of Malacca.[48] Indeed, we can already get a sense of this, as the European Union has published, and regularly updates, a list of what it calls "critical raw materials," necessary to ensure a strong industrial base to the European economy. The US administration did the same, identifying thirty-five critical minerals. For example, cobalt is featured on both lists, as a fundamental mineral in low-carbon technologies like electric vehicles and batteries, including those used every day in smartphones and laptops. Responsible for almost 60 percent of the world's known reserves and 97 percent of global exports of cobalt is the Democratic Republic of the Congo, suggesting that this country will increasingly enter the spotlight of global power politics.[49]

Any predictions about geoeconomics must also acknowledge that climate change is not the only megatrend at play; it overlaps with others, including digitalization, automation, quantum computing, artificial intelligence, and big data, all of which exhibit important winner-take-all effects, and in all of which countries will compete. A hugely important megatrend is the rise of China and the re-dimensioning of the United States after roughly a century of global economic and political dominance.[50] To the relative "optimists," like political economist Dani Rodrik, the relationship between China and the United States could evolve into a milder, second Cold War or mutual recognition of existence and differences—an *us* and *them* dynamic within a framework of continued reciprocal trade and investment. To the less optimistic, as the global center of economic gravity shifts eastward, the risk of escalation to more open confrontation between superpowers is high.

In his 2017 book *Destined for War,* historian Graham Allison turned his attention to what he called the Thucydides trap.[51] Analyzing the Peloponnesian War that devastated ancient Greece, the Athenian historian Thucydides concluded that "it was the rise of Athens and the fear that this instilled in Sparta that made war inevitable." How will the rise of a new superpower play out in the twenty-first century? Allison reviews sixteen instances of swaps in economic supremacy between countries over the past five hundred years and paints a grim picture: war broke out in twelve of them. This does not imply that war is inevitable, but suggests the Sino-American relationship will shape geopolitics in decades to come. Climate mitigation and international cooperation will play out against a backdrop of rising geopolitical tension between at least two superpowers, which will inevitably push other countries to pick sides.

The politics of degrowth

The degrowth international vision to tackle climate change centers on the proposition that wealthy countries, who have more than enough, will shrink their economies. By doing so, they will bring down greenhouse gas emissions, reducing the risk of catastrophic climate change. And they will do so to such an extent that space will be opened up for poorer countries to develop, while the overall world economy and global emissions remain within safe limits.

From a strictly moral standpoint, this might sound reasonable and desirable. In light of what we have seen up to now, however, we can state categorically that there is no scenario under which a small set of countries willingly makes such an act of self-sacrifice for the global common good.[52] While people are primed for altruism and self-sacrifices, these are directed first toward their in-group, and only later toward others, following geographical discounting. Democratic governments are bound by their citizens' preferences, and therefore these basic human tendencies define the set of policies that can be considered feasible. Naturally, citizens of a country constitute only one type of in-group. Groups can also arise based on shared culture, kinship, language, religion, geography, and other affinities, as emphasized by Amartya Sen in *Identity and Violence*.[53] Humanity as a whole, however, cannot act as an in-group in the absence of an out-group, and this will not change in the face of climate change.

The fond hope that some countries will actively shrink their economies ignores the reality that these rich countries themselves will remain in conditions of *relative* scarcity, which will only be felt more acutely as climate change unleashes its negative effects. This implies that the in-group will be in constant need of extra resources, to tackle the needs of the moment, which can be addressing incipient desertification, building infrastructure to prevent flooding, and the like.

In addition, countries themselves, as we have seen, experience relative income theory, and will constantly compare themselves to the past and to other countries. While China's economy expands at a high clip, you cannot expect the United States to look on with indifference. This is not only based on vague predicaments like "perceptions," or sinful feelings like envy, but also on hard realities, like the fact that economic might goes hand in hand with military spending and therefore international political influence. This, as we have seen before, generates a sort of geopolitical growth imperative. Reduced resources imply less capacity to protect (or project) a value system, paving the way to reduced self-determination as a people.[54]

The idea, moreover, that if rich countries shrink poor countries can expand is based on an incomplete understanding of the economics of development. In 2008, a group of nineteen leading policymakers, mostly from developing economies, headed by two economics Nobel laureates, put together *The Growth Report*, analyzing the experience of thirteen countries that had managed to sustain high GDP growth since the 1950s.[55] Drawing on the input of over three hundred distinguished academics, on top of the personal hands-on experiences of the policymakers, the report sifted out common traits among successful cases. These thirteen cases of "miracle" development, all of which featured sharp reductions in extreme poverty, included China, postwar Japan, South Korea, Indonesia, Malaysia, Brazil, and Taiwan—and a critical feature of literally all of them was fast expansion of exports.[56] Note, however, that most exports go to foreign lands that feature consumption aplenty—that is, the rich, growing, industrialized countries.

To recognize this is to see the flaw in the logic of rich countries having to shrink to open up space for poor countries to grow. The whole idea is based on a misreading of the global economy as a zero-sum game. Pursuing a degrowth agenda in the developed world would bring about a collapse of global trade, closing the door to any hope of fast growth in poor countries, turning economic miracles into mirages, and forcing millions to remain in extreme poverty.[57]

The principle that rich countries should bear the brunt of climate mitigation so that poor countries can develop was central to the 1997 Kyoto Protocol—and today, that narrative is widely regarded as the original sin that was the downfall of that early effort to curtail global warming. Kyoto divided the world in two by centering on historical emissions: Rich countries, which have produced the bulk of man-made CO_2 emissions since the Industrial Revolution, should have greenhouse gas emission reduction targets; other countries should be exempted, to give them space to develop. Perhaps unsurprisingly, the United States, which was the largest greenhouse gas emitter in the world at the time, did not even sign the agreement, seeing it as a treaty that would undermine its competitiveness. Canada withdrew from the Kyoto Protocol in 2011, after failing to meet its targets. Meanwhile, global emissions continued to rise due to the strong economic performance of China and India, which were labeled as "developing," and therefore exempted from binding reductions.

The very same principle was reproposed for the follow-up to Kyoto, and effectively sapped the Copenhagen climate summit in 2009, before being finally abandoned, paving the way for an agreement in Paris in 2015.

Managing the common atmosphere

We kicked off this chapter by introducing the tragedy related to an unmanaged common, noting the parallels with the degradation of the environment, and the planet's atmosphere. That very literature inspired the highly influential work of the first woman to win the economics Nobel memorial prize—the late Elinor Ostrom—who published her most famous book, *Governing the Commons,* in 1990. Sharing dozens of case studies from around the world, including inshore fisheries, grazing areas, groundwater basins, irrigation systems, and communal forests, Ostrom sent a message of hope. In many of these instances, people had managed to organize, building on interpersonal trust and self-governing arrangements, to prevent depletion and overcome tragedy. Ostrom's work is frequently referenced in discussions of international relations and climate mitigation as evidence that the problem of coordination between countries can be overcome. There need be no all-powerful global government with coercive powers if countries talk with and trust each other enough. Those advancing this argument might benefit from revisiting Ostrom's work and hearing the counterargument straight from the horse's mouth.

Governing the Commons focuses exclusively on small, local public goods, rigorously placed within one country and affecting groups numbering fifty to fifteen thousand individuals.[58] Ostrom herself, reflecting on climate change in 2009, argued for moving on from the belief that global problems can only be addressed by global solutions, particularly because the chances of producing globally binding treaties are so slim. Given the severity of the threat and the complexity of the issue, rather than waiting for the global-treaty silver bullet, it would be wiser to kick-start action at lower levels of governance, in communities, cities, and regions, and action by nonstate actors, along the lines of what Ostrom calls a "polycentric approach."[59]

The internationalist diehards might at this point pull out the fact that binding international treaties have been successful in addressing global environmental public goods in the past, and point in particular to the case of the fight against the "ozone hole." The story goes as follows.[60] In the 1970s, a group of scientists started to suggest that chemicals known as chlorofluorocarbons (CFCs) were endangering the ozone layer that protects the Earth from solar UV rays that are harmful to living things. In 1987, world leaders came together and agreed on the Montreal Protocol, analogous to the Kyoto Protocol for climate. Industrialized countries would start phasing out CFCs, while more time was given to poorer countries. Three years later, in 1990, the London Amendments adopted even tighter standards, and rapidly the world headed off a major man-made threat. At the end of 2020, it was announced that the largest ozone hole, over the North Pole, had finally closed. While surely some hopeful lessons can be taken from this episode, the parallels with climate change should not be overplayed.

First, the ozone problem was pegged to a specific type of gases—CFCs—which were widespread but mostly used in three products: refrigerators, air conditioning, and hairsprays. Compare this with greenhouse gases, which are emitted by basically every means of transport, by land, air, or sea, implicated in agricultural fertilization, the bulk of electricity production, home heating, construction materials, and even produced by belching cattle. In other words, greenhouse gas emissions are much more pervasive.

Second, the ozone problem was more immediately evident and visible. In 1985, NASA released a satellite picture of the gaping ozone hole over the Arctic, which helped raise the immediate attention of citizens to the problem, in turn exerting pressure on governments. Climate change, if left unchecked, might very well be more catastrophic in its consequences than the ozone hole, but it is more related to a chain of events that are still hard to grasp, even through highly sophisticated climate models. Crucially, it relates to an increase in probability of very harmful tail events, along the exponential risk logic we have become familiar with during the Covid-19 pandemic, rather than a linear progressive increase in risks and damages.

Third, from a technological standpoint, addressing the ozone layer problem was way simpler, as a cheap substitute to CFCs was rapidly found. Indeed, after the initial resistance of multinationals like Dupont, it was the private sector itself that pushed governments to tighten standards even further with the London Amendments. CFCs were phased out simply and seamlessly. Hardly the green revolution that will be required to eliminate greenhouse gases.

This mistaken comparison of apples and oranges led to the Montreal Protocol's serving as a model for the Kyoto Protocol on greenhouse gas emissions a few years later.[61] This included the fundamental principle of drawing a clear distinction between industrialized countries with emission targets, and poor countries which were exempted, nipping the agreement in the bud.

The globalization trilemma: Climate edition

Climate change is hard to manage at the global level through binding international treaties, primarily because of the nature of the problem. Given its pervasiveness, it clashes with all sorts of principles, including national sovereignty and basic liberal democratic norms.

In 1997, Dani Rodrik broke ranks with the pro-globalization mainstream economic consensus of the time by publishing a provocative booklet, *Has Globalization Gone Too Far?*[62] Taking his contentious ideas further, in 2011 Rodrik published *The Globalization Paradox,* where he introduced a famous political trilemma. Many nations were trying to deal with the very high

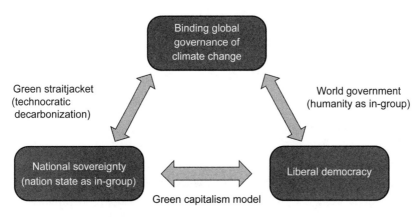

FIGURE 7.2: The global climate trilemma

degree of globalization he called hyper-globalization, while their policy-making was subject to democratic politics, and while insisting on asserting national sovereignty. The three simply cannot be combined, Rodrik concluded. One has to give.

Surely one reason that this articulation of a trilemma is so powerful and has been applied to such a variety of contexts since, is that, however unknowingly, it is firmly rooted in anthropology and evolutionary science. Specifically, it recognizes the interplay of the in-group bias, identity formation, and societal cohesion discussed above. While retaining its basic principles, we can easily apply Rodrik's trilemma to our own challenge, understanding the tensions among international agreements on climate change, liberal democracy, and the nation state (Figure 7.2).

Let us look at the individual corners of the triangle. At the top is a binding global governance agreement to set the required path of decarbonization for humanity, building on the principle that "a global problem requires a global solution." Clearly, this technocratic approach can pose some problems with the other two dimensions, not only in the moment in which some large countries refuse to sign up, like the United States with the Kyoto Protocol, but also dynamically with the vagaries of democracy. We had a taste of it when a democratically elected president decided to withdraw from the Paris Agreement, fulfilling, however, a national campaign promise.

Moving on to the bottom-left corner, the 1648 Treaty of Westphalia established the central principles of modern international law and the standard notion of national sovereignty. First, nations are sovereign and have the fundamental right of political self-determination. Second, states are free to manage their internal affairs without the intervention of other states. The current Westphalian system requires that countries consent to joining international agreements, and all agreements are therefore essentially voluntary (Treaty of Vienna 1969, Article 34). Clearly, the Westphalian system generates a dilemma with respect to global binding climate treaties.

Finally, liberal democracy is a form of government whereby citizens take an active part in decision-making, with a crowning moment being elections, and operating within the limits of some personal freedoms. This mechanism is most likely to operate smoothly within the contours of an in-group, where people agree they are bound by a common destiny and therefore respect the right of alternative political views to compete for leadership and win. Up until now, the nation state has been the largest in-group to be able to sustain the cohesion necessary to allow for common decision-making, compatibly with liberal democracy.

Let us now look at sets of pairs within the trilemma. When proclaiming one-worldism, economist Jeffrey Sachs explicitly states it is time to get rid of the nation state, retaining the other two dimensions, and effectively bringing liberal democracy to the planetary level. When calling for binding treaties, dictated by science alone, without advocating for the abolition of nation states, the ecosocialist planned economy edges toward the demise of standard principles of liberal democracy. This is something we have seen at length in Chapter 3, and is occasionally explicitly recognized by some degrowth advocates themselves.[63]

Clearly, what I have been advocating more or less explicitly throughout this book, and especially in this chapter, is that we stop hoping that a global problem will be solved by a neat, first-best, global-treaty type of solution. Rather, it is better to look for second-best solutions that operate through the nation state and safeguard liberal democracy. This view has been predicated on multiple grounds, generally resting more on practical, rather than moral, considerations. First, liberal democracy has proved a crucial tool to foster creativity and innovation, which under any scenario will be direly needed to stave off climate catastrophe.[64] Second, bringing the concept of in-group at global level, and generating a tight cohesion among all humanity, is unlikely to work, neither over the thirty-year planning horizon that climate mitigation dictates, nor at any point in the foreseeable future. Third, decarbonization and climate mitigation will effectively touch on every aspect of policymaking and people's lifestyles for decades. The Reverend Martin Luther King Jr. once remarked that "budgets are moral documents"—and in the years and decades to come, budgets will be dictated in no minor part by climate mitigation and adaptation. Each nation will need to devise the most appropriate country-specific decarbonization strategy, compatibly with prevalent national value systems, implying that no "green straitjacket" can credibly be imposed top-down.

With the climate trilemma in mind, one should not be surprised that mega agreements such as Kyoto and Paris did not contain explicit binding mechanisms—a cause for vigorous complaints by climate advocates.[65] The Paris Agreement is built around so-called nationally determined contributions. This is a bottom-up process allowing countries to come up with their own carbon reduction plans, with no guarantee that the plans will collectively keep global temperature rise well below two degrees Celsius. Most countries, indeed, are not even on track to meet the self-determined pledges they made under the original Paris agreement (let alone the more ambitious ones presented in Glasgow), and there is little that can be done top-down to coerce them.

Because climate agreements cannot be credibly binding, and will rather remain voluntary, any sort of discussion on the "fair share" of greenhouse gas emissions reductions for each country will inevitably run into a dead end. This could seem a rather depressing and hopeless conclusion. And yet, all it is meant to imply is that any successful scenario of climate mitigation must be made compatible with the iron law of national self-interest. It is for this reason that technology-induced cost reductions, of the type we saw in Chapter 6 for electricity generation, are so important. This dynamic suddenly turns the green transition into a matter of direct and immediate commercial self-interest for a country, shaking off any need to rely on voluntary self-sacrifice for future generations or humanity as a whole.

From national to planetary green transition

Often quoted is this quip by British prime minister Winston Churchill: "Democracy is the worst form of government, except for all the others." Similarly, a green transition carried out in the pursuit of national interest is not optimal, but probably the most feasible and least worst option, taking all dimensions into account: climate, social, and geopolitical. And while this is not a global solution, there are reasons to believe that it can be engineered to ensure propagation across the world.

The master plan would pan out as follows. Industrialized countries with high greenhouse gas emissions would power through with the green transition, and this is more and more likely to happen now that, as we saw in Chapter 6, this is becoming a profitable investment opportunity. This includes not only the West, or OECD countries, but also China, which from the standpoint of resources and technological capacity is well positioned to be a green early adopter. These countries have a strong, self-interested incentive to fast-track the adoption of these new technologies to gain an edge in what will be the energy and production system of the future. By so doing they will be capturing network effects, setting regulatory and technological standards, and effectively shaping the very course of the green industrial revolution.[66]

In the process, early adopters will show the way to the rest of the world, through two channels. First, their actions will prove that the transition is feasible, and even beneficial, given local benefits such as a better air quality and positive effects on citizens' health and well-being. Second, they will demonstrate how such a complex transformation can be achieved in practice.

When this happens, we will fall back onto a well-known dynamic: the *international demonstration effect*. First postulated by Estonian economist Ragnar Nurkse in 1957, and later empirically verified by a variety of studies, this is the process by which people in poorer countries tend to adopt the products, technologies, and trends of richer nations. We see this process of emulation within countries, as discussed in the context of the innovation treadmill in Chapter 5. Nurkse, formulating his theory in the aftermath of the introduction of a powerful early information and communication technology—the television—argued that this also applies at the global level. The propagation effect he described is probably even stronger today thanks to the internet and, more specifically, social media.

As green technologies become a permanent fixture in advanced economies, these production methods and consumption patterns are likely to expand throughout the world, just like blue jeans, Coca-Cola, and Facebook have in the past. Initially, through the highly interconnected cosmopolitan rich elites emulating green icons of success and modernity. Later, through the standard innovation treadmill within countries. This, incidentally, is what Lant Pritchett was noting in his example of how advanced-country technologies get propagated to the developing world, to an extent irrespectively of local relative prices (see Chapter 6). Of course, this will leave ample space for local experimentation and adaptation.

Within this context, having abandoned the logic of self-sacrifice, we can then easily imagine rich countries helping out less developed ones in the energy transition, including through financial transfers, for two reasons. First, by pursuing a model based on a rapid green transition that fosters economic growth and jobs, and technological innovation for climate adaptation, advanced economies will have yet more leeway to devote extra resources to foreign aid, including for a global green transition. Again, abundance rather than scarcity fosters international cooperation. Second, aid will also be powered by a logic of national self-interest, given that the companies of early adopting countries will detain an edge in green technology and infrastructure, and governments will want to push their market and standards also abroad. In other terms, development policy, international climate mitigation objectives, geopolitical interests and national industrial policy will all be closely aligned.

Other motives that could favor large financial transfers to ease the transition abroad could include the protection of some specific global common goods, as for example the Amazon Forest. One could see from this perspective the significant international financing connected to the so-called Great Green Wall, to contain desertification in Africa. Clearly, these targeted financial transfers could also be channeled through international and regional development banks.

To a more limited extent, we are likely to observe important financial transfers connected to the hosting of some green energy generation facilities, such as solar panels or windmills. Based on natural factors, the best locations for these facilities are often found in developing countries, which could then export energy to high-consumption countries.[67] This could be seen as a sort of energy outsourcing, creating local jobs and transferring technological know-how, just like manufacturing outsourcing has over the past decades.

Finally, there are solid reasons to expect that over the coming decades, as the effects of climate change will be more and more visible, a green agenda will rise to the top of the priority list. This will constitute another push factor for a global green transition, even in the absence of binding treaties. This is particularly the case in emerging economies, which currently are the most reticent to prioritize decarbonization, but which will feel the strongest early pinch from climate and environmental stress.

At this point, the reader could be left wondering: Is there a role for global international agreements, like Paris, and its Glasgow update, or are they useless in light of the above? To be clear, they are not useless at all. Climate treaties play a crucial role by creating forums where countries come together, use standard frameworks to assess their progress, share items of concern, exert some useful peer pressure, and even arrive at some very targeted solutions.[68] As with the example of the ozone hole, it is perfectly possible that successful agreements will be signed to adopt some joint policies and fix specific problems. For instance, it is increasingly clear that averting depletion of fish stocks and catastrophic biodiversity collapse calls for, aside from avoiding extreme climate change, establishing no-fishing zones at sea and large natural reserves on land.[69] Alliances might similarly be required to stop the damage thawing permafrost and retreating ice shelves in the Arctic and Antarctica.

Multilateral, all-encompassing international treaties should be seen for what they are: culminations of aligned national greening political agendas, rather than stepping stones toward top-down, technocratic decarbonization, or silver bullets to defeat climate change. They should not exclude other binding international agreements that are smaller both in scope (focused on selected environmental policies) and in size (involving fewer parties). Recent proposals for a joint EU–US, or possibly G7, initiative for a common tax on carbon, would fall into the latter category.[70] Especially among countries that are relatively close in terms of their value systems, levels of development, industrial structures, and political priorities, agreements like this are within reach.

In this chapter we have gained more reason to doubt that, faced with the threat of catastrophic climate change, the right move is to curtail economic growth. That recommendation is predicated on the delusional principle that, under the pressure of bounded resources, people will come together and share peacefully what is available. In the history of past human experience with climate and environmental stress, and in the principles of evolutionary anthropology and cognitive science, there is no evidence that world peace and humanity-wide sharing would be the most likely outcome.

Abandoning economic growth would weaken the scope for innovation, exacerbate perceived scarcity, and reduce resilience—and do so just as climate change was beginning to unleash its powerful effects. It would, in the face of scarcity and heightened uncertainty, set off a scramble for limited resources as people retreated for protection to their nation state in-groups.

We can chastise all this as offensive to morality, but ignoring these human traits and historical lessons will most likely take us where we do not want to go. On the other hand, in my view, our moral compass should guide us all to try and avoid at all costs a return to a zero-sum world and a hyper-conflictual foreign relations scenario. History shows how, when the climate and environment start changing, what we should be most afraid of is in-tergroup conflict. And the only difference today, with respect to the past, is that human weapons are much more deadly.

As societies find themselves between a rock and a hard place—facing catastrophic climate change on one side, and catastrophic conflict over bounded resources on the other—the escape route is narrow but, given the alternative, must be pursued.[71] In abstract terms, it involves using techno-logical innovation and human ingenuity to shift away from the logic that salvation can only come with national sacrifice for the global good, or fair-share valuations, to the logic that casts decarbonization efforts as invest-ment opportunities.

In practical terms, under this new logic, industrialized countries can be expected to act enthusiastically as early adopters for green technologies, at-tracted by the prizes of being at the forefront of the next industrial revolu-tion and able to set its standards for years to come. In turn, this industrial revolution will reverberate through the world, following the standard paths of capitalism and globalization, propelled by the deep-rooted mechanics of emulation, and financed in part by foreign aid.

This is a race against the clock, and the stakes are as high as world peace. The full power of capitalism must be harnessed and driven toward achieving speedy decarbonization, and it is evident that this will require an extensive set of policies. In the words of evolutionary biologist David Sloan Wilson, the mission is to "direct the process of cultural evolution toward planetary sustainability."[72] In practice, this means seizing the viable middle ground between ineffective laissez-faire policies and illiberal command-and-control solutions. Chapter 8 lays out an economic strategy to jump-start the greening of capitalism, and the societal overhaul that will go with it.

8

A BLUEPRINT FOR GREEN CAPITALISM

It always seems impossible, until it's done.
—NELSON MANDELA

In 1781, in his *Critique of Pure Reason,* Immanuel Kant cautioned that "experience without theory is blind, but theory without experience is mere intellectual play." Averting catastrophic climate change is a very concrete challenge that requires practical policy solutions. Many books on the topic, however, including ones by consummate policymakers, jump head-first into the problem without exploring its socioeconomic foundations. As a result, they often read like long laundry lists of policies, technical fixes, and other to-dos. By failing to explain the socioeconomic mechanics involved, these books undermine the credibility of their own recommendations.

This book has taken a different path—some might say it has taken the long way—by first exploring the theory underlying the human origins of the quest for growth, and the mechanics of capitalism and innovation. Now, we switch gears to translate into practice the principles laid down and to assemble a credible policy agenda to tackle climate change. After all, theory without practice is just as incomplete as practice without theory.[1] Building on the previous chapters, a credible strategy of climate mitigation will rest on six defining principles:

Growth is imperative. Progress against climate catastrophe will take place within a capitalist system and proceed in parallel with the continued quest for economic growth.

The price mechanism shapes supply and demand. A successful decarbonization strategy will shift (relative) prices to reorient demand, supply, and the powerful innovation and diffusion machinery of capitalism toward zero-carbon technology across sectors.

Government action accelerates change. Governments will need to take active steps to jumpstart and accelerate the green transition, which will otherwise proceed too slowly.

Social cohesion must be maintained. Avoiding intense political backlash to the climate agenda will require policy measures to protect social cohesion in the midst of accelerating technological innovation and economic transformation.

Early adopters pave the way. Rich countries, which can afford the upfront costs of the transition, should take the lead in climate mitigation, while working to promote green transitions in the rest of the world.

International cooperation can achieve loose coordination. While climate mitigation efforts must play out primarily through national and subnational solutions, they can be loosely coordinated at the international level, as through the 2015 Paris Agreement and the 2021 Glasgow Climate Pact.

This counts as a credible strategy; it is devised to be compatible with individual, group, and national incentives, and also takes into account the current power relations in society. These characteristics in turn maximize the strategy's overall political feasibility and therefore viability. By contrast, any utopian vision built on the premise that "the future can be whatever we want it to be" is not credible. In all likelihood it will fail to garner political backing, prove illusory, or rapidly turn into an undesirable dystopia.[2]

In line with the principles listed above, the most credible policy approach will aim to harness the forces of capitalism. For this efficient machinery to be reoriented to the societal priority of the moment, however, strong government action is needed, even if its role is time-limited. This strategy, then, takes a middle ground between those advocating radical laissez-faire and those promoting command-and-control solutions. In the midst of Covid-19 and its economic fallout, extraordinary policy measures were taken to shore up society, cutting through entrenched ideological positions regarding the proper role of government in the economy. Measures were taken that would previously have been thought radical or impossible. Having swung the political pendulum in favor of government intervention, the Covid-19 crisis has created a unique opportunity to jump-start the green transition.

With that level of intervention, would our system still be considered a free market? The model proposed here is definitely one that moves away from the neoliberal tenets, such as low taxation, embedded in the so-called Washington Consensus. It also moves away from unfettered free trade, complete capital liberalization, and minimal regulation. At the same time, it remains solidly anchored in capitalism. To define this economic model, geared toward addressing a pressing societal problem, some have used colorful terms, calling it a *moonshot* or *mission economy.* These are inspired by the historical experience of the Apollo programs, where an activist government worked in tight cooperation with the private sector toward a common goal.[3] In that case, the vision was to put a man on the moon and win the space race against the Soviet Union. Now, the mission is to reach zero greenhouse gas emissions by mid-century and, more broadly, to reduce the environmental impact of economic activity.

While some parallels are evident, this narrative carries some risks. To be sure, the Apollo program marked a high point in public-private partnership. At its peak, over four hundred thousand people were working for it, and over twenty thousand businesses and universities were contributing in some way. Still, while all this was happening, American citizens could carry on with their lives. Decarbonizing the entire economy at the necessary speed will be different, in that all actors in society—not only governments and some green-innovation firms, but all businesses, the financial sector, and every individual—will be involved and affected. It is impossible to imagine that people will simply carry on with their lives as though nothing has changed. As reflected in Figure 8.1, the green transition will touch virtually every aspect of their current lifestyles, from how they produce and consume goods, to their choices in transportation, diet, and housing. As unpleasant as it may sound to some, full decarbonization will require a "whole-of-nation approach" akin to a wartime economy. While this might seem like an abstract discussion of labeling, it matters. It is important for people to recognize that liberal democracy and capitalism have been subject to government playing a stronger role at times in the past and have proved compatible with it.[4]

Before we turn to the practical elements of a blueprint to achieve a zero-carbon society, a caveat is in order. The challenges of climate mitigation and adaptation are common to all countries in the world, but clearly the mix of national policy solutions to address them will need to vary from country to country, adjusting to the local context. Most of the practical policy examples

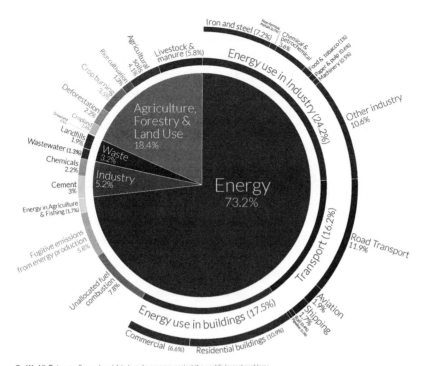

FIGURE 8.1: Global greenhouse gas emissions by sector

Data source: Our World in Data.

Note: Based on most recent data, from 2016.

in these pages will come from Europe, a choice dictated by two factors. First, Europe is a global leader in fighting climate change and can therefore serve as a useful laboratory for other countries that are only now turning to the problem. Second, it is the context I know best, especially having contributed to reflections on the broad economic strategy underpinning the European Green Deal. This focus on Europe, however, is not meant to imply that the solutions chosen in this setting are necessarily the only ones, or the best possible.

The overall strategy

As we begin with the fundamental elements of a green strategy, it must be noted that a whole-of-nation approach requires a vast variety of policies and steps. Over the past few years, these have been generally clustered under the term *Green Deal*.[5] Green Deals differ in their formal names and their exact structures, but they all start with clear, time-bound, and measurable targets. The European Green Deal, for instance, aims to achieve zero net greenhouse gas emissions by 2050—the popular idea of *net zero*. Several other countries have adopted net zero targets. In Japan in October 2020, Prime Minister Suga Yoshihide used the occasion of his first policy address to parliament (equivalent to a State of the Union speech by a US president) to announce the goal of achieving climate neutrality by 2050. Others adopting net zero targets include the United Kingdom, South Korea, and New Zealand, to name a few. President Xi Jinping has announced China's ambition to reach climate neutrality by 2060. Most recently, on Earth Day, April 22, 2021, Joe Biden pledged to cut US net emissions in half (from a 2005 baseline) by 2030, and drive them to zero by 2050.

To the skeptics, all this might sound like rhetoric—in the eloquent prose of activist Greta Thunberg, a lot of "blah, blah, blah."[6] If such targets are set credibly, however, and perhaps even enshrined in law (such as Europe's Climate Law), they can be powerful. By signaling to all actors in society, including businesses, that the green transition is not political hype of the moment but rather an inevitable development, they reframe the debate from *whether* the transition will happen, to *how* it will be achieved.[7] Even powerful corporations are prodded toward a position of "if you can't beat 'em, join 'em"—dialing back their combative lobbying and thinking harder about how to reinvent their polluting operations.[8] Setting quantifiable, time-bound targets can also be a good way to direct efforts across society toward a common goal, as the Apollo program's clearly stated objective did.

Targets can be further articulated at the sectoral level. The most frequent focus is on the automotive and energy sectors; for instance, the UK has committed to ban the sale of gas- and diesel-powered cars by 2030. Norway, an oil-producing country, announced in 2016 that by 2025 only e-cars would be allowed to be registered. In December 2019, France put into law its objective to have nothing but zero-emission vehicles on the road by 2040. On the campaign trail to the 2020 US presidential election, Joe Biden promised policies to achieve 100 percent renewable electricity by 2035.

Focusing targets on energy production and transport is logical because, as Figure 8.1 shows, these sectors are implicated in the majority of greenhouse gas emissions. Visualizations like this not only show that Green Deals will touch basically all aspects of daily life, but can also serve as a compass for action, once national carbon-reduction targets are in place and it comes time to operationalize them. They highlight the sectorial priorities, where technologies and policies are most needed. A "carbon compass" like this makes clear why Green Deals generally prioritize four actions: electrification of transport, renewables to produce electricity, new low-carbon energy processes for industry, and insulation to reduce buildings' energy needs.

While breaking down the problem of greenhouse gas emissions into sector components has evident advantages, including making the problem more manageable, it should also be evident that realistic Green Deal policies cannot be contained to silos—in part because they will need to exploit synergies. For instance, we know that some degree of climate change will take place no matter how proactive mitigation efforts are, so there will have to be adaptation policies alongside efforts to bring down greenhouse gas emissions. Where choices are made to incentivize the retrofitting of buildings to reduce their heat dispersion, or to build new renewable-energy infrastructure, these structures, from roads and sewers to the electricity grid, will also have to be made resistant to climate-induced weather shocks.[9] The same kind of integrated thinking must go into reducing environmental damage. New methods are being explored to reduce the use of chemical fertilizers and pesticides in agriculture without reducing yields—but in parallel, attention must be paid to the crops themselves, selecting ones that are more resistant to drought and flooding.

Capitalism and the decarbonization challenge

Again, the central feature of capitalism is the price mechanism. Demand, supply, finance, and innovation are all coordinated and determined in a decentralized way by relative prices. This also explains why capitalism, left to its own devices, will not fix the problem of global warming: carbon emissions are not priced. Thus an atmosphere that is everybody's property—and therefore nobody's—is overloaded with them. Social ills like this that fall through the cracks of the price mechanism are what economists call negative externalities—costs that are not covered through the market system by the entities that create them. To put capitalism in service of the green transition, this failure must be fixed.

Carbon pricing

The economics textbook solution to externalities is to internalize them. If we could devise a way to price CO_2 emissions, which account for the lion's share of greenhouse gases, they could be inserted into the market mechanism. Broadly speaking, there are two ways to price carbon emissions. The first, and most intuitive, is to introduce a Pigouvian tax on carbon emissions.[10] In plain language: you pollute, you pay. As of April 2021, twenty-seven countries in the world have implemented such forms of carbon taxation.

As an alternative, it is possible to exploit the tight link between demand, supply, and prices by implementing a "cap-and-trade" system, building on the seminal work of economist Ronald Coase. The idea is to set a maximum level of CO_2 emissions allowed per year for the entire system (the "cap"), then auction or dispense emissions permits (called allowances) among polluting firms, who can then proceed to trade these permits among themselves. Under cap-and-trade, a company must buy a permit to emit CO_2, and can do so either from the government or from other firms. This might sound like the kind of abstract system that could only exist in economics textbooks but it is already widely used around the world. The EU has a cap-and-trade solution known as the Emissions Trading System (EU ETS), and so do New Zealand, Switzerland, Australia, South Korea, Kazakhstan, and China, to name just a few.[11] Such schemes also exist at subnational level, as for example in California, Ontario, and the Tokyo region of Japan.

These two alternatives, cap-and-trade (covering the entire economy) and carbon tax, are broadly equivalent. Either can achieve the same carbon prices and raise the same amount of revenue for a government. There are, however, some differences. One complication with cap-and-trade is price fluctuation; given that permits are auctioned and firms are allowed to trade them, the carbon price is subject to market dynamics, no less than a stock trading on Wall Street. For firms trying to decide whether to make large, long-term investments in green production methods, a nonfixed price introduces uncertainty regarding returns, weakening the incentive to proceed. Administratively, a cap-and-trade system is also more cumbersome to set up, especially for developing countries with limited administrative capacities. Once such a system is up, however, it is relatively easy to operate. In several countries, political considerations have steered policymakers toward cap-and-trade, given that any new tax was bound to be unpopular. To the public, cap-and-trade is a less visible mechanism than a carbon tax.[12]

For a good example of the difficult politics of carbon pricing, we need only look to the United States—the second-largest CO_2 emitter in the world, after China, and conspicuously absent from the above list of carbon-pricing jurisdictions (at least at the national level). In 2009, the United States came close to adopting the federal cap-and-trade program specified by the American Clean Energy and Security Act (also known as the Waxman-Markey Bill), but it failed to pass in Congress. There is still some cause for hope given who currently leads the US Treasury. Secretary Janet Yellen spearheaded a petition supporting a carbon tax in 2019 which was signed by more than thirty-six hundred economists, including all living former chairs of the Federal Reserve, twenty-seven Nobel laureates, and fifteen former chairs of the Council of Economic Advisers.[13] The problem, however, is that climate mitigation has become a strongly partisan issue in the United States. Given that filibuster rules in the Senate call for super-majorities, and the moderate democrats known as "blue dogs" remain coldhearted toward carbon pricing, it is hard to believe that Congress will pass carbon-pricing legislation in the near future.

In Chapter 1, we saw that greenhouse gas emissions have been abating in advanced economies over the past few decades, but at a pace that is not sufficient to achieve net zero by mid-century. The problem goes well beyond the failures of politics in Washington. In 2017, a high-level commission on carbon pricing, co-chaired by Joseph Stiglitz and Lord Nicholas Stern, concluded that to keep global warming well below two degrees Celsius, in line with the Paris Agreement, a carbon price of $40 to $80 per ton of CO_2 would be needed by 2020, and $50 to $100 per ton by 2030. In 2019, the IMF estimated the global average carbon price at around $2 per ton. It is simply not sufficient to put in place carbon tax schemes or cap-and-trade mechanisms. What matters is that the level of carbon pricing be set high enough and applied to wide parts of the economy—ideally, all of them. In mid-2021, the carbon price in the EU ETS broke the $50 mark—in principle, not far from the Stiglitz-Stern recommended levels.

The EU ETS, however, currently covers only emissions from industry, the power sector, and intra-EU flights, which add up to roughly 45 percent of total EU emissions. On top of this, roughly half the allowances are given out for free. The degrowth school interprets this as capitalism's inherent inability to shift away from greenhouse gas emission. In reality, there has

simply been too low a price incentive to do so, reflecting too low of a political priority placed on this goal. Going forward, even where carbon
pricing mechanisms are in place, the screws will need to be significantly
tightened to accelerate decarbonization sufficiently to reach carbon neutrality by mid-century.

Another policy action required to accelerate the green transition is the
elimination of fossil-fuel subsidies, and related tax exemptions, as rapidly
as possible. These subsidies amplify the negative externalities associated
with CO_2 emissions, and even create implicit incentives to produce them.
According to recent estimates, global fossil-fuel subsidies remain large, at
$5.9 trillion (6.8 percent of global GDP) in 2020, with the subsidizer-in-chief
being China ($2.2 trillion), followed by the United States ($660 billion) and
Russia ($520 billion).[14] Allowing carbon pricing and fossil-fuel subsidies to
coexist is essentially taking two steps forward and one step back in the
march toward decarbonization. This is something that ambitious net-zero
targets cannot afford. Abolishing fossil-fuel subsidies would not only lift
the drag on decarbonization, it would free up precious financial resources
to further accelerate the green transition.

Regulation

When introducing or ramping up carbon pricing is not an option, perhaps
due to a toxic, polarized political discourse, there is another indirect way
to increase the price of carbon, and that is through regulation. If appropriately crafted, regulation can reduce the appeal of some carbon-intensive
products and advance their green alternatives, rebalancing relative prices
just as carbon pricing does. On this front, the sky is the limit of creativity,
but several examples can be given. Regarding cars, governments might impose energy-efficiency requirements on producers, or introduce local bans
on city center access for more polluting vehicles. The Norwegian government granted privileges to electric vehicles, including free parking and per-
missions to use bus lanes. Such interventions can increase the value of
electric vehicles relative to fossil-fuel alternatives, just as carbon pricing
would. This mechanism extends beyond cars: in electricity generation, some
governments (including several US states) have regulatory provisions
requiring certain percentages of renewable energy sources.[15] Given the
stealth mechanics through which regulation links to the price mechanism,

economists refer to this as introducing a "shadow price" for carbon. For countries with dim prospects of achieving carbon pricing, such as the United States, regulation can be seen as a viable alternative.

While perhaps easier to agree on politically, regulation is a significantly suboptimal policy lever to achieve decarbonization. To begin with, regulatory approaches provide less flexibility at the individual and company level, reducing the space for experimentation with decarbonization solutions. This means that achieving a given climate target through regulation can be much more expensive than if carbon pricing were used. Moreover, regulation often requires governments to make strong assumptions on the future evolution of technologies, or current production methods.[16] Finally, because regulation links to carbon pricing only indirectly, and targets very specific polluting activities or products, it is heavily exposed to the rebound effects described in Chapter 1. If, for instance, a regulation successfully increases the energy efficiency of cars, it could be that consumers will drive more, exploiting the fact they can travel more miles on a full tank. Alternatively, they could use the extra cash they saved on fuel to buy plane tickets, producing even more CO_2 emissions.[17]

In sum, exploiting the price mechanism of capitalism provides for an environment that encourages firms to reduce pollution at the lowest cost. This is why economists tend to see carbon pricing as the "greener in chief"—the decentralized solution to the problem of climate change. It is worth reiterating, however, that no single solution can be enough of a response to the climate challenge.[18] Even the Stiglitz-Stern Commission tasked with determining the optimal level of carbon prices concluded that "carbon pricing by itself may not be sufficient to induce change at the pace and on the scale required."[19] This is because—even focusing on climate change alone, rather than wider environmental degradation—there are distortions and market imperfections in the economic system. The negative externality of CO_2 emissions is only one of these, as we will soon see.

Finance

Just like any other sector in a capitalist system, finance responds to the announcement of targets and price signals. As carbon prices are stepped up, companies tied to old polluting production methods will see shrinking profitability, while businesses important to the green transition will look like good investment opportunities. As a result of credible Green Deal strategies, we are already observing stock markets neatly rewarding green-tech companies.[20]

Financial markets, however, are likely to be afflicted by a shortfall of information in this early stage of the green transition. For fund managers to direct resources toward green activities, there should be a clear understanding of what a green activity, technology, or company is. This might on the surface seem intuitive, but there is no common understanding of what *green* means. If the definition of a green activity is simply "less carbon-intensive" than a current alternative, then burning methane, which is a fossil fuel and a greenhouse gas, can be classified as a green activity, because it produces less CO_2 than coal or gas. Moreover, if the green tag is limited to climate impact, then nuclear power can be considered green, given that it does not emit greenhouse gases. Considering environmental impact more broadly, however, makes this less likely since nuclear reactors have issues related to radioactive waste disposal. Finally, the definition of green cannot be too draconian—specifying, for example, zero emissions throughout the supply chain—because as things stand, basically all human production carries some emissions, including the manufacture of electric vehicles, solar panels, and wind rotors.

Clearly, this uncertainty and proliferation of green indices creates ample scope for so-called greenwashing—the charge leveled at companies who make superficial changes to their activities and messaging so that investors and consumers will think they are concerned with the environment. Aside from the fact that it does not support the Green Deal targets but rather diverts financial resources from more innovative businesses, greenwashing risks fueling a financial market bubble. The precedent of the "dot-com" bubble of the early 2000s, when companies were adding ".com" to their name only to benefit from the hype related to the internet, should serve as a cautionary tale.

This situation of uncertainty calls for a government intervention: policymakers need to produce a clear classification system for sustainable economic activities, which can serve as a guide for investors who want to support the green transition. Just such a classification system, establishing a list of environmentally sustainable economic activities, was recently adopted in Europe. By creating an official standard for financial markets, the EU taxonomy also paves the way for the "green bonds" increasingly demanded by private and institutional investors—essentially, loans issued to finance environmentally sustainable activities.

Mobilizing financial markets early on in the transition is of crucial importance because climate mitigation toward ambitious net-zero targets will require large investments, which cannot realistically be shouldered by only, or even mostly, public finances. More will be said about investment needs below, but here is the point to acknowledge the debate raging now over the role of central banks. Can they, with all their financial firepower, help finance the green transition? Probably in the near future they will tweak their operations to some extent to align with the current green zeitgeist and overall societal efforts toward net-zero targets. As noted by economic historian Barry Eichengreen, in a society taking an all-hands-on-deck approach to an emergency, central banks cannot expect to snooze quietly, hiding behind their operational independence.[21] One advanced economy, where the mandate of the central bank can be more easily amended, is showing the way: In the United Kingdom in early 2021, Her Majesty's government instructed the Bank of England to support efforts toward a green transition. Still, from a broad Green Deal perspective, it is important to be clear that central banks will not do the heavy lifting in financing, given that they must stay tightly focused on their main objective—namely, keeping inflation within reasonable levels. Once again, in the idea of greening central bank balance sheets to finance the transition, there is not a silver-bullet solution.

The role of government

As we saw in Chapter 6, many technologies exhibit high degrees of path dependency and network effects, and innovations can often take a long time to spread across society. As the clock ticks on climate mitigation, a strong jolt is needed to break the locked-in hold of polluting technology and to give low-carbon solutions a healthy push along their "experience curves," to the point that they become cost-effective. Once that has happened, capitalism, through the price mechanism, can run its course.[22] This "push agenda" has a direct connection to carbon pricing. By facilitating learning-by-doing in low-carbon technologies, and making them cheaper, governments also reduce the need for very high carbon prices to achieve climate targets. Investments aimed at rolling out green technologies upfront can make the overall Green Deals more politically acceptable, while also reducing their societal costs.

To break the lock-in effects of current polluting technologies, infrastructure to support a climate-neutral society will need to be put in place. This will likely consist of large, upfront public investments in things like charging points for electric vehicles, renewable power plants, new smart electricity grids, and public transport solutions, such as train lines. Clearly, in line with the idea of exploiting synergies, these projects will need to proceed in parallel with investments in climate adaptation infrastructure, such as flood banks to cope with rising sea levels.

In a complementary way, tax incentives can also be put in place to prod the private sector to invest in green infrastructure. Tax credits can also go to private individuals to incentivize the retrofitting and insulation of residential properties, reducing energy needs and emissions. In principle, tax credits could also cover the installation of solar panels on residential property. Note, however, that infant green industries should take priority over technologies such as solar panels that are already rather cost-effective.[23]

Financial incentives to individuals can also include "cash for clunkers" schemes to foster, for instance, the purchase of electric vehicles. This approach was widely adopted in Europe during the Covid-19 crisis, and helped turn the region into the world's fastest-growing electric vehicle market. The electric vehicle case is instructive as it shows that government policy can significantly speed up technological innovation. For instance, in 2011, virtually *all* cars sold in Norway were powered by gasoline, diesel, and hybrid engines. In 2020, less than ten years later, following a variety of target announcements and measures adopted by the government, a majority (54 percent) of new cars were fully electric. In general, electric vehicles are currently expected to reach upfront price parity with fossil-fuel vehicles before 2025, spurring faster adoption thereafter, and confirming the time-bound rationale for these types of government incentives.

Incentives can also be imagined for digital investments such as broadband access and 5G networks, which, by facilitating greater use of productivity-enhancing digital technologies by companies and individuals, can reduce the need for transport, lowering greenhouse gas emissions. Such incentives may be a legacy of the pervasive teleworking forced by the Covid-19 pandemic, which normalized teleconferencing and remote-work arrangements. The green transition and digital transformation have some areas of overlap.

Beyond discretionary public investment plans and changes in the tax system, governments have another important policy lever to support green technologies: public procurement.[24] Governments are very large-scale buyers of goods and services in the economy and their choices can powerfully orient production toward low-carbon solutions. For example, in Europe in 2019, purchases made by the public sector, including through state-owned enterprises, accounted for roughly 16 percent of GDP. Significant acts of green public procurement might include retrofitting government buildings to reduce their energy needs, or purchasing electric vehicle fleets not only for local public transport, but for use by postal services, police, and certain military activities.

Given that different public sector entities are responsible for each of these spending items, a clarification is in order. The term *government* is often used as if it referred to a single entity. To the contrary, the public sector is organized on multiple, varied levels which respond to often contrasting incentives. For many of the policies needed for Green Deals to succeed, local authorities, and in particular cities, will have to take the lead. In line with Elinor Ostrom's advocacy for a "polycentric (governance) approach" to climate action (Chapter 7), the reality is that Green Deals cannot operate exclusively at the highest level of government. They will rely on actions by all actors in society.

Public investments to speed adoption will need to go beyond a focus on solutions to reduce greenhouse gas emissions. It is also urgent to strengthen natural carbon sinks—the systems in the natural world that absorb and store CO_2. Important investments in afforestation and reforestation, including peatland restoration, could, to begin with, involve dedications of publicly owned land at the local, regional, and national level.

All these activities clearly add up to a large-scale public investment program. Suggesting the order of magnitude involved, the investment plan accompanying Europe's Green Deal aims to mobilize €1 trillion of green investment over the next ten years just for this purpose. Likewise, in the €750 billion European recovery program to relaunch the economy after the Covid-19 pandemic, roughly a third of resources were earmarked for investments in climate change mitigation and adaptation, including clean technologies and renewables, energy efficiency of buildings, sustainable transport and charging stations, and the rollout of high-speed broad-band services.

Green industrial policy

For several sectors, low-carbon solutions are available, and government action is needed mostly to facilitate scaling up so that these technologies can become cheap enough to be self-sustaining. For instance, electricity can be produced without emitting CO_2 by using solar panels or wind turbines in place of carbon-intensive options like coal and gas. Electrification can take us a long way in many fields, including personal transport. For some high-energy-intensity, high-pollution activities, however, we simply do not yet have carbon-neutral solutions. This is the case, for instance, for two fundamental building blocks of our civilization: cement and steel. The same applies to airplanes. One potential solution being explored would use "green hydrogen," produced with renewable energies, but this technology is only at a pilot stage. Likewise, current climate strategies call for augmenting natural carbon sinks with artificial ones—new technologies that can capture and store CO_2, preventing it from contributing to global warming—but experiments with these are only now being done, and at small scale. Yet another example comes from high-yield, intensive agriculture, which relies on nitrogen for soil fertilization. As noted in Chapter 6, alternative bio-chemical processes are being explored to avoid producing this nitrogen through the CO_2-intensive Haber-Bosch process—but solutions are still far from market-ready. Even housing will be revolutionized, in terms of both lower-carbon construction materials and more reliance on circular economy principles, but in ways that are not yet clear.[25] In these and several other realms, governments will most likely need to intervene more actively in the economy, enacting what is generally known as "green industrial policy" to support targeted innovations and achieve breakthrough green technology solutions.

For a long time, the idea of industrial policy has been anathema within the mainstream economic profession. Yet Dani Rodrik, always ready to challenge conventional wisdom, believes that industrial policies are an indispensable part of putting the global economy on a green path.[26] Even the International Monetary Fund—for a long time, the bastion of "Washington Consensus" neoliberal principles—has lately warmed to the idea that, under certain conditions, industrial policy might prove useful.[27] Meanwhile, studies show that green innovation is just as important as carbon pricing in fighting climate change, and should be front-loaded in an efficient climate policy, even with respect to carbon pricing.[28] An active green industrial policy to foster innovation reduces the reliance of the overall Green Deal on carbon taxes which, as we saw, are likely to face speed bumps in certain countries. Clearly, this proactive government involvement in the economy should be seen as temporary, and also limited to the very high-risk early stages of green technology development.[29] Once breakthroughs are made and clean technologies are sufficiently advanced, private sector research would focus on these solutions under an appropriate carbon-pricing regime.

While industrial policy is often presented as a Copernican revolution in terms of economic policymaking, it is important to realize that governments, even in capitalist market economies, have long been very active in fostering technological breakthroughs in specific fields. In 1958, for example, President Eisenhower created the Defense Advanced Research Projects Agency (DARPA) as a reaction to the Soviet launch of the Sputnik 1 satellite the year before. Since then, DARPA has focused on high-risk, defense-related technologies, pioneering innovations like the internet, GPS, and Siri, which later became Apple's voice assistant. In 2009, under the Obama administration, DARPA was complemented by ARPA-E, focused on disruptive energy technology.

Beyond these two agencies, the US government has been in the business of injecting money into promising innovative companies through a variety of policy programs. Perhaps most conspicuously, in January 2010, the Department of Energy, through its Advanced Technology Vehicles Manufacturing Loan Program, lent $465 million to Tesla, which at the time was on the brink of bankruptcy. With the benefit of hindsight, we can safely say that, without this industrial policy bailout, the rollout of electric vehicles would not be where it is today and prospects for the rapid electrification of the transport sector would be much bleaker.

On the other side of the Atlantic the situation is not much different. The European Commission has a variety of targeted R&D policies, such as its €100 billion investment in "Horizon Europe," the world's largest multinational research and innovation program. Beyond this, the Commission can authorize national governments, acting in concert, to fund specific innovation projects carried out by private companies, in the case of so-called Important Projects of Common European Interest, and in deviation from otherwise stringent rules on state aid. Up to now, this exemption has been used for large investments in microelectronics and electric batteries; another is currently being explored for research on hydrogen as a potential breakthrough in low-carbon energy.

If we expect the green transition to be comparable to an industrial revolution in its far-reaching consequences, we should also recognize that, in the early stages of industrial revolutions, governments have generally taken active roles in developing their homegrown industries and ensuring their edge in the technology of the future.[30] In *Empire of Cotton,* Sven Beckert painstakingly assembles a detailed world history of the cotton manufacturing industry.[31] Contrary to the myth of untrammeled free enterprise (and without contemplating here the implications for slavery), Beckert shows that this leading sector of the First Industrial Revolution was fueled at every stage by government intervention. Governments from Denmark to Mexico to Russia lent large sums to early clothing manufacturers. Moreover, they jumped in to build and finance infrastructure demanded by big cotton growers and mills, from canals and railways in Europe to levees on the Mississippi River. The parallels with the public infrastructure program for a climate-neutral economy should be evident.

Above and beyond public spending, during the First Industrial Revolution, the British government forced Egypt, the Ottoman Empire, and other territories to lower or eliminate their import duties on British cotton, to ensure large trading markets. During the Napoleonic Wars, the British set up a naval blockade to prevent goods from reaching Continental Europe. This intriguing historical occurrence allowed Columbia University economist Réka Juhász to explore the link between trade protectionist measures, which represent the "defensive side" of industrial policy, and the spread of innovative production technology in France.[32] Complementing Beckert's insights with detailed empirical analyses, Juhász shows that, by setting up a blockade, the British inadvertently made it possible for a nascent mechanical cotton-spinning industry to develop and spread in France. The fact that positive innovation effects persisted after the blockade was lifted suggests that moves along the learning-by-doing curve allowed France to develop a stable competitive edge in one of the key technologies of the First Industrial Revolution.[33]

When we fast-forward to the twenty-first century, there are good reasons to expect that the green transition will replicate some of these historical patterns, including on the defensive side of industrial policy. At the onset of the green transition, we are likely to see national governments work hard to prod domestic firms to develop competitive advantages in technologies and sectors expected to experience extremely rapid growth. This can be expected based on the sheer scale of the transformation in the global economy expected over the coming years. To get a sense of it, consider some projections issued recently by the International Energy Agency. Its 2020 World Energy Outlook forecasts that reaching the Paris Agreement targets will require the global stock of electric vehicles to reach 245 million by 2030—a thirtyfold increase over 2019 levels. More than half of passenger cars sold in 2030 will need to be electric, up from 2.5 percent in 2019. Accompanying this surge in electric vehicles, global battery manufacturing capacity will need to grow exponentially, doubling every two years over the next decade. By some estimates, the global electric vehicle battery market could be worth as much as $67.2 billion already by 2025—three times its value five years earlier.[34] Investments in the power sector will also need to nearly triple globally, from $760 billion in 2019 to $2.2 trillion in 2030, with more than one-third spent to expand, modernize, and digitalize electricity networks. Onshore wind capacity will need to expand every year by 10 percent between now and 2030, and solar capacity must grow by 15 percent per year. Close to half the existing building stock in advanced economies will need to be retrofitted by 2030, and one-third elsewhere. Seen as a potential game changer for decarbonization in a variety of energy-intensive industries, the global green hydrogen market for the utilities industry alone could be worth as much as €10 trillion by 2050—roughly half the size of the entire US economy today.[35]

As a consequence of these tectonic shifts in the economic system, with some green sectors likely to experience sustained exponential growth, it is rather unlikely that governments will stick religiously to the principles of free trade and capital mobility. Once large public investments are made to fund domestic innovation, we can very much expect governments to take actions to protect their green technology firms from takeovers by foreign actors.[36] Likewise, measures will undoubtedly be taken to prevent forced technological transfer for innovative industries considered to be of strategic national interest. An activist green industrial policy will effectively plant the seeds for its defensive counterpart.

While again, this turn in favor of defensive industrial policies flies in the face of textbook neoliberal economic practice, it is hardly a novelty under capitalism. In 1975, for instance, as concern about Japanese investments ran high in the United States, particularly with regard to semiconductors, President Gerald Ford established the Committee on Foreign Investment in the United States to scrutinize foreign investments on national security grounds. Sectors, technologies, and businesses considered of national interest have ranged from aluminum and steel to banking, automotive, energy, defense, semiconductors, biotechnology, aerospace, and information and communications technology. Today, as similar calls for scrutiny of foreign investment resurface in Western countries, the discourse favors terms like "economic sovereignty" and "strategic autonomy."

Two remaining issues should be mentioned with regard to industrial policy, beginning with the elephant in the room: the question of how to finance such an activist government agenda. At a time when interest rates for many developed countries are below or close to zero, much of the public spending for the green transition is likely to be financed through public borrowing, especially in the first years.[37] Going forward, part of the financing will come from policies set up through the Green Deals, including taxes on carbon and nonrecycled plastic, savings on fossil-fuel subsidies, and allowances auctioned under cap-and-trade.[38] These policies are known for their "double dividend," both accelerating the green transition and raising revenues, in contrast to coarse regulations. Taking the long-term view, taxes will probably have to edge structurally higher, as the role of government increases. Recall the argument from Chapter 7 that, because climate change represents a protracted uncertainty shock to societies, citizens can be expected to favor greater in-group risk-sharing, implying that nation states will need more resources for collective action. This bolsters the case, on efficiency grounds, for taxes on those who can shoulder them: high income earners, high wealth holders, and digital companies. The search for additional public revenues will also, clearly, lead to crackdowns on the long-standing problem of tax havens and international tax evasion. The greater perceived urgency of this is evident in the recent breakthroughs on the issue at the OECD, G20, and global levels.

The second remaining issue is the question of how to ensure that such an activist government agenda does not spiral into a triumph of wasteful spending, entirely captured by interest groups. Rodrik notes with respect to industrial policy that effective government involvement in the economy does not require that officials have perfect knowledge of which sectors will prove crucial for decarbonization and which investments will prove most productive.[39] His observation is applicable to the broader discussion here.

Government action under a Green Deal will rely on a portfolio approach, supporting a wide variety of projects, and pulling in private investment as soon as solutions become profitable. At the same time, government action must remain highly transparent to allow for a vigilant press (the fourth estate) and public engagement to keep interest groups in check, while pushing for investment strategies to be adjusted, as scientific and technological facts on the ground evolve.

The role of business

In principle, in a perfectly competitive capitalist marketplace, the role of firms should be simply to maximize profits, while abiding by the laws of society. This principle is often referred to as the Friedman doctrine, after the famous free-market economist who in the 1960s provocatively proclaimed, "The only business of business is business." In other words, the only social responsibility of a firm was to operate profitably.[40] Given the central role of the price mechanism as a decentralized organizing principle under capitalism, there is some truth to the Friedman doctrine. It is also consistent with the call for governments to twist relative prices and set up regulation to enroll the forces of capitalism in the fight against climate change. At the same time, we know that the realm of climate and environmental impact is likely to be fraught with various market imperfections, externalities, and lack of information.[41] Under such conditions, with the proverbial invisible hand partially impaired, it is not unreasonable to expect that firms should join societal calls for action, going above and beyond mere maximization of the bottom line.

As noted by Bill Gates in *How to Avoid a Climate Disaster*, many CEOs are indeed rising to the challenge and asking what they and their firms can do to accelerate the green transition. This is a particularly welcome development at a time when some corporations wield so much economic power—and so many resources that their plans and investment decisions can have as much impact as those of some nation states. The big tech companies sometimes referred to as GAFAM—Google, Apple, Facebook, Amazon, and Microsoft—dominate the US stock market based on their market capitalizations, and in combination consume as much electricity as New Zealand, a rich OECD country of almost five million people. Today's corporate behemoths have a lot of financial firepower, even as some nation states are increasingly hampered by high debt, low growth, and gridlocked parliaments.[42] If Amazon were a country, its GDP in 2020 (based on revenues of $386 billion) would be greater than that of Israel or even South Africa—a G20 country. Walmart's GDP would almost outstrip that of Sweden.[43] With cash reserves of $137 billion in 2019, Microsoft alone could launch an investment program the size (in today's dollars) of the Marshall Plan—the largest foreign-aid program in history. Moreover, because many of today's problems, including climate change, transcend national borders, multinationals are well placed to be early agents of change who can spread the green transition across the globe.

Realizing that with great powers come great responsibilities, several of these corporate behemoths, such as Facebook and Apple, already run their direct operations only on renewables. Others, such as Microsoft and Google, have adopted ambitious climate targets, such as to become carbon-neutral or even carbon-negative by 2030. Energy-hungry Amazon has pledged to reach net-zero by 2040, knowing that doing so will require a combination of large clean energy purchases, heavy investments in electric vehicles, and the likely purchase of carbon "offsets" or credits to compensate for any remaining emissions. Corporations are also acting in consortia, such as the Renewable Energy Buyers Alliance in the United States, to pool expertise and financial firepower to create a deep and efficient renewable energy market.

All this voluntary corporate green ambition early on in the transition helps in two ways. First, it helps directly, by speeding up the transformation of the electricity system, contributing to making green technologies more cost-effective for all.[44] Second, it helps indirectly, as large corporations implicitly set standards and show the way for other smaller companies, activating a sort of "corporate innovation treadmill," analogous to the classic one discussed in Chapter 5. This is particularly the case if green early action meets the demands of citizens, and especially high-skilled workers' demands for jobs that give them purpose on top of a salary. In a competitive hiring environment, firms cannot be seen dragging their feet on climate mitigation or they risk missing out on top talent.

Extended producer responsibility

Beyond the climate agenda, the private sector plays an important role regarding the environmental impact of production and consumption of material goods. It is often said that 80 percent of the environmental impact of a product is determined in the design phase.[45] The design determines, for example, whether a product can be easily disassembled for component reuse, and what materials will be used. Some choices, like plastic, deteriorate in quality when recycled. Others, like glass, copper, or aluminum, do not. Some materials are biodegradable, while others take decades if not centuries to decompose in the environment. These considerations matter greatly, but firms at the moment have rather weak incentives to factor them in, given that environmental responsibility for a product generally ends once it is sold to the final consumer.[46] After all, a firm cannot be held liable if its product (say, a knife) is used improperly (say, for a robbery). This understandable, limited-liability principle will probably have to be weakened, however, with respect to environmental liability if firms are to have a strong financial incentive to factor environmental costs into their product designs. A green capitalist society, organized around a circular economy, will need to progressively shift toward what is known as the principle of *extended producer responsibility* (EPR), or "product stewardship." This concept, introduced by Thomas Lindhqvist in 1990, involves a shift in administrative, financial, or physical responsibility from governments or municipalities—and therefore from taxpayers, who are usually responsible for products once they become waste—to the entities that make and market the products. The producer becomes responsible for the post-consumer stage of a product's life cycle.

Once again, like many of the policy proposals discussed in this chapter, EPR will sound to some like a Copernican revolution that would revolutionize the mechanics of capitalism. Yet already, basically all rich OECD countries employ some form of EPR for selected products, including tires, car batteries, and domestic appliances.[47] Part of the intent is to create positive dynamics by which the private sector would innovate more to reduce products' environmental footprints and waste management costs—but incidentally, EPR also creates a monetary disincentive against the "planned obsolescence" that degrowth advocates associate strongly with capitalism.

Corporations wanting to show leadership on the environmental agenda should therefore voluntarily embrace the principles of EPR, and some are indeed showing the way on this front. For instance, the Swedish multinational IKEA offers to buy back its furniture, and US-based Patagonia provides free repairs for most of its outdoor items, on top of buying back used garments.[48] As on the climate dimension, these initial, voluntary moves by large corporations will help to create standards in their industries and spur positive emulation, especially to the extent they are appreciated by consumers.

Progressively, voluntary actions should be accompanied by binding legislation, extending EPR to more and more products. Eventually, the list could cover most electronics, including mobile phones, tablets, laptops, and printers, but also plastic, textiles, furniture, and even B2B products like steel, cement, and chemicals.[49] Given the exponential speed at which electric vehicles are expected to grow over the coming years, electric vehicle batteries should be another high-priority target for EPR. Once implementation problems have been ironed out, it should be possible to tighten the screws of the circular economy.

The role of citizens

Firms are not the only parties that lack formal responsibilities for their impacts in a decentralized system like capitalism, under perfect market conditions. Individuals are also able to consume and live their lives according to their personal preferences, responding to price incentives and obliged only to respect the law. Individuals are also citizens, however, and within a liberal democracy their list of duties must be extended. It should be evident by now that Green Deals will rest on a wide set of policies, regulations, and public investments, which will need to have democratic backing from wide majorities of citizens. First and foremost, citizens will need to be supportive of the green transition because, while top-down policies can accomplish much, these measures will not ultimately be successful unless they have bottom-up support.[50] The role of citizens in supporting the green transition, however, does not stop at voting.

The blueprint proposed here to address climate change within capitalism and liberal democracy does not lay the entire burden of mitigation on citizens, or hinge on a complete turnaround in instincts, psychological needs, and preferences.[51] The fact remains, however, that citizens are directly responsible for a large share of total greenhouse gas emissions (over a fifth, for example, in Europe). The more green behaviors they are willing the engage in, the less the Green Deal will cost in terms of investment, and the faster it can yield results. Behavioral change can in principle happen rapidly, whereas building renewable energy infrastructure or reinventing carbon-free production systems will inevitably take several years. As a practical example from the European context, achieving the Green Deal's 2030 intermediate emissions and renewable energy targets will require an estimated total investment, public and private, in the neighborhood of €336 billion per year. Green lifestyle choices by citizens, however, such as enhancing energy efficiency and facilitating a rapid transition toward a circular economy, can bring this number down by roughly a third.[52] This is why it makes sense to say that every single actor in society—including not only firms, financial sector participants, and government at all levels, but also citizens—has an important role to play in meeting net-zero goals.[53]

What can people voluntarily do to reduce their carbon footprint, and directly help society's decarbonization effort? When it comes to household direct emissions, three components account for over 80 percent of the total across advanced economies.[54] These are transport, notably by car and airplane, housing, and diet. Regarding the transport sector, the International Energy Agency estimates that its CO_2 emissions could be cut by more than 20 percent by behavioral changes by consumers—such as replacing flights under one hour with low-carbon alternatives, walking or cycling instead of driving for trips under three kilometers, and reducing average driving speed by seven kilometers per hour. All the better if people choose to forego some long-haul flights, which are particularly carbon-intensive. It is estimated that one person flying from London to New York and back is responsible for roughly the same CO_2 emissions as those produced by heating a typical European home for a year.[55] As a somewhat inferior alternative, it is also possible to purchase carbon offsets for flight segments. Regarding personal diet, we know that meat production—especially beef, but also lamb and goat—contributes heavily to greenhouse gas emissions. Bill Gates has observed that, if cows made up their own country, it would be the world's third-largest emitter of greenhouse gases.[56] Even short of embracing full vegetarian or vegan habits, reductions in meat or dairy consumption can go a long way in reducing personal direct carbon emissions. Finally, emissions related to housing tend to be mostly associated with heating and cooling. As we saw above, investments in insulation and renovation are fundamental to a successful decarbonization strategy, and citizens with the means to afford these can clearly support this rollout.

Proactive green consumers can also apply the "three Rs": reduce, reuse, and recycle. As we saw in detail in Chapter 5, what counts as a personal need versus a capricious want is in the eyes of the individual, and impossible to declare at centralized level. This means that individuals are in the best position to say what is waste and can be easily reduced. By extension, they are well placed to know when practical considerations do not make it a problem to share, rather than personally own, a good such as a car or home appliance.[57]

Aside from behavioral changes, citizens, in their roles as environmentally conscious consumers, can demand more from companies and products. This includes favoring products with less packaging, more recycled and recyclable materials, greater application of EPR principles, and clear zero-carbon plans. Moreover, and perhaps even more powerfully, citizens can demand more of their employers, urging change from within organizations.

Before wrapping up this section, it is worth recalling an earlier discussion of the role of the upper class and its special responsibility toward society. In a capitalist economy organized around prices, the rich, by definition, wield great power through their expanded financial possibilities. In Chapter 5 we saw how, through the innovation treadmill, the upper class's early-adopting behaviors reverberate through the rest of society for years, influencing the path of innovation. While this section has used the word citizen generically, it should go without saying that voluntary adherence to the green societal mission applies particularly to the rich and famous, who can set an example and contribute beyond their individual footprint through emulation effects, especially early on in the green transition.[58]

The green transition beyond national borders

Up to here, the green agenda being proposed has focused mainly on advanced economies, on the principle that rich countries should take the lead in the transition. Still, as we saw in Chapter 7, the green transition cannot be shouldered by advanced economies alone. To say so is not only to push back against the narrow degrowth proposal to mitigate climate change by shrinking rich countries' economies. Apart from sociopolitical considerations, strictly from a technical standpoint advanced economies cannot keep global temperatures within safe levels on their own.

Figure 8.2 displays CO_2 emissions by country, and shows that climate neutrality by rich nations would perhaps have been sufficient in the 1950s. Today, however, the United States, Europe, and Japan combined directly account for less than a third of global carbon emissions. Moreover, most of

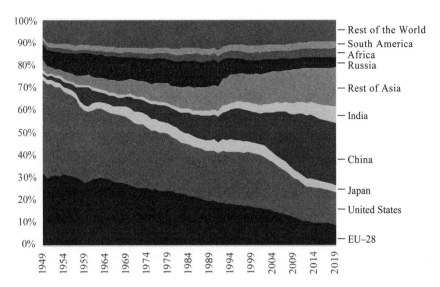

FIGURE 8.2: CO$_2$ emissions by country or grouping as a share of world emissions
Data source: Our World in Data.

the economic growth going forward is set to happen in developing countries, which are currently powered mostly by highly polluting technologies like coal and gas. In simple terms, most of the current and future CO$_2$ emissions are produced in developing countries, and any credible green strategy to fight climate change must take this into account.

To be fair, leaders in most developing countries are well aware of the dangers associated with unfettered greenhouse gas emissions and the catastrophic consequences of climate change. They are likely, however, to be exposed to constant choices between well-known development options that are polluting but cheaper, and a more expensive and uncertain green transition which may pay off only in the longer run. Perhaps counterintuitively, this dilemma is likely to be made indirectly worse by the climate activism of advanced economies, through a sort of global "rebound effect." In simple terms, as rich countries phase out highly polluting technologies such as coal and gas from their energy and transport mix, the global prices for these commodities will drop. While carbon pricing is designed to avoid a rebound at home, the price collapse might very well encourage developing countries to take advantage of these cheap resources, frustrating the overall global climate effort.

The question could be asked: Given that developed countries used these resources in the past, why should the Global South hold back? To put it bluntly, there is no realistic scenario in which developing countries, which account for over 80 percent of the world population, can emulate the carbon-intensive development path of rich countries like the United States yet avoid global climate catastrophe. As they move up the income ladder, developing countries must instead aim for "leapfrogging," skipping the journeys taken by rich countries with all their intermediate technological steps and proceeding directly to innovative, low-carbon solutions.

This "green development" policy might sound outlandish, but technological leapfrogging has happened frequently, including in recent history. For instance, in the field of information and communications technology, large parts of sub-Saharan Africa have skipped phone landlines altogether, avoiding all the costs that setting up that infrastructure would entail. Instead, mobile phones have become widespread. Soon, these became the favored platform for basic payment services through SMS, bypassing another costly infrastructure: a physical banking system. Mobile payments effectively leapfrogged credit card technologies. Several emerging economies, including China and India, rely on QR codes and mobile phones for payments to a larger extent than rich countries do. Once high-speed internet is provided through space by low-orbiting satellites, such as SpaceX's Starlink, which recently became operational, creating a broadband infrastructure on the ground might no longer be needed, and could be leapfrogged. Specifically on the energy dimension, it makes sense in rural parts of sub-Saharan Africa not to build capillary electricity grids from scratch; solutions such as mini-grid systems based on solar photovoltaic, small hydropower, and small wind are not only cheaper but offer more resiliency in conflict-ridden areas.[59]

Green development based on leapfrogging will be particularly important given the path dependency of technology. If a developing country builds coal power plants today, these will remain in operation for decades. With the investment considered locked-in, a switch to low-carbon solutions will be feasible only in the very long term.[60]

How could this international conundrum be solved? To many internationalists, the solution is simply an international treaty setting binding carbon reduction targets, and perhaps even introducing a global carbon price. We saw in Chapter 7 why this is extremely unlikely to happen. Because broad-based decarbonization treaties cannot realistically be binding, they will inevitably run into trouble, or lead to noncompliance every time that short-term economic (and therefore political) incentives are at odds with longer-term green objectives.

Green Deals in rich countries can be designed in a way to shift short-term incentives, also making green development more pressing, palatable, and incentive-compliant in poorer nations.[61] The key is for green pioneers—the advanced economies—to leverage their large markets. Combined, the OECD countries command two-thirds of total world demand. This number goes down to half if the set is limited to G7 members, or 60 percent if it includes both the G7 and China, which has all the credentials to be an ambitious early adopter in climate mitigation.[62] The EU is the biggest export market for goods coming from roughly sixty countries, and the United States is that biggest market for another thirty. Credible Green Deal plans should therefore be flanked by carbon border taxes, which levy taxes on imports based on how much CO_2 was emitted in producing them, effectively acting as carbon pricing at the border.[63] A carbon border tax serves as a strong economic incentive: it nudges other countries to embrace a green agenda or else risk losing revenues from their most important export markets.

Aside from internationalizing greening efforts, carbon border taxes will also serve other purposes, including avoiding what is known as carbon leakage.[64] In simple terms, the tax provides a disincentive for rich-country firms to shift production abroad to avoid tight environmental regulations at home, only to later reimport products for domestic consumption. This, as we saw in Chapter 1 (Figure 1.1), has been a great concern for degrowth advocates, who point to the deceptive gap between production- and consumption-based emissions.[65] Finally, carbon border taxes will add to the environmental revenue stream, helping to finance Green Deal investment plans, and activist government agendas more broadly.[66]

Beyond carbon border taxes, there are other ways to internationalize Green Deals while leveraging the strength of large domestic markets. These include regional and bilateral trade deals, which going forward should include climate and environment chapters and targeted, binding green provisions.

There are, however, countries which simply are not in a position to shoulder the initial investment costs of a green transition. This remains true even when accounting for the fact that, by moving first, advanced economies will reduce the cost of low-carbon technologies, and therefore the overall costs of climate mitigation for others. For these "least developed countries," green pioneers should earmark parts of their foreign-aid spending to finance the investment in climate mitigation and adaptation. This can potentially offer a lot of leverage. At roughly €75 billion in 2019, Europe is collectively the largest donor of international aid in the world, providing over half the total global official development assistance—and that sum does not include financing through (regional) development banks, which likewise should progressively tie their loans to green requirements.

Along similar reasoning, it is very likely that in the aftermath of the Covid-19 pandemic, rich countries will need to engage in substantial debt relief or cancellation for least developed countries, going beyond the debt freeze decided by the G20 at the height of the pandemic. Some of this financial relief could be made conditional on climate action, once again tilting short-term economic incentives in favor of green development.[67]

One last potential avenue of criticism remains to be addressed before this section comes to a close: the argument that the proposed internationalization of Green Deals will hamper much-deserved growth and well-being in countries that desperately need it. The fear is that rich countries will metaphorically "pull up the ladder" of fossil-based development, after having climbed it themselves. Two points should be made on this subject. First, while climate change is surely a global problem, its consequences are expected to be particularly dire in developing countries, which are projected to see droughts, flooding, famine, pests, and forced migration. Policies designed for climate mitigation, then, and especially adaptation, are perfectly aligned with the interests of citizens in the Global South, and this will become very evident in a not-so-distant future. Developing countries realistically have only two paths ahead of them: green development or climate catastrophe. Carbon border taxes and related measures simply make this binary choice more evident. Second, old polluting technologies like coal, charcoal, wood burning, and kerosene are directly detrimental to human health and well-being. People in rich countries learned this the hard way, but there is no reason that developing countries should not benefit from the scientific knowledge that has advanced since then. Unsurprisingly, in all climate mitigation models, poor countries are those where co-benefits from the green transition are the largest, in terms of reduced health conditions and deaths from air pollution. Nobody expects poor countries to walk the entire historical path, or "ladder," of Western medicine, going through ancient nonsensical practices like bloodletting or trepanning. Rather, international organizations and NGOs strive on a daily basis to make widely available the products of modern medical science, such as antibiotics and basic vaccines. Likewise, foreign aid does not fund steam-powered cotton-spinning projects in rural communities in Africa just because those were steps taken by the West in its technological development. Since 2016, the United States bans the imports of products if child labor is used to make them—surely a good thing, even though this was a common practice in the West during the early days of the Industrial Revolution. Concerning climate change and potential catastrophe, along similar lines of reasoning, it should seem reasonable to encourage adoption of green technologies in the developing world, including through carbon border taxes, on the basis of the latest advancements in climate science research.

Ensuring social cohesion

Going back to the national dimension of climate mitigation, we saw in Chapter 6 that innovation, when left to run its natural course, produces winners and losers. With Green Deals, governments will effectively use a variety of policy tools to step on the innovation gas pedal and fast-track technical change. During this process, it will be crucial to ensure social cohesion, not only from a moral standpoint in the name of transitional or ecological justice, but also from a more practical perspective. If a green transition is perceived by some as aggravating poverty and deepening the sense of being left behind, it will be opposed, no matter how compelling the climate science. This is what economic geographer Andrés Rodríguez-Pose has provocatively called the "revenge of places that don't matter." Regions experiencing poverty, economic decay, lack of opportunity, and a sense of being left behind will vote against the established status quo for populist political parties, even if this comes to the detriment of all, themselves included.[68] In all likelihood, such resentments were instrumental to the Brexit referendum and President Trump's election in the United States.

In the context of the green transition, and perhaps in contrast to stronger forces like globalization or automation, direct losses from rapid decarbonization are expected to be rather limited to certain sectors and regions. The classic examples are people employed in coal mining, oil and gas extraction, and fossil-fuel power generation. Some of these jobs are concentrated in specific geographies, and typically these regions are already poor compared to the rest of the country.

Perhaps more concerning, there are several indirect channels through which the green transition might prove cumbersome in terms of its distributional effects. We know that carbon pricing, for instance, will be central to any credible green strategy—but also that pricing carbon will likely have regressive effects. Without compensatory measures, costs will fall disproportionately on the shoulders of low-income citizens, because energy typically represents a larger share of the poor's consumption.

Broadening the scope, it is quite likely that each individual policy measure of Green Deal strategies will have some implications for redistribution, often with contrasting signs and depending on the exact design or even characteristics of the country.[69] For example, taxing aviation sector emissions is likely to be a progressive measure, creating costs that fall disproportionately on the upper classes who fly more. A tax on car fuel will probably fall most heavily on the middle class. This is all, however, very country-specific; in developing countries, road-fuel taxes fall more on the richer part of the population that can afford a car. Each country will need to conduct a detailed analysis on the distributional effect of its greening policies, tailoring them to its own situation.

Concerning the indirect regressive effect of green policies, there are various ways to cushion the impact of the transition for the poorest in society. The standard economics textbook solution recommends using carbon pricing revenues, or environmental taxation more broadly, to decrease labor taxes, particularly on low-income households. Upon closer inspection, however, the challenge is likely to be larger. As Rodríguez-Pose argues, people left behind do not necessarily want income support, but rather renewed opportunities to belong to society through their work. These legitimate demands from people left behind by innovation call for more complex policy solutions, in the form of skillful territorial development, rather than simply throwing money at the problem. As we saw in Chapter 6, the green and social agenda will need to meet and work in tandem for climate mitigation to succeed.

In part in response to these concerns, the EU has set up a "Just Transition Mechanism" as part of its Green Deal, focused on the most affected regions. The objective is to ease the transition with job-search assistance and training support. Moreover, against the submission of a bottom-up regional development strategy, the Just Transition Mechanism can help finance projects, including renovation of buildings, renewable energy, district heating networks, and sustainable transport.

A chapter laying out a blueprint for a complete overhaul of production and consumption is bound to feel rather encyclopedic. This one has detailed what governments at various levels, firms, and citizens can do to help the green transition, and examined the evolving commercial relations among countries of the world. It may be useful to underscore three main takeaways.

First, achieving "net zero" will be a huge effort requiring active involvement by everyone in society. No single action, whether carbon pricing or government regulation, or single technological solution can serve as a silver bullet. Citizens will undoubtedly help the green transition in many ways as they become more conscious of the impact of their choices on the climate and environment. At the same time, it should not be expected that citizens will shoulder the full cost of the green transition, or that a complete turnaround in consumer choices, toward a voluntary "less is more" paradigm, will solve the climate crisis.

Second, a green transition is possible. The daunting complexity of the challenge should not lead to disengagement or fatalism. On the international level, even if no binding global treaty is reached on greenhouse gas reductions, national plans can be successful in arresting catastrophic global warming.

Finally, and perhaps most importantly, the green transition will be a huge and protracted societal challenge, and will require a high degree of cohesion within society. To the extent this cohesion exists, a self-reinforcing process can be established by which government takes on a greater role in the economy to help those in need, and that support fosters even greater unity. In this light, calls for large government need not be interpreted as desires to reject capitalism, undermine liberal democracy, or break the individualistic spirit of Western societies. When a challenge looms large, as even those most avid adherents of the Enlightenment, the US founding fathers, knew well, the simple reality can be that "united we stand, divided we fall."[70]

CONCLUSION

The Future of Growth in a Green World

When written in Chinese, the word "crisis"
is composed of two characters. One represents danger
and the other represents opportunity.
—JOHN F. KENNEDY

As our intellectual journey approaches its end, it is worth looking back at the long way we have come. We started off by identifying the origins of the current disenchantment with capitalism and, by extension, with the benefits of economic growth. Over several chapters, we saw how the concept of growth is tightly intertwined with progress, well-being, liberal democracy, science and innovation, human cultural evolution, relations among groups in society, and relations among countries. In Chapter 8, we explored how the pressing need to decarbonize society, dictated by climate science, can be made compatible with the meta-institutional holy trinity of Western societies: capitalism, democracy, and national sovereignty.

In the process, we have seen that the required green structural transformation will lead to an upheaval in production, consumption, transportation, energy generation, food choices, trade patterns, and geopolitics. Inevitably, this innovation wave will generate winners and losers. Some products and sectors will experience exponential growth, such as renewable energies, electric vehicles, battery manufacturers, and exports of critical raw materials. At the same time, if some other employment categories and regions (such as coal-mining country) are not supported, there are real risks of economic depression and heightened resentment among affected workers toward the rest of society.

One fundamental question remains unanswered: What will a green transition imply for jobs, and the economy more broadly, over the coming decade? Could it be that a policy agenda designed to make capitalism green would inadvertently precipitate the end of economic growth? To conclude this book, we will look to the future of growth in a world of climate change, mitigation, and adaptation.

Growing green

In the early 1990s, around the same time of the UN's first Earth Summit in Rio de Janeiro, William Nordhaus was the first to sketch an integrated model of the climate and the economy: economic activity produces greenhouse gases, these cause climate change, and the damage from that change exerts a drag on economic activity, closing the circle. Nordhaus's Dynamic Integrated Climate-Economy Model quickly became the workhorse model for many policymakers, including the UN's Intergovernmental Panel on Climate Change and, in the United States, the Obama administration. Work with the original version of the model suggested that climate mitigation would damage the economy in the short term, while the costs of climate change would materialize only in the long term and be somewhat contained. This led to the conclusion that an efficient climate policy should tolerate global warming of up to 3.5 degrees Celsius.[1] This early, overly optimistic analysis would later be significantly revised, but its ramifications were far-reaching.

Nordhaus's model set the tone for an economic conversation that, since then, has focused squarely on the cost side of the equation. Because the overall costs of climate change would reveal themselves only in the very long term, scores of macroeconomists then zoomed in on the short term, analyzing in isolation the impacts of specific climate mitigation policies on, for example, gross domestic product or employment. What would the introduction of carbon pricing do to GDP? Crudely, in a basic economic model, carbon pricing is a tax on production and therefore likely to be recessionary, especially in the short run, just like a hike in value-added tax. In a more sophisticated model—one that allows for, say, a zero-carbon innovative sector to emerge and assumes that extra tax revenues generated by carbon pricing will be fully used for some growth-enhancing purpose like lowering income taxes—the impact is more nuanced.

Integrating these various policy measures in their large-scale, macro models, several international organizations have concluded that, if appropriately crafted, Green Deal packages would at best have muted impacts on GDP and employment. Estimates of such impacts edge only slightly above or slightly below zero.[2] The main variable that differs between the marginally positive and marginally negative scenarios has been the assumption of what other countries would do—in particular, whether or not they would also simultaneously implement carbon pricing. If the assumption is that they wouldn't, climate mitigation policies are projected to yield worse economic outcomes, because those companies that were environmentally regulated would lose international competitiveness. As carbon pricing started to be implemented in certain jurisdictions, skillful econometricians worked to tease out the effect of these policies and concluded that the impact on GDP and employment was generally muted and occasionally marginally positive.[3]

On employment, estimates generally tend to be slightly more benign, mostly for two reasons. First, greenhouse gas intensity in an economy is quite concentrated in a small number of industries. For instance, the ten most carbon-intensive industries in the EU-25, while accounting for only 14 percent of total employment, emit almost 90 percent of the CO_2.[4] Taking these facts into account, modeling exercises predict that the direct job losses from taxing CO_2 emissions will be rather modest for the wider economy. Second, renewable activities are more labor-intensive at the moment than carbon-intensive ones. Studies conducted in the United States, for instance, show that the expansion of clean energy creates three jobs for each job lost in the fossil-fuel sector. For each $1 million shift from fossil fuels to clean energy, five additional jobs are created.[5] Bringing findings like these together, the UN's International Labor Organization concluded in 2018 that, if the goal of the Paris Agreement were fully respected, fighting climate change would create more jobs than it destroyed.[6] A circular economy designed for extensive reuse, recycle, and repair of goods could also create many new jobs worldwide.

Evidence like this persuades many that the green transition will be broadly economically neutral. As climate-policy expert Simone Tagliapietra likes to say, the green transition should be characterized as a shift from fossil-fuel production to carbon neutrality—no more, no less.[7] This might rein in the political tendency on both sides of the Atlantic to present Green Deals as a new growth strategy—a claim that should be recognized as analytically unsound, bound to raise citizens' expectations only for them to be disappointed.

The Green Deal economy

All these modeling approaches take too narrow a perspective to answer questions on the future of economic growth. For starters, as hinted earlier, most of the economics literature has concentrated its efforts on carbon pricing, reducing climate mitigation packages to a mere shift in taxation, away from labor income and toward carbon emissions. Chapter 8, one hopes, left you convinced that Green Deal packages are much more encompassing than this single policy element. Measures to accelerate climate mitigation will catalyze forces in the economy across the board to launch a large investment wave, designed to kick-start a total turnaround of economic production.[8] At least in the short and medium term, such a policy package can be expected to boost jobs and domestic demand. Green Deal policies are also highly targeted and sectorial, however, which brings us to a main point of contention.

Green Deals will push early-adopting countries toward developing comparative advantages in advanced green technologies.[9] Mastering technological know-how in the production methods of the future will inevitably have important rebalancing effects on competitiveness across the globe.[10] And as climate mitigation agendas radically modify production and the very structure of the economy, standard macroeconomic models will prove to be poor tools for assessing their overall impact.[11]

As a result of changing production and consumption patterns, Green Deals will redraw the map of trade and investment relations among countries, influence geopolitics, and redefine economic winners and losers. For instance, countries specializing in fossil-fuel exports, like Russia or OPEC members, will likely be on the losing side, unless they manage to diversify their economic models.[12] The green transition will recast completely the map of know-how, as basically every production process must be adapted or reimagined to be climate-neutral and have low environmental impact, in line with a circular economy. Any countries and companies currently basing their growth strategies on cheap production of textiles and fast fashion, for instance, will have to think again; circular economy principles discourage such practices.[13] As production and consumption changes, comparative advantages will shift, and the wealth of nations will be defined and contended on a new technological terrain, which includes the green and digital dimension.

Another issue with current estimations is that, with few exceptions, macroeconomic and advanced climate models are still not well integrated. Macro models without a climate dimension are fine if we assume that climate change and its catastrophic damage are extremely far off in the future—say, a century away, but more problematic if we are already observing their consequences today. Macroeconomic models operating in a vacuum tend to compare two scenarios: climate mitigation versus a general baseline, meaning a continuation of business as usual. The key, however, is to understand that business as usual is a nonexistent scenario. There are only two scenarios ahead of us: successful climate mitigation, and catastrophic damage from climate change. These are the two scenarios that must be compared. And when that is done, it becomes evident that investment in mitigation, and even more so adaptation, carry huge comparative boosts to future growth. When climate and economic models are appropriately integrated, as they are in the 2020 October World Economic Outlook issued by the International Monetary Fund (IMF), it becomes apparent that the long-term economic growth associated with climate mitigation represents a huge boost over the baseline scenario.[14]

Looking more specifically at advanced economies, a large investment wave, such as the one envisioned by the Green Deal, could permanently jolt rich countries out of the negative equilibrium of low growth, low inflation, and low interest rates that afflicted them before Covid-19. A green exit from secular stagnation is a possibility that the IMF acknowledges—without, however, accounting for it in its recent macro-climate modeling exercise. Once that is taken into account, the GDP boost for green early adopters would likely be magnified.[15]

For emerging economies, the path ahead is typically presented as an odious choice between development and greening. This is a false dichotomy built on a wrong premise that there is a feasible development scenario ahead based on polluting production and exports. Moreover, as advanced economies press ahead with the green transition and globalize climate efforts through measures such as carbon border taxes, the window for development based on polluting products will rapidly close. The real choice now is to catch the green development train, or else risk being locked into technologies and products that will soon be outdated. Making large-scale brown investments today would be like pouring more money into horse carriages in the early twentieth century, at the dawn of the automotive era.

More generally, empirical evaluations—all based on past, moderate attempts at introducing some form of carbon pricing—offer little guidance to those trying to predict the economic consequences of full-fledged Green Deal packages. In *The Power of Creative Destruction,* Philippe Aghion, Céline Antonin, and Simon Bunel lay out the basic course of technological revolutions.[16] Reliably, the revolution begins with a fundamental innovation, or general purpose technology, which displays three characteristics: it has scope for improvement, implying a sustained cost reduction; it is pervasive, spreading to all sectors in the economy; and it spawns successive waves of secondary innovation, as the original technology gets adapted in the various sectors. Considered in this light, the green transition—with its wide-reaching implications for production, consumption, energy, and transport—is more of a technological and industrial revolution than an isolated public policy to be evaluated linearly through marginal effects.

To view the coming transition in this way is to see how mistaken it is to generate empirical estimates by extrapolating from past, mild exercises in carbon pricing, and conclude that the overall economic impacts will be muted. That approach would be the equivalent of projecting, in the second half of the eighteenth century—when firms were only beginning to mechanize cotton spinning and making limited productivity gains—that there would be no Industrial Revolution, but only muted effects on the economy. Indeed, we could call this the "Adam Smith fallacy," given that the founding father of modern economics actually did fail to appreciate the massive repercussions that then-nascent mechanized cotton spinning and steam engine technologies would have for the wealth of nations.

Smith's myopia is understandable; in the early stages of the Industrial Revolution, the economic impact of these novel technologies *was* rather muted. Analyzing that specific historical occurrence, recent research by Réka Juhász, Mara Squicciarini, and Nico Voigtländer shows that, as mechanization in the eighteenth century went through a prolonged process of trial and error, productivity gains were initially low on average and widely dispersed. Only in the subsequent decades was high productivity growth observed, as new entrants adopted improved methods of production and organization.[17] The same applied to other technologies of the time, such as metallurgy and paper milling. When electricity first reached companies during the Second Industrial Revolution, a similarly muted initial economic impact was encountered, until the benefits of electrification could be leveraged fully in the assembly line production method, abolishing the older line shaft system.[18] Fast-forwarding to the twenty-first century, as companies start innovating through the process of technological diffusion, inventing new solutions to adapt their business lines to the changing (green) economic landscape, we are likely to observe significant productivity boosts. This is not least a possibility because, unlike mature technologies such as hydropower and fossil fuels, green technologies are likely to have a long way to go in terms of cost reduction.[19] Renewable energies could easily become by far the cheapest energy source in history, leading to a generalized productivity boost across the economy.[20]

Good green jobs

Chapter 8 noted Rodriguez-Pose's preoccupations regarding "places that don't matter." This is not an isolated concern. Several authors have noted the problems of a rising dualism of high-paid jobs, based on the knowledge economy and concentrated in metropolitan areas, and a host of communities neither contributing to nor benefiting from innovation. Several ills of contemporary societies can be attributed to this toxic dichotomy, including inequality, exclusion, spatial and social segmentation, populist backlash, authoritarian politics, and loss of trust in elites, governments, and experts.[21] Having only islands of productive, high-wage activities within a sea of poor jobs undermines the social structures that underpin the economic prosperity of rich countries. This suggests that innovation going forward should be steered to produce not only growth and jobs, but "good jobs"—meaning jobs less prone to geographical concentration, less characterized by dehumanizing repetitive tasks, and less likely to undergo automation or delocalization. With these socioeconomic considerations in mind, the green transition is not only likely to create some jobs, but the types of jobs that advanced societies need today.

In the initial phase, Green Deals will be characterized by large investment programs focused on infrastructure, climate adaptation, renewable energy installations, and renovations throughout national territory. These could include the installation of solar panels and wind turbines, but also investments in recycling and waste management, pollution management and control, reforestation, and public transport.[22] Most of these activities are expected to boost the construction sector—a highly labor-intensive industry. At the same time, some of the most capital-intensive sectors, involving fossil-fuel energy extraction and production, will shrink. Following the early construction wave, renewable energy sources will require regular maintenance services, which clearly need to be carried out in situ and cannot be relocated abroad.

While some reskilling will be needed, studies suggest that most green professions will not require completely different levels of schooling or skills. Much of the retooling of vocational skills could probably happen through on-the-job retraining, making the odds of a seamless job transition much higher than for other economic transformations, such as the digital revolution.[23]

Regarding the characteristics of green jobs, studies suggest that they tend to feature less repetitive task work and call for more interpersonal skills— implying that they are at low risk of automation.[24] Early evidence from green pioneering countries such as the United Kingdom also suggests that the job-creation potential is rather distributed, extending beyond cities and including poorer regions.[25] Aside from the extreme cases of carbon-intensive regions, a green transition can contribute to development in locales that have been left behind, perhaps avoiding the semi-permanent income-support and transfer schemes that hardly solve the problems of "places that don't matter."

The twists and turns of an incipient industrial revolution are hard to foresee in its early stages. With this caveat in mind, however, current evidence suggests that the green transition will contribute to the creation of not only jobs but specifically "good jobs." Green Deals can therefore help to balance out the effects of another current megatrend—the digital transition, with its threats of "winner-take-all" dynamics, strong production agglomerations, disproportionate increases in shares of national income accruing to capital holders versus workers, and global job relocations.[26]

The Flourishing Twenties

As these final lines are drafted, Covid-19 vaccines and antiviral drugs are being rolled out across advanced economies and, as they progressively spread throughout the world, they may soon bring to an end the worst global public health crisis since 1919. Now, as then, exhausted citizens are reemerging from personal loss, economic hardship, and social deprivation, and wondering what the future might hold. When the First World War and the "Spanish Flu" had run their courses, the Western world (the United States most notably) embarked on a decade of social, economic, and cultural progress (Figure C.1). A question therefore springs to mind: Could the West experience a second edition of the roaring twenties? Could we see a "flourishing twenties," in which an improving relation between capitalism and the natural environment yielded a decade of socioeconomic progress?

Mark Twain used to say that history does not repeat itself, but it often rhymes. It is easy to detect the assonance between the first few verses of the

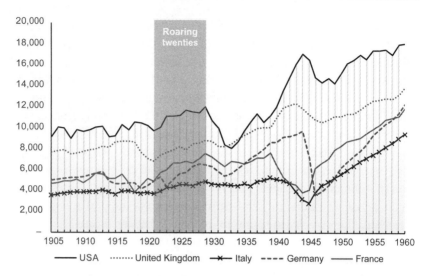

FIGURE C.1: Real GDP per capita, 1905–1960

Data source: Maddison Project (Bolt and Van Zanden 2020).

Note: In 2011 constant prices.

1920s and the 2020s, particularly in the United States.[27] Just like the early 2000s, the early 1900s were characterized by dismal economic growth. Corrected for inflation, US income per capita in 1921, at the onset of the roaring twenties, was basically at the same level as in 1906. Likewise, the early 2000s, and in particular the slow recovery from the 2008 Great Financial Crisis, led economist Larry Summers to resurrect the term "secular stagnation," seen as the malaise of the rich world. In the 1900s, a World War and an epidemic had led people to defer spending, which was then unleashed, fueling the roaring twenties. In 2020–2021, closures of economic activities to contain the Covid-19 pandemic, combined with strong income support measures put in place by governments, have created ample scope for a consumption boom. By some estimates, forced savings today could be worth as much as $1.5 trillion (or 7.2 percent of GDP) in the United States, $300 billion in Japan (5.9 percent of GDP), and $142 billion in Germany (4.3 percent of GDP).[28]

In *Apollo's Arrow,* Nicholas Christakis shows that pandemics are typically followed by periods of increased risk-taking, intemperance, and joie de vivre. This happened after the Black Death in the mid-fourteenth century, for example, and again after the 1919 Spanish Flu. Christakis predicts we will see similar technological, artistic, and social innovations after this pandemic.[29]

The list of striking parallels extends further. Then, like now, the stock market was experiencing a protracted boom phase. Then, like now, the geopolitical map was in the process of being redrawn. While today we are observing the stellar economic rise of China, World War I saw a shift in power from Great Britain to the United States, which then unleashed a wave of investments abroad, including in Europe. The 1920s were also an era of stark contradictions which are paralleled in the 2020s. Specifically, the First World War marked the end of the so-called First Wave of Globalization, as tariffs were ramped up in 1922 in the United States at the onset of the economic boom. In our era, the Trump administration conspicuously raised tariffs, specifically on China, and the Biden administration does not seem in a rush to reverse them. More broadly, pundits speak of a period of deglobalization, with increased calls for strategic autonomy and less reliance on few foreign producers in the name of greater resilience.

Beyond a narrow economic focus, the roaring twenties were characterized by social progress, as women got the right to vote, shortened their hair and dresses, and started driving cars, wearing more cosmetics, smoking cigarettes, and drinking alcohol publicly. Black poets, authors, and jazz musicians found wide audiences, many through what became known as the Harlem Renaissance. With all due caveats, one could view today's #MeToo and Black Lives Matter movements as producing a comparable wave of inclusiveness policies, with perhaps the most conspicuous result being the 2020 election of the first woman and person of color to the office of US vice president.[30]

Even some of the most concerning developments in today's Western societies, such as high inequality, political polarization, and anti-immigration sentiments, were similarly present in the 1920s. President Warren Harding campaigned and won in the 1920 election with the very same motto recently popularized by President Donald Trump: "America First." Inspired by such "nativist" sentiments, the Immigration Act of 1924 led to a strict restriction of immigration into the United States, while the white supremacist Ku Klux Klan experienced a revival. Another element fueling political polarization was the lack of jobs for returning veterans, which led to waves of strikes, feeding anarchist and communist sentiments on one side and the "Red Scare" on the other.[31]

Returning to the economic domain, protracted growth in the roaring twenties was powered by a set of factors, many of which could very well be mimicked in the 2020s, especially with a green transition plan as outlined in Chapter 8. The 1920s were characterized by the spread of new

breakthrough technologies, especially regarding energy with electrification and transport with the advent of combustion-engine cars. These innovations in turn sparked a construction and infrastructure boom. It
should be evident that a Green Deal, by spreading electric vehicles and
other technologies, would also set off a wave of green infrastructure to
support them.

The digital transition, operating in tandem with the green transformation, makes the prospect of a flourishing decade even more real. In the wake
of technological progress in the 1920s, which brought electric home appliances and economic growth more broadly, the time people could devote to
leisurely activities sharply increased, contributing to a sense of liberation
and progress. In 1926, Henry Ford decided to give workers two days off every
week, instead of the usual factory routine of a six-day week. One hundred
years later, in the wake of the Covid-19 pandemic, Spain has started experimenting with a four-day workweek, as are several companies, including
consumer-goods behemoth Unilever.[32] Many companies invested heavily
in information and communications technology to remain operational
when personal contact was to be avoided. Post-Covid-19, these organizations
will capitalize further on these investments, effecting a structural shift
to a new normal in the 2020s.[33] New work practices will likely reduce
hours wasted in commuting and money wasted on excess office space,
and increase time for leisure.[34] Even technology doomster Robert Gordon
is willing to admit that the digitalization push will boost productivity in
the 2020s.[35]

Putting a replication of the roaring twenties entirely within reach is also
the fact that, on closer examination, GDP growth was not even that extraordinary between 1921 and 1929.[36] As evidenced by Figure C.1, growth rates
hardly compared to the period after the Second World War, which might
be impossible to replicate in today's mature advanced economies. Rather,
the defining feature of the roaring twenties was the rekindled spirit of progress, and renewed sense of societal purpose and possibility. These are social psychological feelings that could very well return in the 2020s, not only
due to the shared societal purpose of greening the economy and saving
humanity from climate catastrophe, but also thanks to the possibilities
opened by science and technology, which in the midst of the pandemic received a huge financial boost. Regarding the realm of perceptions, some
conspicuous achievements of research and innovation could set the tone for
the entire decade. For instance, the 2020s appear to mark the beginning of

a new Age of Exploration, this time in outer space rather than around the world. Humans will once again set foot on the moon, with the goal of establishing a permanent outpost by the end of the decade.[37] The same could be said for mRNA vaccine technology, which holds great potential to cure a host of diseases that have dogged humanity for centuries. Perceptions matter, as the tales we tell ourselves about the world spread like a virus. As argued by Nobel laureate Robert Shiller in *Narrative Economics,* they affect decisions to spend and invest, and drive economic outcomes.[38]

It is important to underscore that the flourishing twenties are a possibility, and that history is not destiny. As noted by President Kennedy, moments of large change simultaneously represent risk and opportunity. At times of great economic transformation, some societies manage to emerge triumphant, while others prove incapable and stumble.[39] In the early 1920s, for instance, the Italian economy did not manage to successfully reconvert from a wartime to an open-economy model. Large industrial trusts like Ansaldo and Ilva went bankrupt in 1921, dragging the country into recession while other Western economies boomed. Neither did the country do well on the social progress front, as Benito Mussolini rose to power in 1922. Even for the countries that surfed the economic boom, like the United States, frail fundamentals in the roaring twenties laid the foundations for the 1929 Great Depression, which was promptly exported to the world, setting the scene for a catastrophic Second World War.

The crucial underlying problem of the roaring twenties was income polarization, perhaps best depicted in F. Scott Fitzgerald's *The Great Gatsby.*[40] According to Carlota Perez, an expert in historical economic cycles, the roaring twenties suffered from problems similar to those of the Gilded Age.[41] This was the term Mark Twain used to characterize the uneven prosperity that affected the United States in the late nineteenth century.[42]

This book has advanced the argument that sacrificing economic growth on the altar of inequality would be counterproductive. Reducing inequality, however, is essential to the success of the green transformation, and more broadly to the sustainability of economic prosperity. The coming decade will determine the evolution of climate-related prosperity and well-being, based on how successful societies are in taking their destiny into their hands. In the end, the history of humanity is a succession of such turning points, where civilizations have staved off the crisis of the moment through the instruments made available by technology in the pursuit of self-determination, and powered by curiosity, ingenuity, and perseverance.

Appendix A

GDP and Energy Return on Energy Investment

To calculate Energy Return on Energy Investment (EROEI, sometimes simplified to EROI) is essentially to find the ratio of the amount of energy delivered from a particular energy source to the amount of energy needed to obtain that source. The concept was popularized by system ecologist Charles A. Hall in the 1980s and, while it remains somewhat of a niche interest, the logic behind it is worth exploring. It frequently resurfaces in discussions on the future of economic growth, especially among degrowth advocates and in the work of natural scientists.

Hall first computed EROEI estimates for a variety of energy sources and across time. Then, he showed that the EROEI for each energy source is on a long-term declining trend. This is intuitively clear: the supplies that are most easily accessible are used first. For instance, the EROEI for oil and gas in the United States was greater than 100 in 1919—100 units of energy were delivered for every unit spent accessing it. By the 2010s, this EROEI fell to under 10. Renewables currently have very low EROEI, around 10 for solar PV and 18 for wind.[1]

As EROEI descends to lower levels, more and more resources are needed to extract the same amount of energy. In the end, as EROEI converges toward 1, a point known as the "net energy cliff," the economic system must either stabilize or collapse, as all resources in society (all of GDP, in other words) must be devoted to extracting energy.[2]

Effectively, and linking back to the discussion in Chapter 1, Hall is implying that technological progress will not be large or fast enough to offset the declining quality of energy. In his words, "The decline in EROI among major fossil fuels suggests that in the race between technological advances and depletion, depletion is winning." Considering this to be a fatal blow to the illusions sold by the economic profession, Hall concludes, "Increasing prices, thought by most economists to negate depletion through increasing incentives for exploitation, cannot work as EROI approaches 1:1, and even now has made oil too expensive to support the high economic growth it once did."[3]

To begin with, precise calculations of EROEI build on incredibly heroic assumptions. While the output side is relatively easy to measure, the input side is fraught with complications, especially regarding where the calculation should start and stop in the supply chain. Should the starting point be the energy extraction point, such as the oil well? Of course, much additional energy is used downstream from that point to refine the oil and turn it into usable fuel. Meanwhile, looking upstream, energy is certainly used in the drilling process, but why stop there? Why not take into account the energy used to transport the drill to the well, to manufacture it, and to extract the raw materials used for that? At the extreme, one might even account for the energy embedded in the meals of the workers involved in the process. As energy guru Vaclav Smil writes in *Energy and Civilization,* EROEI "is a revealing measure only if we compare values that have been calculated by identical methods using standard assumptions and clearly identified analytical boundaries."[4] It can then be used to show, for instance, why the rich Middle Eastern oil fields have been preferred to other, lower-productivity wells. It becomes more problematic when the challenge is to make comparisons across energy sources.

EROEI might be a useful indicator for energy and engineering analyses, but from an economics perspective, its use to predict the future of growth looks rather odd. So far the evidence, prima facie, does not seem very convincing. Over the last century, during the "Age of Petrol," the EROEI for oil collapsed by a factor of 10, and yet world real GDP expanded twenty-six-fold.

What is more puzzling, as with the systems dynamics World3 model underlying *The Limits to Growth,* the concept of EROEI assigns no role to (relative) prices, instead focusing narrowly on absolute quantities.[5] To inquire into this choice is typically to be met with sardonic statements mocking economists for lacking understanding of basic thermodynamics. Given the criticism, let us build on the words of quantum physicist Mark Buchanan: "if other factors remain fixed, a temporary drop in EROI, for any reason, means we need to use up more energy in producing energy, and therefore have less to spend on real economic activity—on this logic, economic growth should then falter."[6] The fact is that "other factors" do not remain fixed. Consider that, as the EROEI falls, and more investment is needed to extract resources, *the price of the energy source should increase.*[7] As this happens, if, as Hall suggests, it is not technically feasible to devise more efficient recovery techniques, what will adjust is the economy itself.[8]

Grubb et al (2018) show that, across fifty years and over thirty countries, the long-run energy expenditure relative to GDP remained practically constant, within the range of 8 percent of GDP (plus or minus 2 percent), irrespective of energy prices.[9] When prices increase, as they did during the 1970s oil crises, this leads of course, in the short term, to an increase in the fraction of GDP paid for energy, and recessionary effects. Over the long term, however, significant energy efficiency improvements bring economies back to equilibrium. The effect is symmetric: low prices induce waste, causing energy productivity and innovation to slow down, which leads energy costs back up again.

What matters for economic growth, and ultimately to people, is not the absolute quantities of a specific energy source, but rather the "services" that energy is needed for. Energy expert Roger Fouquet shows that, because of efficiency improvements, the cost of consuming one unit of energy service has declined faster than the price of energy in the long run.[10] Dramatic improvements are possible in energy services: in 2000, lighting in the United Kingdom cost less than one three-thousandth of its 1800 value.[11] This implies that, for a constant level of energy consumption, it is possible to expand the energy services and therefore the size of the economy.

All in all, and linking to *The Limits to Growth,* we should recognize that it is generally problematic to extend the modeling techniques and concepts of system ecology, or even physics, to the realm of economics. While the former are sciences grounded in hard natural laws, where, say, only two atoms of hydrogen and one of oxygen can produce a molecule of water, economics is very much a social science, where flows and relationships are much more adaptable to evolving external conditions.

Appendix B

China's Great Withdrawal and Japan's Seclusion

Generally across history, civilizations that have reached very high levels of complexity and technological sophistication but then resorted to a closed-door policy have fallen in their innovation capacities. China and Japan serve as cautionary tales. China was a beacon of know-how for centuries; it is not clear that Europe had managed to surpass it even by the early 1700s. Japan was an advanced feudal system which rapidly absorbed and adapted European technology following the arrival of the first Europeans. Uninterested in the outside world and wary of religious influences, both countries adopted closed-door policies at different points. When they reopened to the world, their governments found themselves with sharply curtailed capabilities vis-à-vis the West.

China

One of the great subjects for economic historian has been the question of why the Great Divergence took off in Europe rather than China. For almost a millennium before the Industrial Revolution, Chinese technological and social development had been much more advanced than Europe's. Starting in the fifteenth century, however, while the Europeans inaugurated the Age of Exploration, the Chinese moved in the opposite direction.

Initially, the Ming dynasty oversaw epic expeditions, such as the ones led by Zheng He, admiral of a huge armada. The master ship he sailed to Java, Thailand, India, and Arabia in the early 1400s measured perhaps 135 meters—ample room for treasure. Before long, however, such voyages were halted. Imperial mandarins felt that expeditions were not only contrary to Confucian principles but a waste of money, given that China was already the "all-perfect center of the universe."[1] Funds, they believed, would be more wisely spent domestically on the likes of irrigation canals, roads, and granaries to head off famine.

After 1433, a series of edicts prohibited Chinese from traveling abroad, marking the start of a period known as the Great Withdrawal. Forty years later, the Imperial fleet had shrunk from 400 to 140 vessels, and by 1500 it was a crime to build a ship with more than two masts. Eventually, the technology and expertise to build large ships and navigate them was lost. These insular sentiments would culminate a few centuries later, in 1757, in a full-blown closed-door policy toward the Western world, enacted by Emperor Qianlong of the Qing dynasty. Merchants came to Guangzhou as the only trading port in China, but were not allowed to land on Chinese soil and could trade or stay shortly at port only under strict supervision.

With the outset of the First Opium War with Great Britain, in 1839, the Chinese would discover how far their technological and military capabilities had fallen behind. In the Second Opium War, in 1860, France and Britain were able to capture Beijing and decisively defeat the Qing emperor with as few as twenty thousand troops. With its image of imperial power thoroughly tarnished, China embarked on what is known now as the "century of humiliation."[2]

Japan

Europeans, or more specifically Portuguese explorers, first arrived in Japan in the mid-sixteenth century.[3] Thus began the Nanban trade period, characterized by extensive trade, technological transfer, and deep cultural exchanges. Japanese merchants exported silver, diamonds, copper, swords, and other artifacts, and imported Chinese silk and some Southeast Asian products, including sugar. European manufactured items were also in demand, such as Venetian glass. Guns also inevitably attracted attention, especially because the Japanese were embroiled at that time in civil war.

Japanese manufacturers rapidly learned to replicate the European firearms and ship designs, also modifying the latter to incorporate Japanese navigation know-how. By the early seventeenth century, Japan commenced a Great Exploration period of its own, not only spreading across Southeast

Asia with its smaller commercial "red seal" ships (the symbol of the Shogun), but also crossing the Pacific and reaching the coast of Mexico.

As civil war came to an end, however, and the country was pacified by the Tokugawa shogun, Japan became highly wary of an unwanted European cultural transfer: Christianity. This led Japanese rulers to enact a series of isolationist policies. By 1650, with the exception of a few strictly regulated trade posts, contact with the outside world ceased. Foreigners caught on Japanese soil were subject to the death penalty. Guns were banned, to encourage a return to the more civilized sword. Travel abroad was prohibited, as was the construction of large ships.

Westerners reapproached the country on several occasions over the subsequent two hundred years, without success. It was only when US Commodore Matthew Perry sailed into Edo Bay (today, Tokyo Bay) in 1854 that Japan saw the power of the Industrial Revolution, steam engines, and heavy Paixhans naval artillery.[4] Forced to trade again, it saw how far behind it had fallen technologically. The subsequent overthrow of the Tokugawa shogunate by the Meji government kicked off a major westernization of the country, and one of the greatest episodes of technological catch-up in human history.

Notes

Introduction

1. Since the turn of the century, China has experienced an unprecedented growth acceleration. The result is that growth is actually "lifting all boats"—generating billionaires and increasing within-country inequality (along what economists call a *Kuznets curve*) but also bringing unprecedented material improvements to poorer cohorts. Even as the contrast grows between persistent situations of misery and bastions of massive power and wealth, we do not see the same effects on popular sentiment as in advanced economies. A 2014 global survey by Pew Research Center showed support for the "free market economy" to be higher in China than in the United States and all other advanced economies with the exception of South Korea. Pew Research Center, "Emerging and Developing Economies Much More Optimistic than Rich Countries about the Future," October 9, 2014, https://www.pewresearch.org/global/2014/10/09/emerging-and-developing-economies-much-more-optimistic-than-rich-countries-about-the-future/. The current push by President Xi Jinping toward "shared prosperity" can be seen as an effort to keep negative sentiments toward Chinese state capitalism from emerging.
2. Cohen (2018, 1).
3. Milanovic (2019).
4. Saez and Zucman (2016).
5. Cohen (2018).
6. At the extreme, in certain static socioeconomic contexts, the effect of wealth inheritance and privileged access to elite occupations can reverberate for centuries. In a beautiful recent paper, Barone and Mocetti (2021) show that top-earning families today in Florence, Italy, were already at the top of the socioeconomic ladder six centuries ago.
7. OECD (2018). The channels through which this happens can be several but, according to OECD chief economist Laurence Boone, the lack of mobility is mainly due to the organization of the French educational system.

8. Gabriel Zucman has done pioneering work to quantify the extent to which corporations and individuals take advantage of tax havens. The estimate mentioned above is specifically for the Scandinavian nations of Norway, Sweden, and Denmark. Alstadsaeter et al. (2019).

9. In 1889, Andrew Carnegie condemned the upper classes' extravagance, irresponsible spending, and self-indulgence, and insisted that philanthropy was their duty. Matching deeds to words, Carnegie donated over 90 percent of his own wealth. A. Carnegie, "Wealth," *North American Review,* June 1889, 653–665.

10. IPCC (2018).

11. Oreopoulos et al. (2012).

12. Dave Lee, "The Recessionals: Why Coronavirus is Another Cruel Setback for Millennials," *Financial Times,* July 9, 2020; Andrew Van Dam, "The Unluckiest Generation in U.S. History," *Washington Post,* June 5, 2020.

13. Bundick and Pollard (2019).

14. The Pew Research Center compiles interesting statistics on Millennials and previous US generations. See, for example, Kristen Bialik and Richard Fry, "Millennial Life: How Young Adulthood Today Compares with Previous Generations," Pew Research Center, February 14, 2019, https://www.pewsocialtrends.org/essay/millennial-life-how-young-adulthood-today-compares-with-prior-generations/.

15. While it is too early to say, polling suggests that Gen Z and Millennial opinions are not far apart. Pew Research Center, "Generation Z Looks a Lot Like Millennials on Key Social and Political Issues," January 17, 2109, https://www.pewresearch.org/social-trends/2019/01/17/generation-z-looks-a-lot-like-millennials-on-key-social-and-political-issues/.

16. *Economist,* cover, February 14, 2019.

17. "Greta Thunberg's Speech at the U.N. Climate Action Summit," September 23, 2019, transcript, NPR, https://www.npr.org/2019/09/23/763452863/transcript-greta-thunbergs-speech-at-the-u-n-climate-action-summit?

18. For instance, Sachs (2008, 30) dismisses in a few lines "environmentalist" calls to reduce the "income and consumption of the rich world," simply stating that it is better to focus on sustainable technologies. Wilfred Beckerman (1974) may still stand as the most popular deep inquiry into the relationship between economic growth and the environment. A more recent attempt is Andrew McAfee (2019). Economic history books tend to be very clear on the epochal transformation for humanity associated with sustained growth. See, for example, Acemoglu and Robinson (2012). These books, however, remain exposed to the criticism that, while growth might have been important in the past, rich countries have by now reached a level of affluence where it is no longer needed. A badge of honor is owed to economist Max Roser, who frequently writes short blog posts to explain in accessible language why eco-

nomic growth is important and how it can be perfectly reconciled with nature. See, for instance, Max Roser, "What Is Economic Growth? And Why Is It So Important?" *Our World in Data,* May 13, 2021, https://ourworldindata.org/what-is-economic-growth.

19. Chen et al. (2018).

20. Fringe and anti-establishment candidates and parties have disproportionately attracted the youth vote in America (as Bernie Sanders did in the 2016 US Democratic primary), in France (where far-left Unbowed France and far-right National Front candidates performed well in the 2017 presidential election), in Spain (where anti-establishment party Podemos gained ground in the 2015 general election), in Greece (as far-left Syriza regained power in the 2015 general election) and Italy (where, for example, the anti-establishment Five Star Movement and far-right Lega pulled many votes away from the center in the 2018 general election).

21. *Financial Times,* cover, September 18, 2019.

22. Sandbu (2020); Milanovic (2019); Piketty (2013). Also see the collection of essays by the Economics for Inclusive Prosperity network, all aimed at reducing economic and social inequalities: https://econfip.org/about/. Building on this initiative, Dani Rodrik and Olivier Blanchard amplify the voices of economists and former policymakers arguing that economic inequality is the defining issue of our time and proposing policies to reduce it. Blanchard and Rodrik (2021).

23. One recent poll finds that 85 percent of young Americans favor some reform of the student loan program, including 94 percent of Democrats and 76 percent of Republicans. Harvard Kennedy School Institute of Politics, Harvard Youth Poll: Spring 2020, April 23, 2020, https://iop.harvard.edu/youth-poll/spring-2020-poll.

24. The mechanics underlying this point are inspected by political economists Torben Iversen and David Soskice (2019).

25. Boldizzoni (2020).

26. "Southern Europe's Millennials Suffer Two Huge Crises by Their Mid-30s," *Economist,* April 18, 2020.

27. John F. Kennedy, "Address at Rice University on the Nation's Space Effort," September 12, 1962, Houston, https://www.jfklibrary.org/archives/other-resources/john-f-kennedy-speeches/rice-university-19620912.

1. The Limits of Economic Growth

1. McAfee (2019, 54).

2. Meadows et al. (1972) sold over twelve million copies worldwide, was translated in thirty languages, and was reissued in updated versions twenty and thirty years later.

3. Latouche (1989).

4. Demaria et al. (2013) present "degrowth" as an umbrella term for a set of radical academic, political, and social movements all emphasizing the need to reduce production and consumption and define goals other than economic growth. For Buch-Hansen (2018), "the main advocates of the degrowth project are grassroots, small fractions of left-wing parties and labor unions as well as academics and other citizens who are concerned about social injustice and the environmentally unsustainable nature of societies in the rich parts of the world" (162). According to Paul Krugman, "degrowth is a marginal left-wing European position," as reported by Weiss and Cattaneo (2017, 229).

5. This multidisciplinary and holistic approach owed much to Aurelio Paccei, the mastermind behind the Club of Rome. He strongly believed that humanity's biggest problems—environmental deterioration, poverty, endemic ill-health, urban blight, criminality—should not be approached in isolation or as "problems capable of being solved in their own terms." Solutions would only come from a systems-level, multifaceted engagement with a complex "problematique." This perspective runs through Meadows et al. (1972).

6. The degrowth literature is wide, mostly because it spans several decades and several disciplines. Moreover, it encompasses various views, including degrowth *tout court,* but also "a-growth," also known as the pursuit of a steady-state economy, or an agnostic stance that simply rejects growth as a policy objective. To present an organic overview of the movement's basic tenets, I draw from many texts, but with no pretense of completeness: Easterlin et al. (2010), Schor (2010), Wiedmann (2020), Daly (2014), Göpel (2016), Hickel (2020a, 2020b), Van den Bergh (2017), Lipsey (2019), Kallis et al. (2018), Kallis et al. (2012), Kallis (2019), Jakob et al. (2020), Demaria et al. (2013), Jakob and Edenhofer (2014), Büchs and Koch (2019), Buch-Hansen (2018), Hoffmann (2015), Gough (2015), Newell and Paterson (2011), Newell (2011), Victor and Rosenbluth (2007), Victor (2012), Raworth (2017), Park (2015), Jackson and Victor (2015), Binswanger (2009), Klitgaard and Krall (2012), Nordhaus (1973), Ehrlich (1968), Fotopoulos (2010), Hardt and O'Neill (2017), Weiss and Cattaneo (2017), Klein (2019, 2015), and Trainer (2012).

7. The planet already has natural sinks—that is, systems that absorb more carbon than they emit. These include soil, forests, and oceans. Together, they are estimated to remove between 9.5 and 11 Gt of CO_2 per year. To put this into perspective, global CO_2 emissions reached 37.1 Gt in 2017. This means, first, that the carbon reduction still needed is formidable, and second, that while a target of zero emissions by 2050 is a simplification, it gives a useful sense of how much is needed.

8. Jevons (1865) observed that England's consumption of coal soared after James Watt introduced his steam engine, which greatly improved on the efficiency

of Thomas Newcomen's earlier design. Watt's innovations made coal a more cost-effective power source, and the steam engine rapidly spread to a wide range of industries. Thus, total coal consumption rose even as the amount of coal required for any particular application fell. This led Jevons to predict that coal would quickly run out—something that clearly did not happen.

9. Jackson (2009, 102).

10. See, for instance, Gillingham et al. (2013, 2016).

11. Hall et al. (2014).

12. One extension of this argument relates to the concept of Energy Return on (Energy) Investment, popularized by Charles A. S. Hall in the 1980s; EROI (or EROEI) is the fundamental mechanism assumed in the World3 computer model used by Meadows et al. (1972). See Appendix A for more on the link between EROI and economic growth.

13. Rockström et al. (2009).

14. Thomas Jefferson to the Republicans of Washington County, Maryland, March 31, 1809, Founders Online, National Archives, https://founders.archives .gov/documents/Jefferson/03-01-02-0088.

15. Kuznets (1934, 7).

16. Jackson (2009, 151).

17. This critique has deep historical and philosophical roots, dating at least to the time of Jeremy Bentham (whose utilitarian philosophy is fundamental to modern welfare economics) and John Stuart Mill (a theorist of a steady-state economy and point of reference for erudite degrowth scholars). Bentham declared pursuits to have equal utility if they produced equal pleasure. Mill argued that some pleasures are qualitatively better than others; it must, for example, be of greater utility to enjoy learning than to enjoy playing games of dice. Yet Mill's is a difficult proposition to prove (Westacott 2016, 69–71). As I argue elsewhere (Terzi 2021a), the limited nature of GDP measurement is exactly why this indicator was so broadly embraced and has endured. Other composite indicators of economic welfare, such as Herman Daly's Genuine Progress Indicator, inevitably reflect value judgments as they assign relative weights to different dimensions. This makes it extremely hard for them to survive changes in political leadership (Chancel, Thiry, and Demailly 2014). Thus, one of the harshest critiques of GDP—that it does not distinguish between "moral" and "immoral" production, or attach value judgments—may point to more of a feature than a bug.

18. For a recent review of issues related to gross domestic product measurement, see OECD (2020).

19. Robert F. Kennedy, "Remarks at the University of Kansas," March 18, 1968, John F. Kennedy Presidential Library and Museum, https://www.jfklibrary.org /learn/about-jfk/the-kennedy-family/robert-f-kennedy/robert-f-kennedy -speeches/remarks-at-the-university-of-kansas-march-18-1968.

20. Some scholars attempting to estimate the point of flattening happiness in the empirical distribution conclude that the turning point could be around $10,000. To put this into perspective, US GDP per capita in 2017 was almost six times that. Countries at the "optimal subsistence level" would include Namibia, Dominica, and Armenia.

21. Smith (1776), Book IV, Chapter II, paragraph IX.

22. Bregman (2020, 19).

23. This point is taken up by *The Limits to Growth* in its thirty-year update (Meadows, Randers, and Meadows 2004).

24. John Maynard Keynes (1930) predicted that within a hundred years we would reach a fifteen-hour work week. This assumed both that labor productivity (and therefore earnings) would greatly increase, thanks to innovation and technology, and that people would want to work less as a result. Not long after, in 1932, philosopher Bertrand Russell proposed a four-hour work day and attacked the Western work ethic, asserting that "a great deal of harm is being done in the modern world by the belief in the virtuousness of work," and that "the road to happiness and prosperity lies in an organized diminution of work." B. Russell, "In Praise of Idleness," *Harper's Magazine*, October 1932.

25. Raworth (2017, 282).

26. Graeber (2019, 8).

27. For a few examples of the growing coverage of degrowth, especially in the aftermath of Covid-19 and lockdown policies, see Sam Meredith, "A Guide to Degrowth: The Movement Prioritizing Well-Being in a Bid to Avoid Climate Cataclysm," *CNBC*, February 19, 2021; John Cassidy, "Can We Have Prosperity without Growth?" *New Yorker*, February 10, 2020; Shannon Osaka, "Can We Learn to Live with Less? Covid-19 and the Debate on 'Degrowing' the Economy," *Rolling Stone*, August 12, 2020; Akshat Rathi, "How 'Degrowth' Pushes Climate and Well-Being over GDP," *Bloomberg*, September 18, 2020; Ruby Russell, "Five Things You Need to Know about Degrowth," *Deutsche Welle*, November 10, 2020; David Rotman, "Capitalism Is in Crisis. To Save It, We Need to Rethink Economic Growth," *MIT Technology Review*, October 14, 2020.

28. David Wallace-Wells, "Vaclav Smil: We Must Leave Growth Behind," *New York Times Magazine*, September 24, 2019. For energy experts in favor of abandoning economic growth aside from Smil (2019), see also Foxton (2018); Alston (2020); Gaya Herrington, quoted in N. Ahmed, "MIT Predicted in 1972 that Society Will Collapse This Century. New Research Shows We're on Schedule," *Vice*, July 14, 2021; Giorgio Parisi: "The increase in GDP clashes with the fight against global warming," quoted in Francesco Giovannetti, "Clima, il Nobel Parisi: 'Aumento Pil è in contrasto con lotta a riscaldamento globale,'" *La Repubblica*, October 8, 2021; Steven Chu: "You have to design an economy based on no growth or even shrinking growth," quoted in Laurie Goering, "John Kerry

Calls on Scientists to Lead Fight against Climate Change Denial," *Reuters*, April 27, 2021; Michael D. Higgins, president of Ireland, "Climate Action and the Role of Engineers," speech at the Engineers Ireland Annual Conference, October 21, 2020; European Environment Agency, "Growth without Economic Growth," Briefing no. 28 / 2020, EEA, Copenhagen, January 11, 2021, https://www .eea.europa.eu/publications/growth-without-economic-growth. Pope Francis (2015) writes that "Human beings and material objects no longer extend a friendly hand to one another; the relationship has become confrontational. This has made it easy to accept the idea of infinite or unlimited growth, which proves so attractive to economists, financiers and experts in technology. It is based on the lie that there is an infinite supply of the earth's goods, and this leads to the planet being squeezed dry beyond every limit (106). [T]he time has come to accept decreased growth in some parts of the world, in order to provide resources for other places to experience healthy growth" (193).

29. These early American degrowth academics were also influenced by the experience of the Soviet Union, seen as a failed system which, to the extent that it grew, did so with appalling environmental impact.

30. Jackson (2009, 202). Professor Jackson seems to have since reconsidered the question, as his latest book on degrowth is titled *Post Growth: Life after Capitalism* (2021).

2. Growth and the Mechanics of Capitalism

1. Genesis 26:12–13: "Now Isaac sowed in that land and reaped in the same year a hundredfold. And the Lord blessed him, and the man became rich, and continued to grow richer until he became very wealthy."

2. Isaiah 60:5.

3. It is important to note that GDP is a metric that responds to a set of statistical conventions, designed to measure and therefore operationalize the more abstract concept of national income. For more on this, see Coyle (2014) and Terzi (2021a).

4. Notoriously, and perhaps more formally, Lionel Robbins defined economics as "the science which studies human behavior as a relationship between ends and scarce means which have alternative uses" (Robbins 1932, 15).

5. Note that under special conditions, the air we breathe acquires value—especially where it is scarcest, under water. Oxygen tanks for scuba diving have a price and producing them indeed contributes to economic growth.

6. Coyle (2019).

7. GDP computations rest on a wide set of assumptions. Note, for example, that aside from free digital services, services performed by the public sector, such as defense and education, also lack explicit market prices; assumptions must

be made to indirectly estimate their value. The international standard for measuring GDP is contained in the System of National Accounts (1993), compiled by the International Monetary Fund, the European Commission, the Organisation for Economic Co-operation and Development, the United Nations, and the World Bank. The next update of the System of National Accounts is slated for 2025, and will focus on improving how data is accounted for, and factoring in natural capital (meaning the value of so-called ecosystem services, the functions performed by natural processes to the benefit of the economy).

8. Wiedmann et al. (2020).

9. While the question of whether capitalism has a built-in "growth imperative" has provoked rather wide discussion, contributions to the literature come mainly from Marxist and post-Keynesian social scientists. See, for example, Binswanger (2009); Gordon and Rosenthal (2003); Klitgaard and Krall (2012); Fotopoulos (2010); Richters and Siemoneit (2017a, 2017b); and Jackson and Victor (2015).

10. In the words of degrowth economist Herman Daly, "the market, of course, functions only within the economic subsystem, where it does one thing: it solves the allocation problem by providing the necessary information and incentive. It does that one thing very well." Reported by Klitgaard and Krall (2012).

11. For more on the potential clash between post-growth and the political economy of welfare state institutions, see Bailey (2015).

12. This point is acknowledged even by Serge Latouche (2009, 16): "jobs, retirement pensions and increased public spending (education, law and order, justice, culture, transport, health, etc.) all presuppose a constant rise in Gross Domestic Product."

13. Doomsayer sociologist Wolfgang Streeck (2016) notes that in the strong growth years of the 1960s, governments made enormous social security promises to the wider public. With Baby Boomers now retiring, these promises are a ticking time bomb under capitalism itself, as the commitments are unsustainable.

14. Across OECD countries, from 2007 to 2017, the only two government expenditure lines to expand (as a share of GDP) were Health (up by 1.1 percentage point) and Social Protection (up by 1.5 percentage point). Every other line item, including Defense, Culture, and Education, held steady or declined (OECD 2019, 71).

15. This is why we generally express public debt as a share of GDP. It gives investors a rough sense of the size of the potential bucket from which government can extract resources (through taxation) to repay its creditors.

16. For more on this topic, see Terzi (2020b).

17. See, for example, Easterly and Fischer (2001).

18. An interesting parallel can be drawn with the current difference between advanced and developing countries' access to financial markets. Focusing on Latin America, Gavin et al. (1996) show that emerging markets are poorly capable of smoothing the business cycle, as they immediately lose credit worthiness at the time of need, namely during a recession. This in turn results in higher exposure to macroeconomic shocks, high volatility, which has been shown to undermine educational attainment, harm the distribution of income, and in particular the poor (Gavin and Hausmann 1998).

19. Ball, Leigh, and Loungani (2020).

20. Dr. Raghuram Rajan, Governor of the Reserve Bank of India, "Going Bust for Growth," remarks to the Economic Club of New York, May 19, 2015, https://www.bis.org/review/r150526g.pdf.

21. Kelton (2020).

22. Boushey (2019).

23. Bailey (2015).

24. In practice, it is very hard to separate increases in capital from innovation. For instance, buying new machines, and therefore modifying production processes, effectively does both.

25. Schumpeter (1942, ch. 3).

26. Marx (1867, vol. 3, ch. 24, sec. 3).

27. Note that high wages can be monetary, but can also take the form of valued benefits: access to training, a well-appointed office environment, a day care center, a nice cafeteria, workout and well-being facilities, and so on. All such elements help firms land at the top of Best Employer rankings, attracting the best workers.

28. Hall and Soskice (2001).

29. Generally, mainstream neoclassical economic models are geared toward explaining the growth process—in particular, since the Second World War, due to data availability—yet they show remarkably little interest in the origin of the pulsion for growth. Regarding fundamental growth drivers, whether the focus is on investment in infrastructure, innovation (total factor productivity), savings rate, or human capital, the shorthand conclusion has been that, if a variable goes to zero, growth dries up, and there is nothing inherent in the model that dictates that growth must continue. The empirical regularity is clear, however: since the Industrial Revolution, the West has experienced growth, occasionally interrupted by recessions. There have not been long periods of flat growth, and this cannot go unexplained. If, as Solow suggests, it is only a matter of preferences (including intertemporal consumption / investment choices) then said preferences should be examined more closely. Solow quoted in Steven Stoll, "Fear of Fallowing: The Specter of a No-Growth World," *Harper's Magazine,* March 2008, 92. Richters and Siemoneit (2017a).

30. This expression is borrowed from Ferguson (2011).
31. Adam Smith did not realize as he wrote *The Wealth of Nations* (1776) that innovation was about to become the central element of economic growth. Note that the opening passage of his book describes a pin factory, rather than a cotton mill, even though the first steam engines were being applied to spinning around that time.
32. This can include better protection of private property, contract enforcement, or access to finance—all elements that allow the price mechanism to function smoothly. Of course, reality might not be so clear-cut, and there might be some points of contact between the two growth models. The key planks of Smithian growth also support technological progress. At the same time, Mokyr (2016) shows that these elements did not make Europe stand out with respect to other areas of the world, such as China, before the Industrial Revolution.
33. This can be seen as meaning "approaching the study of nature through careful measurement, precise formulation, well-designed experiments, empirical testing, mathematization, and above all, the belief that such activities were virtuous, respectable, and could lead to economic and social reward" Mokyr (2016, 223).
34. As Joel Mokyr (2016, 319) writes, "Intellectual sacred cows were increasingly being led to the slaughterhouse of evidence." See also Strevens (2020, 242).
35. In the words of Mokyr (2016, 248): "the belief in progress . . . was a hallmark of Enlightenment Europe, and it provided cultural lubrication to the innovation-creating machinery."
36. Jones (2005, 1065), explaining Paul Romer's distinction between objects and ideas, observes that "At some level, ideas are instructions for arranging the atoms and for using the arrangements to produce utility."
37. Max Weber (1917, 10) noted that "every scientific 'fulfillment' gives birth to new 'questions' and cries out to be surpassed and rendered obsolete. . . . In principle, this progress is infinite." It is a point that Julian Simon liked to make in his debates with first-wave degrowth advocates like Paul Ehrlich.
38. To borrow an observation by Daniel Cohen (2018), growth and technological innovation are but two sides of the same coin.
39. It is worth noting that roughly 70 percent of innovation spending is currently done by business enterprises in both the United States and Europe, therefore responding to a large extent to market mechanisms. Only at its highest point in the 1980s, at the peak of the Cold War, did investment by the US government come close to matching the private sector.
40. Their influence shows up in the words of nineteenth-century philosopher Auguste Comte: "all human progress, political, moral, or intellectual, is inseparable from material progression" (Comte 1896, 266).

41. Together with the concept of progress, ideas about political power as a social contract, formal limits on the executive branch, freedom of expression, religious tolerance, basic human legal rights, and the sanctity of property rights were all crystallized during the Enlightenment and, to varying degrees, remain the bedrock of Western culture. These in turn led to the codification and formalization of the institutions that kept technological momentum going (Mokyr 2016, 315).

42. Mokyr (2016, 19).

43. In Rousseau's own words, after the development of agriculture, "equality disappeared, property was introduced, work became indispensable, and vast forests became smiling fields, which man had to water with the sweat of his brow, and where slavery and misery were soon seen to germinate and grow up with the crops" (Friedman 2005, 30). Note the striking parallels with much of recent degrowth-sympathizing scholarship, such as Suzman (2020) and Bregman (2020). At the time, French philosopher Voltaire struck back at Rousseau's ideas, noting that if human beings had contented themselves with bare necessities, they would have never made any social, material, or artistic progress (Voltaire 1764). Wanting, producing, and enjoying non-basic needs (that is, luxuries) naturally accompanies progress, as both cause and effect (Westacott 2016, 146–147).

44. Friedman (2005, 54).

45. Dickens (1854).

46. Hansen (1932). This has some parallels with Bob Gordon's (2016) argument. Likewise, it was Hansen (1938) who first spoke of "secular stagnation"—the concept resurrected by Larry Summers in the aftermath of the Great Recession. The theory that a growing economy eventually reaches a steady state was recently reproposed (and celebrated as a sign of success) in Vollrath (2020).

47. See, for example, how Ripple et al. (2017, 1026), in their "warning to humanity," casually insert the need to "reassess the role of an economy rooted in growth" into a list of urgent steps that also includes imperatives to incentivize renewable energies, curb pollution, and so on. Similarly, marine biologist Enric Sala (2020) takes a hard line against GDP and economic growth but leaves the impression that he has not stopped to consider the full implications of rejecting growth.

48. Klein (1972) puts it more bluntly: "to talk of aiming at an economic equilibrium without discussing its political and social implications, is to indulge in meaningless rhetoric."

49. See, for example, Newell and Paterson (2011, 2010); Pollin (2019, 2018); and Mann (2019).

3. Post-Growth Utopia

1. More recent scholarship, which approaches degrowth issues from an environmental rather than social perspective, emphasizes the imposition of a regulatory cap on energy and material consumption as the policy lever to slow down growth (Hickel 2020a).

2. According to John Kenneth Galbraith, it is precisely this link between growth and employment, and therefore poverty alleviation, which persuaded liberals and the American left to get on board with the goal of expanding the economy. In his first inaugural address (March 4, 1933), President Franklin Delano Roosevelt declared, "Our greatest primary task is to put people back to work." In the 1960s, the same issue led Galbraith to start advocating for a "basic income."

3. Note that Piketty (2013) reaches this point from a different angle, arguing that accelerations in wealth accumulation and inequality originate from slowdowns in economic growth. Similarly, "capital is back because low growth is back" (Piketty and Zucman 2014, 1260).

4. For an analysis linking poverty and mental health, see Ridley et al. (2020).

5. On the link between the social and environmental agenda, climate activist Chico Mendes is said to have remarked, "ecology without class struggle is just gardening."

6. Throughout the literature, "Marxian degrowthers argue that degrowth would dismantle current models of business ownership and management, and instead introduce a cooperatively managed economy in which decision-making and incomes are more evenly shared" (Büchs and Koch 2019, 158). "Recession and depression are possible within capitalism; degrowth is probably not. Whereas in theory, growth may not be necessary or inevitable within capitalism, in practice, the system generates growth via dynamics of competition, private ownership, and the availability of cheap energy supply" (Kallis et al. 2018, 300). To "subordinate property capitalist expansion to eco-social considerations," van Griethuysen notes, "alternative institutional arrangements to the private property regime, such as state property and common possession, could be elaborated and implemented, with possible articulations within a global, multilevel regime" (van Griethuysen 2012, 268).

7. The idea that technological advancements would lead human beings to spend more time on "their intellectual and moral faculties" is much older than Keynes, dating back at least to eighteenth-century anarchist William Godwin's *Political Justice* (1793).

8. Jackson (2009, 148).

9. This example is borrowed from Aghion et al. (2021, 161–162).

10. Jackson (2009, 147–148); T. Jackson, "Let's Be Less Productive," *New York Times,* May 26, 2012.

11. Hickel (2020a, 249–288).

12. Smith (2016).

13. IMF (2021).

14. As a small but important distinction, note that most companies exploring this option hope that a working-time reduction will actually increase employees' productivity over the remaining hours—which would not be a desirable effect under a degrowth agenda.

15. This government activist agenda in research has been most popularized by the work of economist Mariana Mazzucato (2013).

16. In Utopia, wealth in any case is of little use, given there is no role for prices, if not to buy foreign products. This is an important aspect that also degrowth advocates should bear in mind, namely the international dimension of their vision.

17. Writing four hundred years before Keynes, Sir Thomas More imagined a minimized workday of six hours, twice as many as the twentieth-century economist predicted.

18. Many of Thomas More's ideas have long traditions, dating back at least to the time of Plato. For instance, in *The Republic,* the ideal city is one in which citizens spend their days discussing philosophy, undistracted by desires for anything beyond basic needs (Westacott 2016, 59).

19. See, for France, Crépon and Kramarz (2002); Chemin and Wasmer (2009); and Estevão and Sá (2008). For Germany, see Hunt (1999); for Quebec, see Skuterud (2007); and for Chile, see Sanchez (2013).

20. This point is taken from Bailey (2015). In his words, "The end of growth provokes a point of enormous tension between the left's environmental sympathies (and a broader social critique of GDP as a measurement of well-being) and the left's deep-seated tendency to view the state as an instrument of progressive politics and welfare provision" (802).

21. See, for example, Hickel (2020a, 20, 218).

22. As noted by Wilfred Beckerman (1974, 14), "not only does no economist support maximum economic growth as an objective *now,* but he would never have done so at any time in the past either."

23. Between 1965 and 1995, the average American family gained six hours a day of extra leisure thanks to work-time reductions, all of which were dedicated to watching more television (Cohen 2018, 126).

24. As Wilfred Beckerman (1974, 13) wrote, "This notion of a contrast between the coarse materialism of the pro-growth school and the exquisitely refined sensibilities of the anti-growth school crops up over and over again."

25. Pecchi and Piga (2008). For discussion of the Bloomsbury Group influence, see the editors' introduction, 11–12.

26. Richard Fry and Kim Parker, "Early Benchmarks Show 'Post-Millennials' on Track to Be Most Diverse, Best-Educated Generation Yet," Pew Research Center, November 15, 2018, https://www.pewresearch.org/social-trends/2018/11/15/early -benchmarks-show-post-millennials-on-track-to-be-most-diverse-best -educated-generation-yet/.

27. Thomas More was well aware of this. In *Utopia* (1516), someone who wants to leave the city to go visit a friend must ask for a permit, which is granted only if this does not encroach on the needs of society.

28. The single most impactful individual action to reduce carbon footprint is to have fewer children, at least one order of magnitude above the second best: living car-free (Wynes and Nicholas 2017). Within our context, this makes us wonder whether even this decision, which until now has fallen squarely within the private realm, would need to be somehow regulated in a world where quantitative reductions are the primary tool to curtail human impact on the climate.

29. Even the Achuar tribe, mentioned as an inspiration by Jason Hickel, while at peace with the environment engage in constant warfare with neighboring groups (Chacon and Mendoza 2012).

30. "In the advanced country . . . increased production is an alternative to re-distribution. And, as indicated, it has been the great solvent of the tensions associated with inequality. . . . How much better to concentrate on increasing output, a program on which both rich and poor can agree, since it benefits both" (Galbraith 1958, 78). According to economist Daniel Cohen (2018, 1): "economic growth . . . is the elixir that eases the pain of social conflict."

31. President Franklin Delano Roosevelt passed his New Deal amid a resurgence of xenophobia and anti-Semitism. Making the broader point, Benjamin Friedman (2005) shows that, cyclically, when shared economic growth waned, sentiments of nativism and intolerance in advanced economies grew, such as in the 1880s and into the twentieth century in America.

32. A debate is still open on whether Hobbes should be considered a liberal or il-liberal thinker, given that his worldview mixes elements of both. For our purposes, Hobbes calls for no constraints on government, rules out a division of powers, favors absolutism and authoritarianism, and is anti-constitutionalist and hostile to democracy (Malcolm 2016).

33. Buch-Hansen (2018). This type of societal escapism has a long tradition, dating back at least to the Greek philosopher Epicurus, who advised his disciples to "withdraw from the world; avoid the pains and dangers of involvement; seek your own security and serenity." He went on to isolate in a community (the "Garden") with them and other friends to enjoy the few fundamental basics for a happy life.

34. Using the lexicon of Chapter 2, the Soviet Union proved effective at extensive growth, but suboptimal on the intensive growth front. See, for example, Easterly and Fischer (1995); Ofer (1987).

35. While supporting the principle of frugal living, which can be seen as encouraging self-sufficiency, philosopher Emrys Westacott (2016, 22) notes: "we might cultivate a vegetable garden, learn how to bake bread, and build a bookcase or two. Such activities are not to be despised; apart from saving money they can be intrinsically rewarding. But unless they add up to a definite lifestyle, we should not kid ourselves that we are doing more than playing at self-sufficiency."

36. In 1844, followers of Charles Fourier founded an intentional community called "Utopia" in southwestern Ohio, only to see it washed away by a flood of the Ohio River in 1847. The community was later reorganized, but between the Civil War and rising land prices, the project could not be sustained.

37. Even if emissions were to stop tomorrow, the world would continue to adjust for the next few centuries to the greenhouse gases emitted in the past. Climate adaptation is to some extent inevitable. Dartnell (2015, 31); Sachs (2008, 107); Wadhams (2017).

38. While having become increasingly part of conventional wisdom, this idea is problematic per se. The question we should keep in mind is: Essential for what? During the Covid-19 pandemic, this answered the simple logic of short-term survival. Many activities we surely would not like to see disappear altogether, such as in the cultural and higher-education (including research) sectors, were deemed non-essential in the wake of Covid-19.

39. Medical devices are a good example. By some estimates, 96 percent of medical devices donated to developing countries no longer work after just five years due to lack of maintenance or missing spare parts—and 39 percent never worked from the beginning, due to lack of training, manuals, and accessories (Malkin 2007; Perry and Malkin 2011).

40. For more on how thoroughly intertwined jobs, know-how, and tasks have become, see Dartnell (2015).

41. Ricardo Hausmann, "Secrets of Economic Growth," World Economic Forum, March 10, 2015, https://www.youtube.com/watch?v=2FeugaLv5Bo.

42. One notable exception is Fotopoulos (2010), who recognizes that degrowth is not only a cultural revolution but will also require a "new system . . . beyond the internationalized market economy" (114).

43. Similarly, Milanovic (2019, 187), considers the (remote) possibility of one country abandoning "hypercommercialized capitalism" on its own.

44. Scientist and natural philosopher Joseph Priestley expressed his belief in the late 1700s that commerce "tends greatly to expand the mind and to cure us of many hurtful prejudices." Montesquieu observed that "it is an almost general

rule that everywhere there are gentle mores there is commerce, and that everywhere there is commerce there are gentle mores" (Friedman 2005, 40).

45. Migrants to less equal countries tend to be more highly skilled on average than the non-migrant population (Parey et al. 2017). For example, Mexican immigrants to the United States are more educated and the majority would fall into the upper portion of the wage distribution in Mexico (Chiquiar and Hanson 2005). This aligns with a positive-selection hypothesis. For evidence of migrants' highly entrepreneurial spirit, see Kerr and Kerr (2020). Finally, superstar inventors tend to migrate from high-tax to low-tax, and therefore lower-welfare, jurisdictions (Akcigit et al. 2016).

46. M. Wolf, "Last Chance for the Climate Transition," *Financial Times,* February 18, 2020.

47. We got a sense of this in 2020. In the face of the worst economic contraction since the Great Depression, barely a dent was made in CO_2 emissions at the global level; they came down by 6.4 percent vis-à-vis 2019. This suggests that addressing CO_2 emissions through quantitative reductions in economic activity would require enormous contractions in the global economy and income per capita.

48. Friedman (2005, 17).

49. See, for example, Glaeser (2012).

50. Nowadays, international collaboration accounts for papers that gain high citation impact, while purely domestic research output has plateaued in quantity and quality (Adams 2013; Adams and Gurney 2018).

51. China is an interesting case: at break-neck speed, the country has managed to catch up with advanced economies and is now a technological leader, in part due to massive funding of R&D. At the same time, China's political organization, if left unchanged, is unlikely to prove very attractive to innovation leaders and scientists over the long term. This may prove to be a major roadblock on the transition path to an innovation-led growth model.

52. See, for example, Bloom and Canning (2007); Rosero-Bixby and Dow (2016). For a pioneering paper, see Preston (1975).

53. "Before 1870, health in rich and poor countries was very similar, but after 1870 health improved in rich countries whereas improvements in poor countries only began after 1930. This is consistent with the view that technological advances are employed first in rich countries before eventually diffusing to poorer societies" (Bloom and Canning 2007, 498).

54. For a thorough look at the dimensions (life, health, prosperity, safety, peace, knowledge, and happiness) that have improved across the world over the past decades, not only in advanced economies, see Pinker (2018).

55. This extends even to nonstandard categories which people care about, such as sports. For instance, "real GDP is the single best predictor of a country's

Olympic performance" (Bernard and Busse 2004, 413). The medal count at the 2021 Tokyo Olympics would seem to confirm this pattern.

56. GDP is designed to track only goods and services transacted on the market because most needs in society are fulfilled through specialized labor and exchanges on the market, rather than self-production or spontaneous benevolence of others. While providing occasional reward to those extending and receiving it, benevolence would hardly work as an efficient mechanism to organize society at large, as noted by Adam Smith (1776, 10): "Nobody but a beggar chooses to depend chiefly upon the benevolence of his fellow-citizens."

57. Coyle (2014) notes that various studies have established that commute time is a strong predictor of subjective happiness.

58. Jones and Klenow (2016), using a novel and broad metric of well-being that factors in consumption, leisure, mortality, and inequality, find that the correlation of "welfare" as they define it with GDP per person is near perfect: 0.98.

59. Rodrik (2011, 135–136).

60. The *World Happiness Report* (Helliwell et al. 2021) reports the top ten countries based on data collected in 2020 as follows: Finland, Iceland, Denmark, Switzerland, The Netherlands, Sweden, Germany, Norway, New Zealand, and Austria. The United States appears in fourteenth position. The highest-ranked middle-income country is Saudi Arabia, in twenty-first position.

61. Describing the same interplay, Andrew McAfee (2019, 4) refers to the "four horsemen of the optimist," meaning capitalism and responsive government complemented by "tech progress" and public awareness.

62. Adam Smith begins his inquiry into *The Wealth of Nations* precisely by referring to what will determine whether "the nation will be better or worse supplied with all the necessaries and conveniences for which it has occasion" (Smith 1776, 1).

63. In *The Metaphysics* (Book I, Part 1), Aristotle asserts that "All men by nature desire to know."

64. The preferences of American economic elites have far greater impact in determining policy than those of the average citizen: "When a majority of citizens disagrees with economic elites or with organized interests, they generally lose. Moreover, because of the strong status quo bias built into the US political system, even when fairly large majorities of Americans favor policy change, they generally do not get it" (Gilens and Page 2014, 576). Iversen and Soskice (2019) make a similar point.

65. Alesina, Glaeser, and Sacerdote (2005).

66. The United States and Western Europe have quite similar living standards in terms of longer life expectancy, extra leisure, and lower inequality, despite Europe's having a GDP per capita roughly 70 percent of that of the United States (Jones and Klenow 2016). In the early 2000s, productivity levels in Europe and

the United States were similar, with the European increase in leisure rather than income driving differences in overall GDP (Blanchard 2004).

67. It is perhaps in this sense that one can read Dani Rodrik's call for "good" economic populism (Rodrik 2018).

68. In *The Theory of Moral Sentiments,* Adam Smith (1759, 313) writes that "before we can feel much for others, we must in some measure be at ease ourselves. If our own misery pinches us very severely, we have no leisure to attend to that of our neighbor." For degrowth advocates, citizens *will* be at ease themselves and, because scarcity will be embraced voluntarily, it will not cause a pinch of misery (Kallis 2019).

4. The Italian Canary in the Growthless Coalmine

1. Most information in this paragraph is taken from the OECD's Better Life Index, using the data posted in July 2021: https://www.oecdbetterlifeindex.org /countries/italy/.

2. One statistic known to most Italians, because it is mentioned on every possible occasion, is that the country hosts the largest number of UNESCO World Heritage Sites.

3. According to the Convention on Biological Diversity, Italy has the highest number and density of both animal and plant species within the European Union, as well as a high rate of endemism. Details are available at https://www .cbd.int/countries/profile/?country=it.

4. The Cultural Influence ranking is a derivative of the annual *US News and World Report* "Best Countries" ranking, which uses a survey methodology developed by BAV Group and David Reibstein of the University of Pennsylvania's Wharton School. Italy reliably scores highest in this subranking, including a first-place finish in 2021. On soft power, see for example Jonathan McClory, "The Soft Power 30: A Global Ranking of Soft Power," Portland Communications, 2015, https://portland-communications.com/pdf/The-Soft-Power_30.pdf.

5. For an explanation of how Italy's economy and politics became increasingly detached from the rest of Europe's, see Merler (2020).

6. Italy has not, strictly speaking, experienced a sharp or sustained recession such as seen in Argentina in the early 2000s or Greece during the eurozone crisis. Rather, Italy's economy has been incredibly stagnant over a protracted period of time. This stands in contradiction to those, such as Binswanger (2009), who argue that capitalism is sustainable only with long-term average growth rates of at least 3 percent.

7. Many studies support this point. Particularly useful are Pellegrino and Zingales (2017) and Calligaris et al. (2016). Avid readers might also consult Terzi (2016) on wage bargaining, regional heterogeneity, and productivity.

8. For more, see A. Terzi, "The Great Fiscal Lever: An Italian Economic Obsession," *Bruegel Blog,* August 21, 2018, https://www.bruegel.org/2018/08/the-great-fiscal-lever-an-italian-economic-obsession/.

9. Whether it is an especially Italian trait or a widely shared human characteristic is hard to say, but the impulse to look for external culprits has been frequent—and this search for scapegoats has prevented Italians from correctly identifying the systemic origin of the country's vulnerabilities and the right solutions to get back on its feet.

10. On the poor return on public investment in Italy, see de Jong et al. (2017) and Fiorino et al. (2012). Similar considerations will apply to the new Recovery and Resilience Facility set up at the European level to kick-start post-Covid recovery in the various EU member states. This can either serve as a modern-day Marshall Plan, if funds are used for productive purposes and end up fostering much needed reforms, or as just another short-term burst fueled by public investment, adding to long-term (national and European) debt.

11. A. Terzi, "Clouds Are Forming over Italy's Elections," *Bruegel Blog,* February 28, 2018, https://www.bruegel.org/2018/02/clouds-are-forming-over-italys-elections/.

12. Italy ranks as one of the most pessimistic countries about the future. Only 19 percent of Italians believe their children will be better off in the future, much lower than in emerging markets, but also strikingly low by advanced economy standards (where the median was 34 percent). Bruce Stokes, "Expectations for the Future," part of the Pew Report "A Decade after the Financial Crisis," September 18, 2018, https://www.pewresearch.org/global/2018/09/18/expectations-for-the-future/.

13. The primary balance is the budget balance net of interest payments. From 1995 to 2019, Italy ran the largest average primary surplus in the world.

14. Eichengreen and Panizza (2014).

15. Among rich countries, Italy has the highest ratio of net private wealth to net national income. Pegged at 250 percent in 1970, it exceeded 700 percent in 2015. World Inequality Database, "Income Inequality, Italy, 1980–2019," https://wid.world/country/italy/.

16. On July 10, 1992, the government led by Giuliano Amato introduced overnight a wealth tax on current account holdings to balance the public accounts and ensure Italy's successful candidacy to join the euro. While the tax had a rate of six per thousand, and thirty years have since passed, this episode remains a horrific memory to most Italians, forestalling any attempt to discuss new wealth taxes and undermining people's perception that bank deposits are safe.

17. Data for 2019 from the OECD data warehouse based on average compensation of employees.

18. Data from PayScale, available at https://www.payscale.com/research/IT/Job=Mechanical_Engineer/Salary.

19. High degrees of fragmentation and government instability are permanent fix-tures of Italian politics. Since 1945, the country has had sixty-seven different governments. Surely, that executive powers are so weak—a purposeful limita-tion by the drafters of the constitution after fascism and the Second World War—plays a role.

20. For a more in-depth discussion of the dynamics of a low-skill, bad-job trap, see Snower (1994).

21. This point is made by Milanovic (2019, 186), who cites Florence and Venice as places where native populations are being displaced by rich people of other nationalities.

22. Naturally, this data point leaves the young men of Italy open to easy jokes by those who would characterize them as "mamma's boys" in no hurry to stand on their own feet—a temptation even for finance ministers of high intellectual caliber. In 2007, Tommaso Padoa-Schioppa described a proposed tax break for renters as aimed at the "bamboccioni," the big babies who needed the extra in-ducement to leave home. While there is perhaps a cultural component, the trend suggests there is more to blame. According to Istat, Italy's National In-stitute of Statistics, in 1983, among adults aged eighteen to thirty-four, 49 percent were single and still living with family; that share in 2000 reached 60.2 percent, and in 2020 stood at 65 percent. In 2016, among young people aged twenty to thirty-four, 43.6 percent gave one of three reasons for living with their parents: being a student, the difficulty of finding adequate employment, and the in-ability to meet the cost of housing. The share of those who perceive living with their parents as a comfortable and enjoyable situation almost halved, from 17.4 percent in 2009 to 9.9 percent in 2016; as a motivation, this moved from first to fourth place. ISTAT (2019, ch. 3).

23. Emigration from Italy is draining entrepreneurial talent (Anelli et al. 2020).

24. A similar dynamic played out in London vis-à-vis the rest of the United Kingdom and, directly or indirectly, is likely to have affected many votes in the 2016 Brexit referendum. See A. Terzi, "How to Make the Single Market More Inclusive after Brexit," *Bruegel Blog,* August 18, 2016, https://www.bruegel.org /2016/08/how-to-make-the-single-market-more-inclusive-in-the-aftermath -of-brexit/.

25. This effect might increase thanks to advances in information and communi-cations technology in coming years, with services being more and more ex-posed to delocalization, as argued by Baldwin (2019).

26. The Italian "citizen income" is not truly a universal basic income, first, because it is not distributed to all but only to those of low income, and second, because recipients must document that they are actively seeking and being denied em-ployment to qualify. As such, it is a benefit program targeted at the long-term unemployed. In certain southern provinces, however, where the informal

economy is large, it has such a high level of take-up that it is effectively closer to a time-bound basic income.

27. As women either choose or feel compelled to delay pregnancy by economic instability, the risk of involuntary infertility increases and the nation's fertility rate falls.

28. R&D spending carried out by companies, research institutes, and university and government laboratories accounts for 1.4 percent in Italy, against an OECD average of 2.4 percent, 2.2 percent in France, and 3.2 percent in Germany. OECD data is available at https://data.oecd.org/rd/gross-domestic-spending-on-r-d.htm.

29. For more on this topic, see Terzi (2015).

30. A similar point was made with respect to green politics: "what happens in Italy rarely stays in Italy"—suggesting that the recent shift toward a green political discourse will set the tone for national political debates in Europe over the coming years. L. Bergamaschi, "Italy's Green Moment," *Politico Europe,* March 18, 2021, https://www.politico.eu/article/italy-green-draghi-cop26/.

31. In this respect, and specifically as a reaction to northern feelings of superiority, there is a fascinating resurgence in attachment to the Bourbonic kingdom. According to neo-Bourbonic supporters, the South is poor not because of its own doing, but because, by means of national unification, the North extracted all the resources from it. As such, a national hero like Giuseppe Garibaldi gets turned into a villain. Yet another scapegoat, and yet another dynamic fracturing society and the nation.

32. Clearly, widespread tax evasion and a large shadow economy, both roughly twice the EU average, contribute to undermining the foundations of the state and the social contract.

33. Directorate-General for Communication (2019). Another survey with similar results was conducted in 2013 by Italy's Institute of International Affairs, which concludes: "Italians pay attention to foreign policy, but global issues occupy a subordinate position in the scale of their priorities. The issues become relevant to them only when they affect the country's interests directly, as in the case of immigration and border security. Italians also see their country as a weak actor in the international arena" (CIRCaP / LAPS and IAI 2013, 4). Evidence from March–April 2021 confirms that Italians are among the least interested in foreign affairs within a group of eleven OECD countries (GMF and Bertelsmann 2021).

34. A. Terzi, "Italian Populism Calls for Hard Choices," *Bruegel Blog,* May 31, 2018, https://www.bruegel.org/2018/05/italian-populism-calls-for-hard-choices/.

35. One clear example is access to water and water efficiency. Italy's water infrastructure is old and in need of urgent upgrades. Leaks amount to between 30 percent and 40 percent of water extracted—even hitting peaks of 50 percent

in some regions. Combine this with the likely long-term impact of climate change, such as droughts and potential desertification in certain areas of the country, and it's clear that lack of access to clean water is a disaster waiting to happen. Still, given the huge investment needed to fix the current network, other, more short-term problems constantly take priority.

36. One such example involves earthquakes. Italy is unfortunately located on a major fault line, exposing it to frequent earthquakes, sometimes of great magnitude. In Japan, however, which is similarly exposed, the deaths and damages associated with earthquakes of similar magnitude are much lower. The difference is that Japan has invested extensively in earthquake-proof engineering. For example, a 6.2-magnitude earthquake with a certain depth of epicenter in Kurayoshi (Japan) in 2016 caused some buildings to collapse and seven people were injured, but there were no deaths. A very similar one in Aquila (Italy) in 2009 caused eighty thousand people to lose their homes and injured sixteen hundred. Worst, 309 people lost their lives.

37. References to degrowth appeared in the Five Star Movement's electoral program as early as 2013.

38. "The Experiment" is also the title of an insightful (and highly critical) book by Iacoboni (2018).

39. For instance, "If prosperous degrowth became a national objective how would we know whether we were doing well? We might want to ask also why some countries, such as Japan or Cuba, maintained wellbeing while not (or de) growing" (Kallis et al. 2012, 175). See also Büchs and Koch (2019).

40. On the World Bank's government effectiveness index—which combines measures of the quality of public services, civil service, policy formulation, policy implementation, and credibility of the government's commitment to raising these qualities or keeping them high—Japan ranked fifteenth in the world in 2019. For a thorough treatment of the country's famously high degree of social cohesion (albeit from a critical perspective), see Hirata and Warschauer (2014).

41. Recent anecdotal evidence suggests nonetheless that, operating in a growthless environment, Japanese youth have become highly risk averse, pessimistic about the future, and inclined to cling to what they have. This dynamic is shared with Italy. L. Lewis, "Japanese Youth Search for Stability in a World without Growth," *Financial Times*, November 27, 2020, https://www.ft.com/content/e5b9e75a-a8b8-4899-a0e4-4aea19d97ff6.

5. The True Origins of the Growth Imperative

1. Hickel (2020a, 209, 239). This principle builds on deep historical foundations. For example, eighteenth-century anarchist William Godwin believed that the "evil in man" was merely due to the corrupting influence of social conditions

and that, once these conditions were changed, the true good nature of humans would be revealed (Godwin 1793). Interestingly, Thomas Malthus, another intellectual point of reference for today's degrowth community, regarded Godwin's worldview as irremediably utopian (Malthus 1798).

2. This critique of capitalism, viewed as a totalitarian system generating false needs through the manufactured reality of commercials, was popularized by Marxist philosopher Herbert Marcuse (1964).

3. The classic example is that of light bulbs, which even in the 1920s could be produced (more expensively) to last longer, but were engineered to bring down their maximum life, from around two thousand hours to one thousand, according to the standard set by the "Phoebus cartel" (dissolved in 1939). Another such example from the early 1900s was the refrigerator, which could have been based on absorption rather than compressor technology. That would have cut down the number of moving parts and the risk of breakdown, making the appliances require less maintenance (Dartnell 2015, 92).

4. Advertising is surely important but the idea that it alone can produce sustained sales is clearly false. Business history is full of products that were launched with great fanfare, but then failed to connect with consumers. Examples come even from marketing masters, such as Coca-Cola with its New Coke debacle in 1985, Ford with the Edsel in 1957, Apple's Newton in 1993, Microsoft's Bob software assistant in 1995, McDonald's Arch Deluxe burger in 1996, Amazon's Fire Phone in 2014, and Google glass in 2013. No wonder that, when venture capital and angel investors meet with startup founders, their most frequent question is "what previously untapped need are you trying to fulfill?" and not "what need are you aiming to create?"

5. Because advertising can in part help firms evade the tough logic of "innovate or die," R&D and advertising are to an extent substitutes. Cavenaile and Roldan-Blanco (2021) show that banning advertising, as degrowth advocates recommend to rein in consumption and GDP, would counterintuitively boost US GDP, by as much as 0.22 percentage points, by spurring innovation.

6. Bregman (2020). Many recent books, combining cognitive science, neuroscience, social psychology, anthropology, and evolutionary biology, reject the simplistic dichotomy of good versus bad, and instead inquire into the cultural and evolutionary origins of human behavior. These include Sapolsky (2017) and Wrangham (2019). Also see Buss (2001).

7. This dichotomy has long historical roots. In Ancient Greece, Epicurus distinguished between desires that were "natural"—for example, for food—and ones that were "non-natural," as for music (Westacott 2016, 89). More recently, a similar dichotomy gained traction during the Covid-19 pandemic, particularly during peak lockdown, between essential and nonessential workers. The former are those working to meet "needs."

8. From an evolutionary perspective, what is *needed* is just the resources required to survive and reproduce so that an organism's genes are transmitted to the next generation, ensuring the survival of the individual's genes.

9. The concept of conspicuous consumption was not discovered but rather popularized by Veblen (1899). In ancient Rome, Horace reflected on the topic, as did Alexander Pope in the eighteenth century, and John Rae and John Stuart Mill in the nineteenth. Mill wrote: "I disclaim all asceticism, and by no means wish to see discouraged, either by law or opinion, any indulgence which is sought from a genuine inclination for, any enjoyment of, the thing itself; but a great portion of the expenses of the higher and middle classes in most countries . . . is not incurred for the sake of the pleasure afforded by the things on which the money is spent, but from regard to opinion, and an idea that certain expenses are expected from them, as an appendage of station" (Mill, 1849, 425–426).

10. Richard Layard treats the societal dissatisfaction created by rich people engaging in conspicuous consumption as a negative externality. Given that it makes others feel less worthy, he suggests it be taxed at a rate as high as 30 percent to restore economic efficiency (Layard 2006). For more theoretical discussion, see also Fleurbaey (2009).

11. This problem corresponds to a situation called "the prisoner's dilemma." Two criminal accomplices are caught but evidence is scant, only enough for them both to receive short prison sentences. The accomplices are held in separate cells, with authorities hoping at least one of them will turn on the other. Each is offered freedom for evidence that proves the other's guilt, so that the accomplice receives a heavy sentence. Clearly, the pair together will benefit most if they both resist this offer. But individually, each is most likely to benefit by implicating the other. By confessing, he gets off free if the accomplice remains silent; if the accomplice confesses, he receives a shorter sentence by confessing than by remaining silent. Not knowing how the other will behave, each provides evidence of their crime. To return to the problem of luxury consumption, people would in principle be better off under a collective agreement not to engage in it. At the individual level, however, there is incentive to deviate.

12. For more formal exploration of this topic, see, for example, Amaldoss and Jain (2005); Corneo and Jeanne (1997); Frank (1985, 2011); and Frank et al. (2014) on "expenditure cascades." For a pioneering paper, see Leibenstein (1950).

13. Even the same act could classify as a need or want depending on the situation. If you live in New York City and fly every weekend to party in Key West, most would call that extravagant luxury. But if you take the same flight every weekend because your mother lives alone in Florida and is having concerning health problems, that could qualify as a need, or at least not be chastised as a needless, self-centered expenditure.

14. Older technologies might be used to provide a sense of occasion, such as horse-drawn carriages for solemn funerals and candles for romantic dinners (Dartnell 2015, 106). Other examples might include fountain pens used by US presidents to sign bills, fireplaces used for atmosphere rather than heat, and even handwritten letters in the era of e-mail and texting.

15. The "innovation treadmill" also operated with respect to energy, with upper classes willing to pay premiums for cleaner and better solutions. For instance, theatres, restaurants, and wealthy homeowners in the United Kingdom in the early twentieth century were willing to pay for the new "luxury" of electric lighting, rather than use gas or kerosene lamps, never mind the much higher price. Electric lighting was twenty-five times more expensive per unit of energy than gas, but only seven times more efficient. By 1920, electric lighting became the cheapest available option, paving the way for widespread adoption (Fouquet 2010). See also Fouquet (2014, 2008).

16. As well explored in the 2020 Netflix documentary *The Social Dilemma*, social media services like Facebook, Twitter, Instagram, and Pinterest deliberately target people's psychological needs to fit in and stand out, generating addictive tendencies and managing to push users to spend on average over two hours per day on their platforms.

17. The same point was made by economist Robert Frank (2007, xx): "No one denies that a car experienced in the 1950 as having brisk acceleration would seem sluggish to most drivers today. Similarly, a house of given size is more likely to be viewed as spacious the larger it is relative to other houses in the same local environment. And an effective interview suit is one that compares favorably with those worn by other applicants for the same job. In short, evaluation depends always and everywhere on context."

18. This example is from Coyle (2014, 65).

19. Langgut (2017).

20. Smith (1776, Book II, p. 352).

21. Sen (1999).

22. In Sen's words, "the levels of capabilities that are accepted as 'minimum' may themselves be upwardly revised as the society becomes richer, and more and more people achieve levels of capabilities not previously reached by many. These variations add further to the need for more income in the richer countries to avoid what is seen as poverty in terms of 'contemporary standards'" (Sen 1985, 26).

23. In the words of Joseph Schumpeter: "as higher standards are attained . . . wants automatically expand and new wants emerge or are created" (Schumpeter 1942, 116). Fifty years earlier, Alfred Marshall (1890, 60) noted that "Human wants and desires are countless in number and very various in kind: but they are generally limited and capable of being satisfied. The uncivilized man indeed has

not many more than the brute animal; but every step in his progress upwards increases the variety of his needs together with the variety in his methods of satisfying them. He desires not merely larger quantities of the things he has been accustomed to consume, but better qualities of those things; he desires a greater choice of things, and things that will satisfy new wants growing up in him." Hirsch (1976) dissects the topic of internal and external utility in great detail. Like others, however, Hirsch fell into the fallacy of separating the world into needs (inherent utility) and wants (external utility). Believing that inherent utility was already fulfilled, he concluded that people's only option was to accumulate what he called "positional goods," characterized solely by external utility. Moreover, he excluded what I call the process of democratization of consumption. This led him to conclude (erroneously) that there were hard social limits to economic growth, which was in the end driven only by social rivals vying for that still scarce resource: status.

24. This is why I refer often in this book to "perceived" scarcity.

25. At the country level, across the world, per-capita national income is the best predictor of average national life satisfaction. Controlling for income, however, changes in life expectancy also increase happiness, no matter whether life expectancy is high or low to begin with. Progress, it seems, fuels happiness (Deaton 2008).

26. At the 1529 wedding of Ercole II d'Este, Duke of Ferrara, and Renée of France, for example, twenty-five statues of sugar depicting the trials of Hercules were crafted for the banquet (Marshall 2015).

27. See Douglas and Isherwood (1979), and reported in Jackson (2009).

28. For a classic argument that status symbolism, often portrayed as a product of the modern age or capitalism, has deep historical roots, see Reinhold (1969).

29. Two archeologists in particular argue that commodity branding did not emerge with contemporary global capitalism, but rather dates back to the beginnings of urban life (Bevan and Wengrow 2010).

30. Braudel (1992, 1:313–315). Braudel argues that while the rich led the pack, the rising bourgeoisie and even the poor were not immune to fashion, following it by imitation, albeit at a distance.

31. One of the inscriptions carved on the Temple of Apollo at Delphi, Greece, was μηδὲν ἄγαν ("nothing to excess"). The principle reappears through the writings of Socrates, Plato, and Aristotle. In a way, the Greek myth of Icarus, who wanted to fly toward the sun but in so doing melted his wax wings and fell from the sky, teaches this basic lesson. On moderation in Ancient Greece, see also Kallis (2019).

32. Westacott (2016, 1).

33. As testament to this cultural transmission process, over a century later, Horace penned his famous line *Graecia capta ferum victorem cepit:* Captive Greece conquered her savage victor.

34. Thomas More's *Utopia* (1516) was produced as a critique of the time. On the back of disproportionate inequality in income and wealth, the European social order was characterized as a "conspiracy of the rich," perceived as greedy unscrupulous gluttons.

35. This is called the "social-rank hypothesis of income inequality." See, for example, Bricker et al. (2020) on luxury car purchases or Walasek and Brown (2015) on internet searches. The same point is made in Wilkinson and Pickett (2010).

36. Social media platforms and reality TV shows play a role in this by breaking down previous visual barriers and showing, for entertainment purposes, how rich people and their offspring burn through cash (Cohen 2018).

37. I would argue that the ranking should see as a first-best reducing the market tendency toward income and wealth concentration including, for example, strong competition policy measures vis-à-vis digital technology companies. Second-best measures include redistribution through taxes and subsidies; these are slightly less preferred because they are likely to generate more tensions within society than the first-best solution. As a less preferred solution is, in my view, the taxation of luxury items, because demand for luxury items among the rich is likely to increase (or at best remain constant) as the price increases. For "Veblen goods," demand is fueled by high price and scarcity. In a high-inequality society, taxing luxury goods will not increase democratization of consumption or decrease the sense of imbalance within society. Coming somewhere in between redistribution and taxing luxury goods is a progressive consumption tax, championed by Frank (2011), which is, however, likely to run into implementation problems.

38. See in particular Chancel (2020, 95–108).

39. Milanovic (2019, 180–184).

40. In the wise words of Chinese billionaire Jack Ma, "When you have one million dollars, you're a lucky person. When you have ten million dollars, you've got trouble, a lot of headaches. When you have more than one billion dollars, or a hundred million dollars, that's a responsibility you have—it's the trust of people on you, because people believe you can spend money better than the others." Maria Tadeo, "Jack Ma: The Richest Man in China on How to Be Happy and Successful in Business," *Independent,* September 24, 2014. This very concept is, however, antique, dating back even to the Bible (Luke 12:48): "To whomever much is given, of him will much be required; and to whom much was entrusted, of him more will be asked."

41. Love for possession, wealth, power, and luxury could raise fear of loss, making individuals more selfish, less generous, and less brave. Thus frugality would complement the suite of traits that promote cooperative human morality (Westacott 2016, 55–56).

42. The presence of large-scale philanthropists, such as Bill Gates, George Soros, or Warren Buffet, somewhat diminishes the standing of Branko Milanovic's

(2019) pessimism about single individuals taking moral actions, coupled with seemingly unfeasible systemic intervention. Instead, John Kenneth Galbraith (1958, 76) argues that a return to morality has happened before, following the incredible lavish excesses of first-stage capitalism in America.

43. This raises the long-standing debate about the role of philanthropy, particularly in the United States. Philanthropy contributes positively to society but cannot substitute for redistribution through taxation because individual philanthropists typically focus on causes close to their heart, without assessing societal priorities. For example, a neighborhood could end up with a top-quality philanthropy-financed library but lack food, sanitation, or shelter.

44. Individuals "express a preference for wage profiles which rise over time, even though these have lower present discounted values than alternative profiles with constant or decreasing wages. Such an observation is very hard, if not impossible, to square with a fixed utility function that does not depend on past incomes" Clark et al. (2008, 102).

45. If you read only one paper on this subject, I would advise the review by Clark et al. (2008). Other papers supporting some of the statements I make on relative income theory are Luttmer (2005), Perez-Truglia (2019), and Neumark and Postlewaite (1998).

46. This example comes from Solnick and Hemenway (1998). Similar experiments and surveys report similar results. See Cohen (2018, 134–135).

47. Max Weber (1905, xxxi) expressed the same conviction, using very powerful words: "The impulse to acquisition, pursuit of gain, of money, of the greatest possible amount of money, has in itself nothing to do with capitalism. This impulse exists and has existed among waiters, physicians, coachmen, artists, prostitutes, dishonest officials, soldiers, nobles, crusaders, gamblers, and beggars. One may say that it has been common to all sorts and conditions of men at all times and in all countries of the earth, wherever the objective possibility of it is or has been given. It should be taught in the kindergarten of cultural history that this naïve idea of capitalism must be given up once and for all."

48. Matthew 20:1–16.

49. "Two Monkeys Were Paid Unequally: Excerpt from Frans de Waal's TED Talk," https://www.youtube.com/watch?v=meiU6TxysCg. For more formal evidence see Chen et al. (2006).

50. While not using this specific analytical framework, this was already hinted at by Galbraith (1958, 80).

51. William K. Vanderbilt gave a ball in 1883 costing today's equivalent of $6.4 million (Galbraith 1958, 50). In 1897, in the midst of a nationwide economic depression, Bradley Martins threw what came to be known as "the most ostentatious party in US history," transforming the ballroom of the Waldorf Hotel into a replica of Versailles Palace for today's equivalent of $12.5 million. One

of the guests supposedly showed up in a gold-inlaid armor valued $314 thousand in today's prices. M. Galante and G. Lubin, "The Most Ostentatious Party in US History," *Business Insider*, November 6, 2012, https://www.businessinsider.com/bradley-martin-ball-of-1897-2012-11.

52. Ricardo (1817).

53. Hobbes (1651, 89). One element that could reinforce the stability of a "bottom-right" society could be a premodern value system whereby one's social standing (aristocrat or peasant, master or serf) is determined at birth, and is largely outside one's control.

54. These examples are from Henrich (2016, 22–33).

55. Why did the Great Divergence kick off in Europe rather than in China, which had a comparable level of technological development, property rights, and functioning public administration? Mokyr (2016) posits a number of reasons, including that Europe was made up of close-by neighbors in constant competition, and that it had shifted toward an Enlightenment, scientific mentality. The first of these builds on Landes (1998), who notes that competition favored freedom of thought, given that a man of science could easily find service in a rival court. See also Appendix B.

56. This point is made forcefully in Fouquet and Broadberry (2015), drawing on detailed data for European countries.

57. The famous Su Sung clock, possibly the most advanced water clock ever built, took eight years to make during the eleventh century. By the time the Jesuits arrived in China four hundred years later, know-how of sophisticated clock-making technology had disappeared (Mokyr 2016, 319). The Ancient Romans invented hydraulic cement, which allowed them to build quays, watersheds, lighthouse foundations, and ports all across the northern coast of Africa (which lacked natural harbors), and therefore to secure military and economic control of the Mediterranean. With the fall of the Roman Empire, however, this know-how was almost lost. No medieval sources mention it and the great Gothic cathedrals were built using only lime mortar (Dartnell 2015, 126). Damascus steel of the third and fourth centuries was renowned for its quality but produced by a method that has been lost to history. Walter Sullivan, "The Mystery of Damascus Steel Appears Solved," *New York Times*, September 29, 1981.

58. Morris (2013) looks at the West and the East separately, given that his argument hinges on evidence that these two broad civilizations developed in parallel. I am reporting the figures for the West. The two regions, in any case, display rather similar trends until the Industrial Revolution.

59. Steady growth of population led to a mechanical increase in the number of ideas in circulation. According to Jones (1992, 2), "the number of new ideas produced in a year rises by a factor of 110,000 . . . between 25,000 BCE and the twentieth century. A factor of 108 of this increase is due to the fact that the

twentieth century has a larger population base from which inventors are drawn." This leads Jones to the conclusion that some type of Industrial Revolution was inevitable—it was only a question of time and place.

60. This argument was first made by economist Esther Boserup (1965). More recently, Nobel laureate Michael Kremer supported Boserup's theory with empirical evidence. This work complements Malthus's theory, which looked at the other side of the coin: how population growth depended on available food production technology (Kremer 1993).

61. In addition to the stories of the Mayans and Easter Islanders, Diamond (2004) tells those of the ancient Cambodian Khmer civilization at Angkor, the pre-Inca Tiwanaku around Lake Titicaca, the great urban center of Tenochtitlan in ancient central Mexico, and the Anasazi in the American Southwest. Each collapsed due to a change in climate, prolonged or sustained droughts, or deforestation. On a smaller scale, the Greenland Norse disappeared after failing to adapt to the Little Ice Age. See also Boccaletti (2021) and Fagan and Durrani (2021).

62. "Growth, in different forms, has been a feature of mankind from the very beginning. . . . It is true, a number of civilizations stalled, some even disappeared, but the strength of the overall process was such that world population kept growing" (Cohen 2018, 62).

63. A set of recent papers question the conventional wisdom that Millennials display very different preferences in practice than previous generations—for example, in their car ownership and vehicle miles traveled. The differences that have been observed tend to relate to Millennials' delays in forming families and lower incomes—the result of having entered the labor market during the Great Recession (Leard, Linn, and Munnings 2019; Kurz, Li, and Vine 2018; Knittel and Murphy 2019). No clear generational effect on carbon emissions emerges in the United States, but a generational effect is observed in France, largely due to economic constraints with respect to Baby Boomers rather than to greater fears of ecological crisis (Chancel 2020, 98–101).

64. Terzi (2020a).

65. Galbraith (1958, 91).

66. Karl Marx in 1846 related the standard of living to "replacing the domination of circumstances and chance over individuals by the domination of individuals over chance and circumstances" (Sen 1984, 87).

67. This element takes on a negative connotation, specifically linking back to cruel acts under colonialism. And yet, rich democratic countries are constantly urged to chastise violations of democracy and human rights across the globe, and take action. This similarly belongs to the class of projecting value systems, and its effectiveness likewise rests on political and economic clout.

68. Beyond cases of complete civilizational collapse, the crucial difference in economic growth patterns between advanced and emerging countries is not the

absolute growth rate but rather the volatility. Poor countries alternate periods of extremely high growth (generally unseen in the developed world) with sharp collapses, while rich countries display slower growth rates that are sustained with low volatility year after year (Pritchett 2000). This is yet another indicator of how economic growth and technical progress contribute to resilience. For my own work on this topic, see Peruzzi and Terzi (2021).

69. "The ultimate, hidden truth of the world, is that it is something that we make, and could just as easily make differently" (Graeber 2015, 89). Tim Jackson (2009, 183–184) echoes these words, claiming that "economics is an artefact of human society. . . . We devise the rules of the game and establish its mores. We build and regulate the institutions that serve it. Its gatekeepers are the characters in a drama of our own making. Rewriting their role lies entirely within our remit."

70. A similar line of argument is advanced by evolutionary biologist Lonnie Aarssen (2015).

6. Understanding the Winds of Change

1. Rifkin (2019).
2. Nordhaus (2021).
3. Robert Solow famously remarked: "You can see the computer age everywhere but in the productivity statistics." Solow, "We'd Better Watch Out," *New York Times Book Review*, July 12, 1987. Three decades later, his remark holds up well.
4. Every technology takes time to spread, but big data and AI are particularly hard to apply to business processes that were created for the analogic economy (see Coyle 2014, 84; Brynjolfsson and Hitt 2000). As a result, the technology spreads slowly, and we are seeing a few innovation leaders pull further away from the many laggards, with a poor overall effect on aggregate productivity (see also Andrews et al. 2019; Aghion et al. 2021, 118–123).
5. Diamandis and Kotler (2012).
6. Wind power dates back at least to 5000 BCE and the first sailboats on the Nile River. See also Tagliapietra (2020a).
7. In the language of evolutionary biology and cultural evolution, knowledge transmission occurs in three ways: vertical transmission, from parents to offspring (or from ancestral societies to present societies); horizontal transmission, between individuals of the same generation (or across contemporary societies); and oblique transmission, from non-parent individuals in the parental generation (or from non-ancestral past societies). See Gould (2011) or Christakis (2019, 367). In perhaps the largest-scale historical example of horizontal / oblique cultural transmission, Japan's Meji Restoration featured much imitation of what was considered the best from the West. From France came its education system and legal code; from Prussia, the structure of the army

and the constitution (the American model was considered too liberal); from the British, the navy; and from the United States, business regulation.

8. The proof of the technological superiority of Western gunpowder weapons can be inferred from the Southeast Asians' preference for buying Western guns (Sharman 2019, 74). Hoffman (2015, 13) confirms this and notes that European firearm experts were in such high demand in the Middle East and Asia that "in seventeenth-century China, even Jesuit missionaries were pressed into service to help the Chinese emperor produce better cannons."

9. Capitalism can be seen as one such idea, as described in its irresistible global spread by Milanovic (2019) and, adapted to local context, as often discussed by Dani Rodrik regarding the Asian "tigers" or the Chinese growth miracle (Rodrik 2009). See Mokyr (2016, 22–33) on biological evolutionary principles applied to cultural accumulation processes. See also Cavalli-Sforza and Feldman (1981) for pioneering work on the topic.

10. This is true also in policymaking. As I once heard remarked, "today's bad policies are typically successful policies of the past that failed to be adapted in time." This is why there is no end to policymaking, which is just the constant adjustment of laws and practices to current needs and a changing reality.

11. Another reason for ammonia's importance in the run-up to World War I was increasing interest in producing explosive weapons at scale.

12. On the handiwork in Sulawesi forty-five thousand years ago as the product of cognitively "modern" members of our species, see Brumm et al. (2021). For the more general argument, see Gould (2011).

13. Given the increasing nostalgia toward the often idealized past, ranging from food habits to medical treatments, it is worth underlining that the opposite is also unlikely to be true. Far from proclaiming that modern is always better than what came before, it is worth reflecting each time on why a specific innovation came about and spread widely. In most instances, this was to address a specific problem created by the previous practice.

14. For an overview of technological diffusion, see Hall (2006). On the slow spread of mechanized spinning in France, see Juhász et al. (2020). In this case it is important to recognize that diffusion was slow until the advent of cotton mills and organization of processes for spinning and weaving at large scale. The green transition will face challenges, but at least will leverage previously developed production structures.

15. For more on the evolutionary basis of hyperbolic discounting see, for example, Gowdy et al. (2013).

16. In the language of economics, all these features are useful in that they allow capitalism to reduce "misallocation" of resources, bringing the economy closer to its possibility frontier. This includes also the improved allocation of talent (Jones 2016). Economists interested in macroeconomics, policy and long-term

growth devote large part of their time to improving these mechanisms of optimal resource allocation, often falling under the term "structural reforms." For more on this, see Terzi and Marrazzo (2020).

17. When factory targets were just quantitative, which is realistically what could be monitored through central planning, the result was that it no longer mattered whether shoes were durable, comfortable, or even the right size for the majority of wearers, nor whether TV sets actually worked after a few months. More broadly, centralized planning left little room for quick adjustments to errors in judgment or external factors beyond the state's control, not to mention tinkering and constant learning-by-doing. When one industry failed, the other industries followed suit (Coyle 2014, 69). In an all-time classic, Fukuyama (1992) made similar points on the Soviet central-planning impossibility of managing the amount of information needed for innovation.

18. This is true even though, as a comparatively efficient allocative mechanism, a capitalist system increases the chances that geniuses end up in positions allowing them to express their talent. For discussion of the misallocation of geniuses in developing countries, see Agarwal and Gaule (2020). Beyond capitalism, discrimination poses a challenge to this mechanism, as argued inter alia by Jones (2016).

19. Building on a vast literature on long-term growth, Jones (1999, 41) suggests that "The rise and decline of institutions such as property rights could be responsible for the rise and decline of great civilizations in the past." Certainly, throughout history, changes in property rights have been a fundamental determinant of economic growth. The Industrial Revolution is no exception.

20. This might be particularly important as there is some evidence that new ideas are becoming harder to find, meaning that technological and economic progress can only be sustained through increased research efforts. See Bloom et al. (2020).

21. Top countries for R&D expenditure, as a share of GDP, are these ten: Israel, South Korea, Japan, Sweden, Finland, Taiwan, Austria, Denmark, Switzerland, and Germany. Nine of these are OECD countries and all are high-income.

22. We saw this mechanism at play in 2020 with the development of new Covid-19 vaccines at breakneck speed, supported by almost $40 billion in public, private, and NGO investments. As the virus became the most pressing short-term problem for humanity, resources flowed to fighting it, and research programs were adapted to secure this funding.

23. Regarding industrial policy for the green transition, Rodrik (2014) argues that occasional failure is suggestive of a successful industrial policy, evidence that it takes some risks that would otherwise go unfunded.

24. Capitalism by no means excludes a large public sector involvement, including in R&D. Typically, however, this is smaller than the private sector's, especially

for rich OECD countries (on average, 71 percent of total in 2018). From 1950 to 1980, the federal government was the largest funder of R&D in the United States, and roughly half of its spending fell under the heading of defense. This could only, however, be sustained politically in the context of the Cold War with the Soviet Union. Public R&D expenditure started shrinking after the Moon era of the space program. See Jones (2016).

25. See, for example, Hickel (2020a, 198); Jackson (2009, 151, 166–168).

26. This is the process that Dani Rodrik has in mind when he talks of multiple paths of innovation, the idea being that society, through government, can steer the focus and direction of innovation. Rodrik (2011); D. Rodrik, "Technology for All," *Project Syndicate,* March 6, 2020, https://www.project-syndicate.org /commentary/shaping-technological-innovation-to-serve-society-by-dani -rodrik-2020-03.

27. Petroleum had been known and used already in antiquity. Bitumen was widely used from 4000 BCE to waterproof boats, baths or pottery. In the Book of Genesis (6:14–16), God instructs Noah to build the famous ark and smear it with bitumen. Bitumen was also used as a construction glue, and the Greek historian Herodotus describes how it was used in the construction of the mighty walls of Babylon. In the ninth century, Arab and Persian chemists were already distilling kerosene from petrol, for lamps. In medieval Europe, petroleum was used for supposed medical properties.

28. The modern kerosene lamp, invented in 1854, ultimately created the first large-scale demand for petroleum products. Kerosene first was made from coal, but by the late 1880s, most was derived from crude oil.

29. Morris (2013). This opens up an interesting debate, namely about what the future of energy demand is in a growing planetary economy. Projecting past trends would suggest that future technologies, powering the increasing complexity of society, will eventually continue increasing energy demand, particularly given we do not seem to be approaching supply-side limits (see Fouquet 2008). This debate is open, however, and several energy and climate experts have suggested that per-capita energy demand could very well plateau at some point, as we have seen in OECD countries over the past decades. Aggregate energy demand has remained practically stable across the OECD average from 2000 to 2016, despite a 32 percent increase in GDP (Tagliapietra (2020a, 170). For more on this view, see Smil (2017, 14). All in all, the relationship between economic growth and energy demand seems positive, although not necessarily linear.

30. Surveys across countries gauging voters' top policy priorities show "climate change" has surged only recently as an answer. See, for example, Pew Research Center, "As Economic Concerns Recede, Environmental Protection Rises on the Public's Policy Agenda," February 13, 2020, https://www.pewresearch.org

/politics/2020/02/13/as-economic-concerns-recede-environmental-protection -rises-on-the-publics-policy-agenda/. In no small part, this is because CO_2 emissions have a strong lagging effect, generating huge costs but only in the long run. They are also a global public "bad," in that my own emissions affect not only me but also others around me, even in other countries. Economists refer to a situation like this as a "market failure," because prices cannot convey the information required for efficient allocation.

31. For details on Exxon's treatment of climate-related discoveries, see Hasemyer and Cushman (2015).

32. See Keynes (1936, 384). This echoes nineteenth-century novelist Victor Hugo, credited with the phrase "Nothing is more powerful than an idea whose time has come."

33. For a riveting in-depth recount of the time around the introduction of the car, see Standage (2021). This perception of air pollution as a minor annoyance remained prevalent well into the middle of the twentieth century (McAfee 2019, 55).

34. Outright regulatory bans, such as the EU's ban on single-use plastic in 2021, remain a tool available to governments. But bans would need to be targeted on end products to avoid the challenges of understanding and regulating complex production processes.

35. As documented in the innovation economics literature, an adoption curve for new products and technologies takes an S-shape: it starts slowly, then accelerates exponentially, and finally stabilizes when the market is saturated (Hall 2006). The concept of a tipping point therefore seems relevant.

36. This is discussed in, among others, SDSN (2019). Along similar lines, Systemiq (2020) argues that a green tipping point has been reached.

37. The episode of the wager is taken from Coyle (2014, 71–72). Jones (2016) analyzes the price of a basket of industrial commodities, consisting of aluminum, coal, copper, lead, iron ore, and zinc, corrected for inflation. During the twentieth century, world demand for these industrial commodities exploded with the rise of the automobile, electrification, urbanization, and industrialization in general. The real price of these commodities *declined,* however, by almost 80 percent over the twentieth century.

38. In an abstract sense, recycling can be seen as a way of converting energy into material, which raises a question: If running out of material is not the issue, then can we run out of energy? That, too, seems extremely unlikely. Aside from the obvious pioneering work on nuclear fusion, which would for practical purposes generate an unlimited amount of energy, even just looking at current technology, the supply constraint seems very far. Taking just one among the renewable sources, a total of 173,000 terawatts of solar energy strikes the Earth continuously. That's more than 7,500 times the world's total energy use.

39. See, for example, Quah (2019).

40. In G7 countries, for example, the area covered by forests increased every year between 2001 and 2015. Turning to Europe, in 1990, 28 percent of Spain was forested; by 2020, the proportion was 37 percent. Greece and Italy both moved from 26 percent to 32 percent over the same period. "State of Europe's Forests 2020," Ministerial Conference on the Protection of Forests in Europe, https://foresteurope.org/wp-content/uploads/2016/08/SoEF_2020.pdf.

41. Note that throughout most of human history innovation has taken place more by simple trial-and-error or serendipitous discoveries. The quantum leap in innovation can be traced back to a scientific positivistic approach, however—which is to say, the realization that nature is organized along some fundamental regularities and that one should exploit, or bend, those to one's favor. As per Mokyr (2016, 41), "inventions are mostly not accidental events. The probability of an invention occurring is related to an epistemic base underpinning it, that is to say, to an understanding of the natural regularities and phenomena that make the technique work. Some inventions require a rather extensive understanding of the underlying science: nuclear reactors are not built by accident." The broader principle is applicable also to agriculture. Our ancestors have been increasing crop yields through artificial selection and cross-breeding for millennia, through a trial-and-error process. Only in 1866, when Gregor Mendel proposed the basic rules of intergenerational trait transmission, laying the ground for genetics, did this lead to more effective breeding programs.

42. Stephen Jay Gould notes that human knowledge is exploratory and teleological, so that knowledge accumulates faster and is transmitted more broadly faster than are genetic changes in biological evolutionary processes. In evolutionary terms, knowledge accumulation follows a Lamarckian process rather than Mendelian heredity. Gould asserts that the concept of progress and continuous improvement or increased complexity makes sense in a human cultural context, although not in biological evolution. See Gould (2011).

43. Scientific debate has long raged on this topic. To some, increasing diversity and complexity is a necessary result of the divergence process, implying progress. To others, including Stephen Jay Gould, this is far from necessary and should be imputed to what he calls the "left-wall"—the idea that, starting from unicellular life, you can only move in the direction of increased complexity. Moreover, Gould argues that we are deceived by our focus on the narrow tail of the complexity distribution (meaning us humans) into largely ignoring that the most common organisms on the planet have always been, and still are, simple bacteria (Gould 2011).

44. Romer (2016). For a similar treatment of the problem, see Weitzman (1998, 359), who concludes that "the ultimate limits to growth may lie not so much in our

abilities to generate new ideas, as in our abilities to process to fruition an ever-increasing abundance of potentially fruitful ideas."

45. This concept clarifies the economic motivation for public investment in R&D. Because private-sector investors do not factor in the wider benefit to society from inventions, capitalist firms invest less than what is socially optimal.

46. In the candid words of Nobel laureate physicist Frank Wilczek, "When I was a teen and trying to put it all together, I thought that life was a matter of figuring out the answer to questions and that was that. Now I'm learning that good answers lead to better questions, and that the cycle never ends." Quoted in Claudia Dreifus, "A Prodigy Who Cracked Open the Cosmos," *Quanta Magazine,* January 12, 2021. See also Weber (1917).

47. Technological achievements in medieval Europe, such as heavy plows, mechanical clocks, spectacles, iron casting, fire arms, shipping design, and navigational equipment, were of incredible importance (Mokyr 2016, 143). No "Medieval Industrial Revolution" ensued, however, because they were not associated with an accompanying depth of propositional knowledge explaining why they worked. The kind of self-reinforcing innovation dynamic that, centuries later, led to the Great Divergence was not established.

48. This also explains how it could be that past civilizations full of skillful artisans, in the Middle East, Southern Asia, and East Asia, nevertheless fell into technological stasis. Having such artisans is a necessary but insufficient condition for sustained and accelerating progress.

49. Based on an extensive literature review, Bloom et al. (2019) draws the same conclusion: globalization and trade openness are good for competition and innovation, but likely to exacerbate inequality among people and places.

50. Mokyr (1994) calls this Cardwell's Law, for historian Donald Cardwell, who observed that no nation has been very creative for more than a "historically short period." Among history's most well-known protests against technological development are the Luddite uprising and the Captain Swing Riots in Britain during the early nineteenth century. Recent research establishes a causal link between the adoption of threshing machines and the frequency of local riots during the 1830s (Caprettini and Voth 2020). The same holds for electrification in Sweden, which had labor-saving effects and led to local conflicts and strikes (Molinder et al. 2021). Of course, this resonates with current fears regarding automation, AI, and robots and their effects on jobs and the future of work.

51. Juma (2016). We cannot rule out that the comparatively generous welfare system prevalent in England at the time of the onset of the Industrial Revolution, the so-called Poor Laws, helped contain social tension following the introduction of labor-shedding technology, allowing for faster deployment of the new technology.

52. This has nothing to do with shortsighted politicians seeking reelection. To expect politicians to self-sacrifice and do what "needs to be done" for the better good is to miss the point of what good politics is about—which is building majorities and consensus. Policies adopted solely in the name of self-sacrifice would simply result in social tension, polarization, and a reversal once the incumbent politician was ousted. This is an argument I have made previously in discussion of proposed reforms to change the structure of the economy, especially during structural adjustment programs (Terzi 2017).

53. "Culture affected technology both directly, by changing attitudes toward the natural world, and indirectly, by creating and nurturing institutions that stimulated and supported the accumulation and diffusion of 'useful knowledge'" (Mokyr 2016, 7).

54. H. Ritchie, "Wild Mammals Have Declined by 85 Percent since the Rise of Humans, But There Is a Possible Future Where They Flourish," Our World in Data, April 20, 2021, https://ourworldindata.org/wild-mammal-decline.

55. Recalling the Judeo-Christian tradition, this view represents a modern reinterpretation of the Book of Genesis, with humanity moving from "dominion" to stewardship. Nature is the creation of God, and, as such, should be treated with respect.

56. According to the "Gaia hypothesis" proposed by James Lovelock and Lynn Margulis, nature is unfathomable because everything is interconnected in ways we cannot possibly understand (Lovelock 2000).

57. One such framing has been called "ecologism," an ideology that starts from the position that the non-human world deserves moral consideration, and that social, economic, and political systems should take this into account (Baxter 1999). For ecologists, "the environmental crisis is so great . . . that only a thoroughgoing reorganization of the political, social and economic system would achieve a solution. This would necessitate a massive change in human values" (Harrison and Boyd 2018, 275). They also note that, by contrast, "Environmentalists believe that green issues, however important they are, can be addressed within the existing political and economic structures. To succeed, this would require wise government, appropriate legislation and the voluntary adoption of environmentally sound practices by consumers" (275). These two quotes sum up the divergent ecological and environmental positions that we have explored in this book, especially with respect to ecosocialism and green capitalism.

58. Mokyr (2016, 17).

59. In 2018, while rafting in the Grand Canyon, I had an enlightening conversation on this topic with plant biologist Jiří Friml. Our proximity to the canyon wall, exposing nearly two billion years of Earth's geological history, made it easy for him to express the natural sciences view that, in the grand scheme of things, humans are nothing special. Our presence on the planet for just a tiny

fraction of its history is in no way comparable to other successful animals like the trilobites, with their ocean presence spanning almost 300 million years. As a social scientist I had to push back, pointing to all the marvels of our technological progress, which will make us in all likelihood the first species to expand beyond our planet. The debate was not settled, but two points stick with me: our history is much shorter than we like to imagine it; and it is best understood in the light of continuity, with parallels to other organic forms on Earth. These perspectives have shaped this book.

7. The Global Fight for Planet Earth

1. Galbraith (1958, 55); Boldizzoni (2020, 220). As noted by Max Weber, Marx's *Communist Manifesto* is at the same time a scholarly achievement of the highest order and the founding text of a religion of salvation. The latter makes it impossible to change its followers' convictions by rational argument, as noted by Boldizzoni (2020, 56).

2. The words of Italian revolutionary Antonio Gramsci come to mind. In *The Prison Notebooks* (1930), he writes that a "crisis consists precisely in the fact that the old is dying and the new cannot be born; in this interregnum a great variety of morbid symptoms appear" (Gramsci 1971, 275–276).

3. Hardin (1968).

4. Hardin concludes that appeals to morality alone cannot solve the tragedy of the commons, with regard to the environment or any other shared resource. Decades later, William Nordhaus showed quantitatively, using game theory and his DICE economic and climate model, that voluntary abatement with no sanctions can only lead to a noncooperative, low-abatement equilibrium (Nordhaus 2015).

5. Sachs (2008, 3). Raworth (2017, 198, 201) reaches similar conclusions: "for several centuries we have been encouraged to identify ourselves foremost as nations" but that we should now take the "inevitable" twenty-first-century step of considering ourselves part of a global community, opening up possibilities for globally redistributive design. As a consequence, "in the century of the planetary household, global taxes" will turn out to be inevitable. Perhaps not intentionally, these ideas echo the writings of Marxist revolutionary and political theorist Leon Trotsky, who noted at the outbreak of the First World War, that "the War proclaims the downfall of the national state" which had "outlived itself" (Trotsky 1918, 21–22).

6. For a recent example on the pandemic and vaccine rollout, see Brown et al. (2021). Among their conclusions is that "because a globally cooperative and better coordinated effort is needed, rich countries must stop their infighting and perhaps slow their own consumption of the currently limited stock of vaccines to deploy more to the world's hotspots as soon as possible."

7. As full disclosure, I have at times stood in this camp—for example, in work with Jim O'Neill on reforming the G7 and G20 to make them the economic steering board for the world (O'Neill and Terzi 2014).

8. Goldin (2021, 176).

9. On the origins of the Anthropocene, John Robert McNeill and Peter Engelke conclude, "it is best to return to the firmer ground of the past, to try to see how the present came to be what it is. . . . That will not allow us to know the future—nothing will, not even the most sophisticated modelling exercise, but it may help us to imagine the range of possibilities or, as Saint Paul put it, to see through a glass, darkly" (McNeill and Engelke 2014, 6).

10. For an extensive coverage of the challenges arising from climate adaptation, see Kahn (2021).

11. Burke et al. (2015b) produce a similar map of the economic impact of climate change across countries, at a lower level of granularity and without accounting for adaptation. Cruz Alvarez and Rossi-Hansberg (2021), looking at welfare more broadly defined, confirm these effects, as do Desmet et al. (2021), whose focus is on economic costs of coastal flooding due to climate change. See also Ricke et al. (2018).

12. Noy et al. (2021), examining the fiscal consequences of earthquakes at various levels of government in Japan, find a pattern of short-term increases in disaster-relief spending, but long-term reductions in other spending, at lower (city and town) levels of government. At higher levels of government, such as prefectures (regions), fiscal policymaking is robust enough to prevent unwanted reductions in public services. This illustrates why it is better to join forces against the threat of natural disasters, just as insurers do by risk pooling.

13. Migration has always been one of the main ways in which people respond to environmental stress (Butzer and Endfield 2012; Boccaletti 2021). The advent of agriculture and the creation of permanent settlements made it more costly to move, but the effect was only to tilt the cost-benefit analysis. In situations of need, and when little is left to be lost, humans still migrate away from adversity. Burzyńskia et al. (2019) estimate that the twenty-first century will see between 200 million and 300 million climate migrants, mainly due to rising sea levels.

14. This has been recognized by, among others, Wadhams (2017), Dartnell (2015), Sachs (2008, 107), and Barrett (2007).

15. To protect the city, a flood protection system known as MOSE was put in place, featuring an underwater mobile gate 1.6 kilometers long. As evidence that the consequences of climate change can be contained but the solutions will not come cheap, this piece of climate adaptation infrastructure cost a whopping 5.5 billion euros ($6.7 billion US). For more details, see the website, https://www .mosevenezia.eu/?lang=en.

16. In social psychology, this dynamic has been well researched since early work by Henri Tafjel on biases observable even in "minimal groups"—that is, groups with no prior history together and therefore no psychological or otherwise meaningful interpersonal connections. For more on the experiment involving five-year-olds and T-shirt colors, see Dunham et al. (2011).

17. In the *Descent of Man,* Darwin puts it this way: "There can be no doubt that a tribe including many members who, from possessing in a high degree the spirit of patriotism, fidelity, obedience, courage, and sympathy, were always ready to aid one another, and to sacrifice themselves for the common good, would be victorious over most other tribes; and this would be natural selection" (quoted in Wilson 2019, 77–78).

18. Christakis (2019, 277).

19. As per Huntington (1996, 21): "People use politics not just to advance their interest but also to define their identity. We know who we are only when we know who we are not and often only when we know whom we are against." As grim and unfortunate as this reality might be, it should not be ignored. People who seek to reinforce their identity will look for, or artificially create, foes.

20. Anderson (1983, 6).

21. Indeed, it has been proposed that religions have evolved culturally to exploit this piece of human psychology (Henrich 2016, 204).

22. Analyzing the history of European national anthems, Elgenius (2011, 141) reports that anthems adopted since the Napoleonic War era have been filled with references to war, conflict, death, and brotherhood.

23. Émile Durkheim, a founder of the discipline of sociology, discussed national flags as modern objects of worship, and therefore as extensions of a secular form of divinity. Indeed, their use is regulated by law and acts against them are seen as acts of desecration. Uses of flags by human societies date back at least to 4000 BCE. The Roman Empire emblazoned its standards with images of the emperor and of the eagle—the latter being treated as a sacred animal, given official recognition within the sanctuary of the Roman Parthenon, the divinity of which was transferred to the empire. Note, too, that blood-related symbolism is widespread, with eight in ten flags of the world including the color red (Elgenius 2011, 28–29, 91).

24. For a theoretical treatment of the link between conflict and measures of dispersion such as polarization and inequality, see Esteban and Ray (2011). The key parameters they identify are the "degree of within-group cohesion" and the quantity of public goods to be allocated.

25. Alesina and Glaeser (2004). Important studies in the field include Luttmer (2001) and Alesina and La Ferrara (2005), both of which provide overviews of the literature. Other studies have been carried out at the subnational and city

levels in the United States, reaching similar conclusions, and in Europe (Alesina et al. 2021). In the 1970s, Nobel laureate Thomas Schelling (1978) proposed a "bounded neighborhood model" by which, past a certain tipping point, a racial group moves out of a neighborhood altogether to avoid being in a minority. The clear underlying premise is a "sense of us" that relates to race, which leads to less willingness to share a common good—in this case, a neighborhood.

26. Evolutionary anthropologist Joe Henrich shows that humans evolved to parse ethnicity—becoming highly attuned to a set of cultural factors including language or dialect—rather than race, with its focus on morphological traits including skin color and hair type. In the modern world, however, these sets of cues tend to strongly overlap (Henrich 2016, 204–205).

27. In 2012, the prestigious journal *Proceedings of the National Academy of Sciences of the United States of America* devoted a special issue to the links between collapses of ancient states or civilizations and changes in climate or the environment. In it, Butzer and Endfield (2012) argue that environmental stress rarely dealt a direct blow to a society. Rather, over a decadal timeframe, it generally ignited socioeconomic dynamics that in turn brought about collapse. In their words, in cases of affected societies "with low resilience, cascading devolutionary feedbacks can then destabilize a system through famine, internal conflict, and political simplification, eventually leading to subsistence crises, breakdown of the social order, and civil wars" (3630).

28. Based on historical weather data from medieval and early-modern Europe, Anderson et al. (2017) show that during the fifteenth and sixteenth centuries, in periods of unusually cold temperatures, linked to poor crop yield, the likelihood of a Jewish community's being expelled was greater. This correlation is stronger for those cities known to have had poor-quality soil and weak states. Historians suggest an explanation: Jews were convenient scapegoats for social and economic ills. Along similar lines, rising numbers of witchcraft trials in Europe in the Little Ice Age were strongly associated with the fall in temperatures and consequent economic malaise (Oster 2004). A similar dynamic has been identified in modern Tanzania, linking extreme rainfall conditions (drought and floods) to witch murders (Miguel 2005).

29. Christakis (2019, 276) argues that power-hungry, malevolent leaders can take advantage of these evolutionary in-group identification dynamics to actively foster out-group hatred, escalating expressions of xenophobia, racism, and prejudice to generate scapegoats.

30. See Butzer and Endfield (2012).

31. As noted by Wilson (2019, 222), the novelist Joseph Conrad said that one reason he liked writing sea stories was because life on a ship is so morally simple: everyone on board knows the common imperative is to keep the ship afloat.

The same appeal, Wilson muses, might draw one to space adventures like *Star Trek* and *Star Wars*. When a planetwide catastrophe appears imminent, anyone "will start regarding the whole earth as the appropriate moral circle."

32. Parker (2013). See also Fagan (2001).

33. Iyigun et al. (2017a, 2017b).

34. See Jia (2014) for analysis of data from 267 prefectures in historical China. Using high-resolution paleoclimatic reconstructions to match events with incidences of conflict in China over the past millennium, Zhang et al. (2006) conclude that "the stressed human-nature relationship generated a 'push force,' leading to more frequent wars between states, regions and tribes, which could lead to the collapse of dynasties and collapses of human population size" (459).

35. Miguel et al. (2004).

36. For example, strengthening the role of Egypt's highest-ranking religious authority enabled them to threaten political authorities with revolt (Chaney 2013).

37. Using evidence from the early Holocene in Saudi Arabia, Petraglia et al. (2020) show that, following the extreme aridity of the Last Glacial Maximum, precipitation increased around ten thousand years ago during the Holocene humid period. Records show, however, that after six thousand years there was a return to aridity, culminating in the establishment of the desert environment prevalent today. Because of this great fluctuation it is possible to analyze how human societies adapted to environmental stress. A fascinating aspect of the work is that droughts are documented in rock art (as in the Shuwaymis petroglyphs) depicting starving cattle. Competition for grazing led to a sharp increase in conflict, as documented by a growing share of burials with victims showing marks of fatal trauma. When necessary, humans migrated from the desertic inland to the resource-rich coasts.

38. Burke et al. (2015a), but compare Carleton et al. (2016).

39. Indeed, as several have noted, the G20 works well at times of crisis but poorly in peacetime. Borrowing a classic title from absurdist theater (by playwright Luigi Pirandello), Angeloni and Pisani-Ferry (2012) refer to the G20 as "characters in search of an author." A recent confirmation of this pattern was the debt moratorium for poor countries during the Covid-19 crisis.

40. For economists, these differences within the group relate to the *adverse selection problem* in insurance, by which those at higher risk are more apt to seek coverage. In the context of climate change collaboration it is the poorer countries, with more to lose given their exposure to more frequent negative events of greater magnitude, that are more likely to cooperate willingly, but also to lack the resources to do so. Rich countries have less incentive to join the collaborative effort.

41. Barrett (2007) lays out a spectrum of global problems according to how much global cooperation is required to solve them. On one end are problems like

asteroid deflection, which in principle could be solved by a single, self-interested country. At the other end are challenges like disease eradication. Barrett calls solutions in this part of the spectrum the "weakest links" among global public goods because arriving at them requires literally all countries to cooperate. One country failing to fight a pandemic, for example, can be the source of a mutation and resurgence of the problem globally. Based on Barrett's taxonomy, climate change is closer to the latter end, because many, though not all, countries need to join the collaborative effort. The cooperation of countries with low emissions, whether they are very small (for example, Monaco) or very poor (for example, Afghanistan) is not essential to success in avoiding climate catastrophe.

42. Note that in the United States the policy was implemented by President Trump's administration but retained by President Biden's, despite its more internationalist agenda. The forces dictating these behaviors are deeper and stronger than a single leader.

43. Even if advanced economies were fully vaccinated by the end of 2021, they would still experience a drag on their economic recoveries because of their trade and production links to emerging and developing economies. These costs were predicted to fall somewhere between $400 billion and $1.9 trillion, placing them well above the strict cost of manufacturing and distributing vaccines globally (estimated at $27.2 billion). The authors concluded that global coordination on vaccines would make sense even from just a strict economic standpoint, setting aside moral considerations (Çakmaklı et al. 2021).

44. Unfortunately, when various socioeconomic forces are in play, the fact that a course of action makes sense from a scientific, moral, and organizational perspective is no guarantee that it will be taken. Investment in research and development provides a clear example. In areas from cancer research to space exploration to climate technology—not to mention pandemic vaccine development—joining forces at the global level could achieve faster progress. Countries could pool their resources in the search for breakthrough technologies, and avoid redundant efforts. This, however, is rarely seen. Instead, for example, in the provision of Covid-19 vaccines, China and Russia immediately seized the opportunity to engage in so-called vaccine diplomacy, striking deals for vaccine deliveries based on geopolitical objectives, rather than directing sales and donations to infection-rate hotspots. Anyone engaging in wishful thinking about humanity's capacity to join forces in face of common threats should study these examples as cautionary tales.

45. This seems to echo Francis Fukuyama's (1992) famous predictions. In the aftermath of the Soviet Union's collapse, he believed that Western liberal democracy would spread universally as the final form of human government and

that wars would end at least among the nations traditionally grouped as first-
and second-world powers. The euphoria following the end of the Cold War
generated an illusion of global harmony, which rapidly revealed itself for what
it was. This feeling is typical at the end of large-scale conflicts, including World
War I, which was supposed to be "the war to end all wars" and World War II,
which was supposed to be followed by a universal system of "peace-loving Nations" according to President Franklin D. Roosevelt (as reported in Huntington
1996, 31–32).

46. By historical analogy, even with the rise of petrol at the beginning of the twentieth century, we did not see coal, the energy source of the previous century,
disappear. Rather, to this day, even in advanced economies, coal still has a generally small, and yet relevant, share in the energy mix.

47. See Yergin (2020). We already had a foretaste of this in 2010 when, within the
context of a geopolitical clash over some contested island, China suspended
the export to Japan of rare earth minerals, used to manufacture products like
hybrid cars, wind turbines or guided missiles. While hosting around 36 percent
of world reserves, China currently produces around 90 percent of the world's
total output of rare earths, giving the country a strong lever to gain the upper
hand in tense geopolitical situations (Tagliapietra 2020a, 129).

48. To a certain extent, China's strong presence in Africa, and particularly in the
management of the Democratic Republic of Congo's cobalt and copper reserves,
and more broadly the far-reaching Belt and Road Initiative, can be read also
through these geopolitical lenses.

49. The Democratic Republic of Congo is one of the poorest countries in the world.
As discussed at length by development economist Paul Collier, the discovery
or increased prominence of natural resources can be a double-edged sword for
a poor country. On the one hand, it carries the promise of a sizeable inflow of
money, and with it the potential to lift many from poverty. On the other hand,
the country will be exposed to the risk of rampant corruption and graft. Collier proposes a set of international standards to reduce this risk (Collier 2011).

50. Another trend playing out in parallel is the acceleration of digital technologies, including the use of Big Data and artificial intelligence. This, just like the
green transition, will imply a shift away from ownership of material resources
and enhance the crucial importance of technological dominance.

51. Allison (2017). Earlier scholars have made the same point; see, for example,
Mearsheimer (2001).

52. In a controversial paper, Nobel laureate Thomas Schelling broke with the
prevailing narrative of "leaving a better planet to our children," pointing out
that, from a strictly economic perspective, climate mitigation involves facing
costs today to reap benefits (or avoid catastrophic costs) tomorrow. When we

combine this perspective with calls for the largest costs to be borne by the rich, industrialized countries—whose combined populations today represent less than 20 percent of the global total and in coming decades will represent less—it is easy to see that climate mitigation spending will function as a sort of implicit global redistribution machine (Schelling 1995).

53. Sen (2006).
54. "As Western power declines, the ability of the West to impose Western concepts of human rights, liberalism, and democracy on other civilizations also declines and so does the attractiveness of those values to other civilizations" (Huntington 1996, 92). Cultural imposition is not a uniquely Western habit. Throughout history, increased power has almost always meant assertive projection of values, practices, and institutions to other societies.
55. World Bank (2008).
56. Exports serve a double role in allowing for rapid development. First, they open space in the current account. In plain English, the fact that you export more means you then have the possibility of importing more without destabilizing your currency. This, in turn, makes it possible to import machinery or inputs from abroad, and to move up the value chain. Second, exports imply that you have access to large demand. The alternative is to rely on the domestic market, but, in a developing country, that will by definition be small and can grow only slowly. A. Terzi, "Is the Golden Age of Globalisation Over?" *World Commerce Review,* Spring 2020, https://www.worldcommercereview.com/publications /article_pdf/1828.
57. Another part of the degrowth standard agenda that also carries profound ramifications for rapid development prospects is the idea of fostering local production and breaking down global value chains. Trade openness is no guarantee of rapid development, despite the usual Washington Consensus arguments. Still, without trade openness, economic miracles are unlikely to occur.
58. Ostrom (1990, 26). Discussing the problem of global climate change a decade later, Elinor Ostrom concludes that "It is obviously much easier to craft solutions for collective action problems related to smaller-scale common-pool resources than for the global commons." Further, she argues, "extensive empirical research on collective action . . . has repeatedly identified a necessary central core of trust and reciprocity among those involved that is associated with successful levels of collective action. If the *only* policy related to climate change was adopted at the global scale, it would be particularly difficult to increase the trust that citizens and firms need to have that other citizens and firms located halfway around the globe are taking actions similar to those being taken 'at home'" (Ostrom 2009, 13, 35). Elsewhere (Ostrom 2012, 356) she declares that "relying entirely on international efforts to solve global climate problems needs to be rethought."

59. V. Ostrom, Tiebout, and Warren (1961) is the pioneering paper that first used the term *polycentricity* in the context of metropolitan governance, to describe a system of, as the lead author later summed it up, "(1) many autonomous units formally independent of one another, (2) choosing to act in ways that take account of others, (3) through processes of cooperation, competition, conflict, and conflict resolution" (V. Ostrom 1991, 225). Also see Dorsch and Flachsland (2017). Regarding climate change, this decentralized approach is built on the realization that subnational-level initiatives contribute to a dynamically evolving web of policies, led by businesses, civil society groups, and individuals, which often organize in transnational networks. A polycentric approach can also encourage local experimentation, which is particularly useful when complex problems like complete decarbonization demand experimentation and solutions that can be later scaled up to higher levels of governance.

60. Sachs (2008, 112–114).

61. Montreal imposed trade sanctions on those who refused to comply once a quorum of countries had signed. In contrast, Barrett (2007) emphasizes that Kyoto was designed to apply only to the thirty-eight wealthiest countries. This gave more than 150 other countries an incentive to free ride, and failed to reassure the wealthy countries that their efforts would suffice to solve the problem over the long term.

62. Rodrik (1997).

63. See, for instance, Creutzig (2020).

64. Among others, geographer Karl Butzer reviewed historical cases of civilizations dealing with heightened environmental stress. His conclusion is that authoritarian states make it harder for bottom-up options to emerge, silence dissent, and stifle initiative. Meanwhile, tolerance for new ideas may have helped western European societies avoid collapse over three centuries (between the Medieval warm period and the end of the Little Ice Age) of extreme climatic perturbations (Butzer 2012).

65. The Kyoto protocol was in principle binding, but then again, when Canada failed to meet its obligations and withdrew from the treaty, no sanctions were levied. Moreover, developed countries failing to meet their CO_2 targets could simply pay to offset the extra emissions through the "clean development mechanism" specified under the protocol.

66. For a deeply insightful explanation of how this process operates for the European Union in the world, see Bradford (2020).

67. Morocco and Tunisia have the potential to become important renewable energy hubs as the European Union strives to become carbon neutral by 2050 (Bennis 2021). More broadly, thanks to hydrogen produced using renewable energy like solar and wind, North Africa could become an energy supplier for Europe, and so could vast neighboring countries like Ukraine (Leonard et al. 2021).

68. While cooperation in the realm of climate change, along the lines of the Paris Agreement, creates an important global framework, we should recognize that action will inevitably happen at the national level (Tagliapietra 2020a, 77–78).

69. Regarding fish, see Sala (2020), who argues that 35 percent of world oceans should be protected areas. Regarding land, see Strassburg et al. (2020) on how to select locations for ecosystem protection and restoration.

70. See, for example, Leonard et al. (2021) and Tagliapietra and Wolff (2021). For theoretical foundations of the idea, see Nordhaus (2015), especially on the useful contribution of climate clubs. In the end, "the roots of economic cooperation are in cultural communality" (Huntington 1996, 135).

71. In an impressive display of sensing emerging trends, Martin Wolf, writing in 2007 when green was by no means the name of the game, came to the strikingly similar conclusion that the Great Enrichment had moved humanity out of a zero-sum world and, in the process, transformed international relations. Wolf warned that, should growth come to an end, "dark days will follow." Martin Wolf, "The Dangers of Living in a Zero-Sum World Economy," *Financial Times*, December 19, 2007.

72. Wilson (2019, 111).

8. A Blueprint for Green Capitalism

1. John Maynard Keynes, in an obituary essay on his teacher Alfred Marshall, wrote that "the master-economist must . . . contemplate the particular, in terms of the general, and touch abstract and concrete in the same flight of thought. He must study the present in the light of the past for the purposes of the future. No part of man's nature or his institutions must be entirely outside his regard. He must be purposeful and disinterested in a simultaneous mood, as aloof and incorruptible as an artist, yet sometimes as near to earth as a politician" (Keynes 1924, 322).

2. Francesco Boldizzoni (2020, 232–234, 274), writing on the many mistaken foretellers of the end of capitalism across history, concludes that one trait that has consistently clouded their thinking is their strong utopian tendency.

3. See, for example, Mazzucato (2021), which takes inspiration from "moonshot" programs to advocate for large-scale coordination between the public and private sector to achieve shared societal goals. Indeed, EU Commission president Ursula von der Leyen referred to it as Europe's "man on the moon moment." "Press Remarks by President von der Leyen on the Occasion of the Adoption of the European Green Deal Communication," European Commission, Brussels, December 11, 2019, https://ec.europa.eu/commission/presscorner/detail/en/speech_19_6749.

4. When Friedrich Hayek wrote his critique of state socialism (Hayek 1944) he was also responding to this type of wartime economy, with its strong role for the government in economic planning and decision-making, as seen in the United States and Britain. This also shows that capitalism progresses through cycles, with alternating periods of government activism and government minimalism.

5. The notion of a Green New Deal was introduced by Thomas Friedman in 2007, invoking Franklin D. Roosevelt's transformative set of policies during the Great Depression. Thomas Friedman, "The Power of Green," *New York Times,* April 15, 2007. In the aftermath of the 2008–2009 financial crisis, the United Nations Environmental Program (UNEP) began to promote the launch of the Global Green New Deal Initiative, a vision to foster jobs and growth after the recession while fighting climate change.

6. Damian Carrington, "'Blah, Blah, Blah': Greta Thunberg Lambasts Leaders over Climate Crisis Greta Thunberg." *Guardian,* September 28, 2021, https://www .theguardian.com/environment/2021/sep/28/blah-greta-thunberg-leaders -climate-crisis-co2-emissions.

7. For instance, focusing on the United Kingdom and post-pandemic recovery packages, Martin et al. (2020) point out the equal importance of the long-term targets and vision with respect to short-term mechanisms when trying to prod a shift to new clean technologies. On the importance of a vision to set a new green direction of society's complex system, see Kupers (2020, 122–123).

8. As noted by a recent British report, creating clear and concrete expectations on the climate mitigation path ahead helps the private sector reduce costs associated with decarbonization, notably preventing stranded assets and early capital write-offs. See *Report to the Committee on Climate Change of the Advisory Group on Costs and Benefits of Net Zero,* 2019, https://www.theccc.org.uk/wp-content /uploads/2019/05/Advisory-Group-on-Costs-and-Benefits-of-Net-Zero.pdf.

9. The images of Texas, where many lost electricity and heating for days when the power grid was shocked by an extraordinary "arctic blast" in early 2021, should serve as a cautionary tale.

10. Arthur Cecil Pigou first introduced the concept of a tax to correct an externality problem (Pigou 1920), building on work of his University of Cambridge predecessor Alfred Marshall.

11. Proving the earlier point on the power of a credible announcement, the proposed plans to reach ambitious climate targets are already leading to an increase in the carbon price in the EU ETS. This is because unused permits can be saved to offset future emissions. Firms are therefore setting aside permits in the expectation that emission targets (the cap) will be tightened. Further keeping prices suppressed was also the surplus of emissions allowances that accumulated in the EU ETS between 2009 and 2013, in the aftermath of the

2008 financial crisis. Since 2019, this excess supply has been reduced by the introduction of the so-called Market Stability Reserve.

12. For more on cap-and-trade versus carbon taxes, see Aldy and Stavins (2012), Goulder and Schein (2013), and Fouquet (2010), who adds a historical perspective. See IMF (2019) for an overview of policies to price carbon, and strategies to enhance their political feasibility.

13. https://clcouncil.org/media/EconomistsStatement.pdf.

14. In the EU, fuel subsidies were estimated to be worth roughly $289 billion (Coady et al. 2019). Parry, Black, and Vernon (2021) builds on and updates Coady et al. (2019).

15. Twenty-nine US states have so-called Renewable Portfolio Standards (Stock 2020).

16. Regulatory targets worked to foster corn-based biofuel in the United States. But they failed for other, less-polluting advanced biofuels, because policymakers failed to see the poor state of advancement of these technologies (Gates 2021, 265–266).

17. Note the interaction between environmental regulations and cap-and-trade. If you successfully bring down emissions through regulation, this might push down the carbon price under cap-and-trade, potentially nullifying the benefit through a "rebound effect." This could be prevented by introducing a carbon price floor within the cap-and-trade system. In any case, it shows that regulation can bring us only so far in tackling climate change.

18. "The absence of a price on carbon is but one of the externalities plaguing climate policy, and carbon pricing alone at politically plausible levels is unlikely to be particularly effective in reducing emissions from the oil and gas used in the transportation, commercial, and residential sectors" (Stock 2020, 402). Energy systems have always adjusted in response to price signals, but the adjustment has come with large social and economic dislocation, so "it is doubtful whether any elected political system could drive equivalent transformations using price alone" (Grubb et al. 2018, 3).

19. High-Level Commission on Carbon Prices (2017, 3). See also Stern and Stiglitz (2021).

20. See, for instance, G. Tett, "Wall Street's New Mantra: Green Is Good," *Financial Times,* January 29, 2021. At the end of 2020 the return in global market portfolios over the previous decade on renewable power companies was estimated at 422.7 percent, much higher than the return on fossil-fuel companies at 59.0 percent (IEA 2021).

21. Specifically, given climate change will have sharp repercussions for financial stability, central banks are well positioned to use their powers as regulators to jolt the market toward green investments. Barry Eichengreen, "New-Model Central Banks," *Project Syndicate,* February 9, 2021, https://www.project-syndicate

.org/commentary/central-banks-have-tools-for-climate-change-and-inequality
-by-barry-eichengreen-2021-02.

22. Aghion et al. (2016) find strong evidence of technological and investment path dependency in the auto industry. Firms that have innovated extensively in dirty technologies in the past will find it more profitable to innovate in dirty technologies in the future. This justifies government intervention to redirect technological change. More broadly, reading the economy through complexity science, Kupers (2020) argues that technological lock-ins require an external jolt to move the economy to a new, green equilibrium.

23. For a more technical treatment of this point, see Gerlagh et al. (2009).

24. In a curious turn of events, and in his own words, even tech billionaire Bill Gates moved from seeing government regulation as counterproductive red tape to acknowledging the crucial role the government will need to take, including on public procurement, to successfully stave off climate catastrophe (Gates 2021, 253, 282–283).

25. The need for new ideas and technological solutions for housing has led the European Commission president Ursula von der Leyen to call for a "New European Bauhaus," referring to the famous 1920s German artistic movement, to reimagine living spaces in the face of climate change. Two examples spring to mind in this respect. Architecture and engineering expert Catherine De Wolf has been compiling a global database of the carbon embodied in buildings, illustrating how this can be reduced through reusing structural components and fostering circular economy principles for cement and concrete. Along different lines, space architect and MIT Media Lab affiliate Valentina Sumini is among the few pioneers developing design and architectures to sustain human life in extreme environments beyond Earth. While this is pursued to enable space exploration, the low-resource use and resistance to extreme circumstances can clearly inform Earth-based climate-change-compatible architecture.

26. Similarly, Tagliapietra and Veugelers (2020) conclude that achieving net zero by 2050 will represent an industrial revolution against a deadline, and a green industrial policy will be fundamental to facilitate this shift within the allowed timeframe. See also Gates (2021, 278–279).

27. See, for example, Cherif and Hasanov (2019); but also IMF (2020), arguing that government must take a more active role in climate-change mitigation, and specifically in boosting low-carbon technologies.

28. See, for instance, Acemoglu et al. (2016 and 2012), but also Gerlagh et al. (2009).

29. See also Aghion et al. (2021, 184).

30. For instance, Alexander Hamilton, a US founding father and first secretary of its treasury, was a strong proponent of infant industry protection for the nascent cotton industries in the United States (Juhász 2018).

31. Beckert (2015).

32. Juhász (2018).
33. An important caveat is in order. Juhász herself recognizes that trade protection was effective because France already had the underlying prerequisites for the development of a comparative advantage in mechanical cotton-spinning. Before the invention of this technology, differences between France and Britain were minimal, especially if compared with those between rich and poor countries today. We should not, therefore, jump to the conclusion that a blunt instrument like infant-industry protection will always spur rapid development. As Ricardo Hausmann emphasizes (see Chapter 3), achieving economic development is much more complex.
34. MarketsandMarkets (2021).
35. Goldman Sachs (2020).
36. President Biden's executive order strengthening the "Buy American" provisions for public procurement (Executive Order 14005, signed January 25, 2021) is a victim of the same logic. It seems that, where large public investments are involved, some protectionism is unavoidable.
37. Simulations show that, in the first few years, the climate-mitigation package to net zero will likely need to be debt-financed, as carbon revenues are smaller than the initial spending on infrastructure, subsidies, and compensatory transfers to negatively affected households. As the price on carbon increases, carbon tax revenues are thereafter broadly sufficient to finance the additional green infrastructure and transfers (IMF 2020).
38. The EU is currently considering, among various options, introducing a tax on non-recycled plastic to eventually repay the debt incurred due to its Covid-19 recovery program, known as Next Generation EU. In 2018, the EU ETS generated €14 billion in revenues, 80 percent of which was then used by national governments to combat climate change, at home or abroad. It is estimated that between now and 2050, total accumulated ETS revenues could range between €800 billion and €1.5 trillion (Fuest and Pisani-Ferry 2020).
39. See Rodrik (2014).
40. Milton Friedman expressed his conviction that "there is one and only one social responsibility of business—to use its resources and engage in activities designed to increase its profits so long as it stays within the rules of the game, which is to say, engages in open and free competition without deception or fraud." Milton Friedman, "A Friedman Doctrine: The Social Responsibility of Business Is to Increase Its Profits," *New York Times Magazine,* September 13, 1970, 17. See also Friedman (1962). The point was reiterated, in milder tones, by another Chicago School economist: Raghuram Rajan, "What Should Corporations Do?" *Project Syndicate* commentary, October 6, 2020. https://www.project-syndicate.org/commentary/what-are-corporations-for-stakeholders-or-shareholders-by-raghuram-rajan-2020-10.

41. Dani Rodrik (2015) shows that each economic model explains only partially how the world works. The skillful economist must therefore understand, depending on the setting, which model is to be applied. Within our context, drawing insights from a perfectly competitive market model for climate and the environment, when we know that there are several market imperfections in this specific field, would therefore lead to erroneous policy recommendations.

42. This situation is not without historical precedents, comparing well with the emergence of the English and Dutch East India Companies (EICs) four hundred years ago, which in many ways are the forerunners of modern multinational corporations (Sharman 2019, 65). Under their original charters, European governments granted these companies unique privileges to exploit the opportunities offered by the discovery of the "new world." The EICs had the right to sign treaties with regional powers such as the Mughal Empire and the Maratha Confederacy, build forts, issue money, carry out administrative functions in the region, and even recruit armed forces. By 1803, the English EIC had a force of 260,000 soldiers—twice the size of the British Army. In the absence of regulations, these corporations had turned into private empires. When the British and Dutch governments tried to regain control, they realized that the EICs were buying parliamentary influence to retain their privileges, answering only to their shareholders. By some estimates, the Dutch EIC's peak net worth was around $7.9 trillion in today's dollars: roughly the equivalent of Japan's and the UK's GDP combined. Stefano Marcuzzi, and Alessio Terzi, "Are Multinationals Eclipsing Nation-States?" *Project Syndicate,* February 1, 2019.

43. Data for nominal GDP in 2020 is based on IMF estimates, October 2021.

44. In some segments corporate demand outstrips that of private individuals. For instance, roughly 60 percent of new vehicle purchases in Europe are made by corporations, including company cars then offered as perks to employees. Thus the private sector can be a significant actor in transforming the energy-source landscape within the transportation sector. William Wilkes and Stefan Nicola, "Tesla Needs to Crack Europe's $360 Billion Corporate Car Market," *Bloomberg Businessweek,* March 3, 2021.

45. See, for example, European Commission (2020b, 2).

46. Clearly this is a simplification and broad generalization. Most products are covered by regulations, which typically prohibit the use of some toxic, dangerous, or polluting materials, and define detailed technical standards for production. There are also substantially different legal philosophies on the two sides of the Atlantic regarding product liability. The United States has a tendency to keep product standards limited, while allowing consumers to make ample use of lawsuits against producers. In Europe, technical and production standards tend to be more stringent, applying a sort of ex ante screening to consumer products.

If a company does not violate the regulation, however, it is quite shielded from product liability litigation.

47. Despite EPR being, in theory, an individual obligation, in practice producers often exert this responsibility collectively. In collective schemes, a *producer responsibility organization* (PRO) is set up to implement the EPR principle on behalf of all the adhering companies (the obligated industry). Competition among different PROs is also a possibility. For more information on the European experience with EPR, see, for example, Marques and Ferreira da Cruz (2015).

48. In 2020, the Swedish company, which also plans to become "climate positive" by 2030, pledged to take back its furniture in exchange for vouchers good toward new purchases. Returned items would be resold in the store, donated to charity, or recycled. D. B. Taylor, "Ikea Will Buy Back Some Used Furniture," *New York Times,* October 14, 2020. Patagonia has operated a repair and recycle "Worn Wear" program since 2005. A. Engel, "Inside Patagonia's Operations to Keep Clothing Out of Landfills," *Washington Post,* August 31, 2018.

49. Given the international polluting power of plastic, as evidenced by the omnipresence of microplastics on land, sea, and air, an EPR for plastic could be covered by an international agreement, in line with the targeted treaty approach discussed in Chapter 7. This option was recently considered in "Chemistry Can Help" (2021).

50. A historical parallel occurs with the precursor to the Green Deals—namely, FDR's New Deal—which at the time was dealt a significant blow by the conservative majority on the Supreme Court, on the grounds that it went beyond federal competences. By 1936, the Supreme Court had invalidated eleven of the thirteen New Deal laws that had come before it. After a landslide reelection, in his inaugural speech President Roosevelt proposed to add as many as six new justices to the court, in what became known as the "court-packing plan." To oppose this, two justices effectively changed their positions almost overnight, making the plan no longer necessary. It was a move that came to be known as "the switch in time that saved nine." The key takeaway for our purposes is that policy action can be immensely important in restructuring an economic system, but it is only viable insofar as it is strongly supported by voters through the democratic process. Otherwise, it will face roadblocks and run into inertia.

51. Climate scientist Michael Mann (2021) comes to a similar conclusion: individual behavior matters, especially as it sets an example, but solutions to climate change will need to be systemic, including pricing carbon and ending fossil-fuel subsidies.

52. European Commission (2018, 2020a).

53. Writing a decade ago, Weyant (2011) looks at fundamental policies to accelerate the development and diffusion of new energy technologies and, on top

of carbon pricing and public research and development, places particular importance on consumers and their power to make green choices.

54. Terzi (2020a).

55. More specifically, one round-trip flight from New York to Europe creates a warming effect equivalent to two or three tons of carbon dioxide per person. The average European generates about ten tons of carbon dioxide a year (Carmichael 2019; Terzi 2020a).

56. Tagliapietra (2020a, 212).

57. Cars are a case in point. In most advanced economies, personal cars are used on average between nine and ten hours per week. This means that almost 95 percent of the time, a car sits idle on the street or in a garage, suggesting ample scope for optimization going forward (Kupers 2020, 104–114). Mindful of Larry Summers's frequent warning to students not to make measurable predictions that could be proven wrong, I will not attach a date, but will nonetheless express confidence that, when people of the future look back to photos of our era, they will be puzzled by all those bulky aluminum boxes taking up much of the urban space. Andrew McAfee (2019, 241–242) makes a similar prediction.

58. This is the emulation channel that advertisers use when they feature successful, famous people purchasing their product or engaging in an activity, often wholly unrelated to the talents they are known for. The human tendency for hyper-emulation based on cues of success and prestige can be very powerful, as illustrated by suicide epidemics that have occurred in the aftermath of celebrity suicides (Henrich 2016, 49–50). Marketers' workhorse models assign particular importance to mass media early on in the diffusion process, at the beginning of the so-called S-curve. As time passes, word of mouth becomes more important (Hall 2006).

59. Tagliapietra (2020a, 220).

60. Industrializing economies should avoid locking themselves into antiquated fossil-fuel energy systems, as global environmental pressures will turn these into bad investments (Fouquet 2016).

61. Aghion et al. (2021, 185) reach similar conclusions, noting that "it is not necessary that all countries in the world coordinate from the outset. Unilateral coordination among developed countries to redirect innovation toward green technologies, combined with a resolute policy to disseminate these technologies to less-developed countries, would suffice to successfully combat global warming." See also Mercure et al (2021).

62. An important question is whether liberal advanced democracies will be able to collaborate and coordinate with China on the climate front despite the wider context of souring geopolitical relations. While there are ample reasons for skepticism, hope springs eternal. Even at the height of the Cold War in the

1960s, the United States and the USSR managed to cooperate in financing and distributing a smallpox vaccine, leading to the global eradication of the disease by 1980 (Barrett 2007).

63. Designing these measures in legal compliance with WTO rules, and in an administratively feasible way, is likely to prove possible, but nonetheless very challenging from a methodological standpoint. For a review of the issues, see Cosbey et al. (2019), but compare Horn and Sapir (2013). It is also entirely possible that the long-standing objective of reforming WTO rules might include provisions explicitly allowing a carbon border tax.

64. At the moment, within the EU ETS, free allowances are given to certain sectors as compensation for the environmental compliance costs borne by carbon-intensive companies. Between 2013 and 2020, these were roughly 50 percent of total auctioned / sold allowances. An effective carbon border tax would allow for reduction, and eventually elimination, of these free allowances.

65. There is rather limited evidence of carbon leakage, defined as companies relocating production abroad while trying to avoid costs related to climate policies at home, even in environmentally strict Europe. Nonetheless, as carbon prices rise and regulation becomes tighter, this phenomenon could become more prevalent, and therefore a carbon border tax is justified.

66. In the case of Europe, the estimated revenue from an EU carbon border tax could range from €36 to €83 billion per year (Krenek et al. 2019). While estimates fluctuate, Fuest and Pisani-Ferry (2020) point out that border taxes will not credibly do the heavy lifting to finance the green agenda. Nonetheless, they can contribute to the task.

67. A. Shalal, "World Bank, IMF to Consider Climate Change in Debt Reduction Talks," *Reuters,* February 19, 2021.

68. Rodriguez-Pose (2018).

69. For more, see Zachmann et al. (2018) and Tagliapietra (2020a, 207–208).

70. This statement has been a rhetorical trope, used by Benjamin Franklin and, later, Abraham Lincoln. It was popularized by the "Liberty Song" of John Dickens in 1768, perhaps, but a similar line was already to be found in the ancient Greek writer Aesop's fable of "The Four Oxen and the Lion."

Conclusion

1. Nordhaus (1994). More recent estimates using this model have concluded that optimal climate policy should aim to stabilize global warming below two degrees Celsius, in line with the international consensus of the Paris Agreement (Hänsel et al. 2020), for two reasons. First, climate mitigation has progressed slowly, so the "long-term" damage has become more imminent. Second, climate modeling suggests that, due to non-linearities and tipping

points, the damage of even small temperature increases could be much larger than initially thought.

2. Evidence indicates that "leading macroeconomic modelling analysis tend to find that green policies—or at least energy and climate change policies—do not have large implications on overall employment" (OECD 2017, 11). Leveraging the model developed by Nordhaus, IMF (2020) concludes that climate mitigation will have some, albeit limited, long-term impact on growth by 2050 (about −1 percent of global GDP). Three macroeconomic models point in the direction of marginally positive or marginally negative impacts of achieving stringent climate targets (European Commission 2020a). For another analysis of the impact of climate policies on employment and growth, see Goulder and Hafstead (2017), who show that careful policy design, including the judicious use of the environmentally generated revenues, can achieve desired reductions in CO_2 emissions at low cost, avoid uneven impacts across household income groups, and prevent losses of profit in the most vulnerable US industries. Their estimates suggest an effective climate policy based on carbon pricing would lower GDP by a negligible 1 percent over thirty years, equivalent to three basis points per year.

3. For example, Metcalf and Stock (2020) analyze the thirty-one European countries under the EU's Emission Trading Scheme (ETS), fifteen of which also have carbon taxes, applied to the emissions not covered by the ETS. They find no adverse effect on GDP or employment, and possibly a positive one. OECD (2017) focuses on British Columbia, Canada, which in July 2008 implemented a carbon tax on most industries and residents, covering three-fourths of GHG emissions. The tax was designed to be revenue-neutral, meaning that its entire revenue was used to reduce personal and corporate income taxes, and included direct transfers to low-income households. The tax led to a reduction in GHG emissions with respect to the rest of Canada, and generated approximately ten thousand jobs per year between 2007 and 2013, leading to an overall 4.5 percent increase in employment over the six-year period.

4. See OECD (2012, 2017). European Commission (2019, 171) reports comparable data, notably that "electricity production, (some) transport, manufacturing, agriculture, and mining sectors together produce close to 90 percent of all CO_2 emissions in the EU, while they account for 25 percent of gross value added and less than 25 percent of employment."

5. Garrett-Peltier (2017). Along similar lines, the renewable energy sector tends to require more workers per megawatt of energy generated than fossil-fuel-based energy sectors (Wei et al. 2010).

6. Specifically, eighteen million net jobs created worldwide by 2030 (ILO 2018).

7. Claeys et al. (2019, 7).

8. In an overview of policies to promote innovation, Bloom et al. (2019) say the most important element of mission-oriented innovation policies, or "moon-

shot strategies," might very well be political. A political vision can generate significantly more resources for research and innovation. As a pressing societal challenge, climate change can cause political actors to support much-needed investment and innovation.

9. Climate mitigation policies will create jobs in the short term due to infrastructure and construction investment but also because renewables require labor-intensive maintenance (Blyth et al. 2014). Regarding these policies' long-term impact, however, they note that it goes well beyond considerations of labor intensity. Growth will depend on the scope for efficiency enhancements in these new green sectors, and on whether they are strategic sectors expected to grow.

10. Expanding modeling efforts to capture the comparative advantage dimension of the green transition in an international context, Lutz et al. (2015) note that positive economic and employment effects of Germany's investment in renewable energy are then observed, both in the short and long term.

11. In economics, this type of argument is known as a "Lucas critique," since Nobel laureate Robert Lucas used it to argue in favor of microeconomic foundations for macroeconomic models. "Projections by international organizations primarily rely on energy-specific and computable general equilibrium models that provide useful insights but are ill-suited to macroeconomic analysis. They often minimize the macroeconomic effects of decarbonization." Jean Pisani-Ferry, "A Credible Decarbonization Agenda Can Help Strengthen Europe's Economy," *Peterson Institute for International Economics Blog,* December 9, 2019, https://www.piie.com/blogs/realtime-economic-issues-watch/credible -decarbonization-agenda-can-help-strengthen-europes. For a damning account of how current economic models insufficiently take climate risks into account under Integrated Assessment Modeling, see Stern and Stiglitz (2021).

12. IMF (2020, 100); Gustafson (2021).

13. For a pioneering paper on the tortuous paths countries take to discover their own competitive advantage, see Hausmann and Rodrik (2003).

14. IMF (2020).

15. More technically, according to the secular stagnation hypothesis, private investment has a chronic tendency to be insufficient to absorb private savings—and this leads, in the absence of extraordinary policies, to extremely low interest rates, inflation that is lower than desirable, and sluggish economic growth. The Green Deal investment plan could serve to absorb the large supply of private saving and rekindle private demand, jolting advanced economies out of this negative equilibrium (Rachel and Summers 2019). Some macro-climate models project that this strong public and private investment plan would have a demand-side economic boost effect for the first fifteen years (IMF 2020).

16. Aghion and Antonin (2021, 42).

17. Juhász et al. (2020).
18. Aghion and Antonin (2021, 43–45); McAfee (2019, 27–28). Incidentally, this creates another knowledge externality, as potential adopters of the new technology have an incentive to wait for incumbents to improve their know-how and master production organization and techniques, before entering the market. In other words, if you are a traditional carmaker, you have an incentive to wait it off and see, while Tesla struggles to organize production and overcomes manufacturing issues for years, until it becomes clear that the EV technology will become predominant.
19. In the words of Stern and Stiglitz (2021, 61): "For two hundred years, technologies based on fossil fuel have been explored. Diminishing returns may have set in. Climate change has induced new searches in other parts of the technology frontier. Possible paces of innovation in these relatively unexplored areas can, at least for now, be markedly higher . . . the green economy may usher in a new era of high productivity growth."
20. "Solar projects now offer some of the lowest-cost electricity in history" (IEA 2020).
21. See Rodrik and Sabel (2019), but also Blanchard and Rodrik (2021).
22. In line with this list of growth and employment enhancing activities, Hepburn et al. (2020) build on a survey of 231 central bank officials, finance ministry officials, and other economic experts from G20 countries to identify the five policies with highest potential in terms of both economic multiplier and climate impact metrics. These include clean physical infrastructure, building efficiency retrofits, natural capital investment, clean R&D, and investment in education and training.
23. From a strict skills perspective, the green transition is expected to be much less dramatic than the outsourcing revolution associated with hyper-globalization (Bowen et al. 2018).
24. Consoli et al. (2016); Vona et al. (2018).
25. Unsworth et al. (2020).
26. On the link between information and communications technology and the rising capital income share since 2000, see Nordhaus (2021). Regarding job relocation, see Baldwin (2019), who predicts that this technology will enable white-collar jobs and services to be performed abroad, generating shockwaves through labor markets of advanced economies.
27. On the historical parallel regarding Europe, see Terzi (2021a).
28. Estimates by Bloomberg Economics are based on data from national central banks and statistical agencies. Simon Kennedy, "Consumers Saved $2.9 Trillion during the Pandemic: Their Money Will Drive the Global Recovery," *Bloomberg*, March 3, 2021.
29. Christakis (2020, 283).

30. This trend can be seen beyond the United States, as we also see the first women presidents of the European Commission, World Trade Organization, European Central Bank, and the nation of Greece, to name just a few.

31. On twenty-nine occasions between 1919 and 1920, deployments of US military troops helped to quell labor disputes and other disturbances.

32. Stefan Nicola, "Tell Your Boss the Four-Day Week Is Coming Soon," *Bloomberg*, March 2, 2021.

33. A 2021 survey by the World Economic Forum found that more than 80 percent of employers intend to accelerate plans to digitize their processes and provide more opportunities for remote work, while 50 percent plan to accelerate automation of production tasks. In another global survey, by McKinsey & Co., executives assessed their digitization efforts in 2021 to be, on average, seven years ahead of original plans.

34. Happiness surveys show that long commutes significantly lower personal happiness (Coyle 2014, 116).

35. Economist Robert Gordon noted that "this shift to remote working has got to improve productivity because we're getting the same amount of output without commuting, without office buildings, and without all the goods and services associated with that. We can produce output at home and transmit it to the rest of the economy electronically, whether it's an insurance claim or medical consultation. We're producing what people really care about with a lot less input of things like office buildings and transportation. In a profound sense, the movement to working from home is going to make everyone who is capable of working from home more productive. . . . I think we're going to see a period of considerable growth in overall productivity statistics." Robert Gordon, "The Rise and Fall and Rise Again of American Growth," interview with Leo Feler, UCLA Anderson Forecast Direct, February 2021, https://www.anderson.ucla .edu/about/centers/ucla-anderson-forecast/research-and-reports/forecast -direct-podcast/february-2021.

36. The term "roaring twenties" first appeared in the *Lincoln Journal Star* (Nebraska), August 15, 1923, and did not refer to the social and economic boom at large, but rather developments in the movie industry. "During the war and after the war up to the big movie slump of 1920 . . . new companies were springing up like mushrooms. . . . But the 'roaring twenties' have gone." This nostalgic use of the term suggests that the economic bonanza of the era was far from clear in the first years of the 1920s to those living through them.

37. Ironically commenting on the lack of breakthrough technologies in the early twenty-first century, investor Peter Thiel famously quipped: "We wanted flying cars, instead we got 140 characters." The perception that technological innovation is slowing down may be because much of it has lately occurred behind

the scenes, focused on business-to-business uses and applying things like big data and artificial intelligence to production, rather than to consumer products. Over the next few years, this dismal perception could very well change, not least because there are now promising experiments with flying cars, thanks to advances in battery energy density and material sciences.

38. Shiller (2019).
39. A particular signpost to watch will be the capacity of governments and policymakers to phase out the exceptional support granted to the economy during the Covid-19 pandemic. If poorly timed or managed, this policy withdrawal could easily spark a wave of bankruptcies, or even a financial crisis.
40. First conceived in 1922 and released three years later, F. Scott Fitzgerald's *The Great Gatsby* (1925) proved prescient, describing the precarity of the roaring twenties. The success of the novel was propelled during and after the 1929 stock market crash and the Great Depression.
41. Perez (2019).
42. The coexistence of economic progress with greed and corruption in the United States following the end of the Civil War is described by Mark Twain and Charles Dudley Warner in *The Gilded Age: A Tale of Today* (1873).

Appendix A. GDP and Energy Return on Energy Investment

1. Buchanan (2019).
2. Murphy and Hall (2011).
3. Hall et al. (2014).
4. Smil (2017, 15–16).
5. Nordhaus (1973, 1182) goes further in his critique, noting how World3 "contains no clear concepts of production functions, consumption, or output; nor is there any discernible method of allocating resources over time or between sectors."
6. Buchanan (2019).
7. It is important to separate evaluations of individual energy sources, their prices, and their supplies from speculations about the future of energy prices. Hausman (1996) created a data set of historical energy prices from 1450 to 1988 and found no support for the view that fuel prices were rising over the very long run. When prices of individual fuel sources have risen, consumers have switched away from the more expensive ones.
8. The statement that the efficiency of extraction cannot be improved has yet to be proven right. Throughout history, commentators on wood fuel and coal first (between the seventeenth and nineteenth century) and petrol later have reliably expected that prices would soar and the economy would eventually collapse. Every time, prices temporarily increased, including in the 1910s and

1970s, and then new extraction techniques and infrastructure, or new cheaper energy sources, brought prices back to a fairly stable long-term average price of energy (Fouquet 2011).

9. See Grubb et al. (2018), who test a hypothesis by Bashmakov (2007).

10. See Fouquet (2011), who builds on pioneering work by Nordhaus (1996) for what concerns the "true" price of lighting. The price of the service is the price of the fuel multiplied by the amount of energy required to produce one unit of the service, such as heating, lighting, or transport.

11. Fouquet and Pearson (2006).

Appendix B. China's Great Withdrawal and Japan's Seclusion

1. For more on this topic, see Mokyr (2016); Ferguson (2011).

2. Sharman (2019, 141).

3. For a detailed historical recount, see Lidin (2002), source of much of the information in this section.

4. This episode has captured the imagination of several artists, serving, for example, as background to Giacomo Puccini's *Madama Butterfly*. For cinephiles, it is the period to which *The Last Samurai* (2003), with Tom Cruise, makes nostalgic reference.

Bibliography

Aarssen, Lonnie William. 2015. *What Are We? Exploring the Evolutionary Roots of Our Future.* Kingston, Ontario: Queen's University.

Acemoglu, Daron, Philippe Aghion, Leonardo Bursztyn, and David Hémous. 2012. "The Environment and Directed Technical Change." *American Economic Review* 102 (1): 131–166.

Acemoglu, Daron, Ufuk Akcigit, Douglas Hanley, and William Kerr. 2016. "Transition to Clean Technology." *Journal of Political Economy* 124 (1): 52–104. https://doi.org/10.1086/684511.

Acemoglu, Daron, and James A. Robinson. 2012. *Why Nations Fail: The Origins of Power, Prosperity, and Poverty.* London: Profile Books.

Adams, Jonathan. 2013. "The Fourth Age of Research." *Nature* 497 (7451): 557–560. https://doi.org/10.1038/497557a.

Adams, Jonathan, and Karen A. Gurney. 2018. "Bilateral and Multilateral Coauthorship and Citation Impact: Patterns in UK and US International Collaboration." *Frontiers in Research Metrics and Analytics* 3 (March): 1–10. https://doi.org/10.3389/frma.2018.00012.

Agarwal, Ruchir, and Patrick Gaulé. 2020. "Invisible Geniuses: Could the Knowledge Frontier Advance Faster?" *AER: Insights* 2 (4): 409–424. https://doi.org/10.1257/aeri.20190457.

Aghion, Philippe, and Celine Antonin. 2018. "Technical Progress and Growth since the Crisis." *Revue de l'OFCE* 157: 55–68.

Aghion, Philippe, Céline Antonin, and Simon Bunel. 2021. *The Power of Creative Destruction: Economic Upheaval and the Wealth of Nations.* Cambridge, MA: Belknap Press of Harvard University Press.

Aghion, Philippe, Antoine Dechezleprêtre, David Hemous, Ralf Martin, and John Van Reenen. 2016. "Carbon Taxes, Path Dependency, and Directed Technical Change: Evidence from the Auto Industry." *Journal of Political Economy* 124 (1): 1–51. https://doi.org/10.1086/684581.

Akcigit, Ufuk, Salomé Baslandze, and Stefanie Stantcheva. 2016. "Taxation and the International Mobility of Inventors." *American Economic Review* 106 (10): 2930–2981. https://doi.org/10.1257/aer.20150237.

Aldy, Joseph E., and Robert N. Stavins. 2012. "The Promise and Problems of Pricing Carbon: Theory and Experience." *Journal of Environment and Development* 21 (2): 152–180. https://doi.org/10.1177/1070496512442508.

Alesina, Alberto, and Eliana La Ferrara. 2005. "Ethnic Diversity and Economic Performance." *Journal of Economic Literature* 43 (3): 762–800. https://doi.org/10.1257/002205105774431243.

Alesina, Alberto, and Edward L. Glaeser. 2004. *Fighting Poverty in the US and Europe: A World of Difference.* Oxford: Oxford University Press.

Alesina, Alberto, Edward Glaeser, and Bruce Sacerdote. 2005. "Work and Leisure in the United States and Europe: Why So Different?" *NBER Macroeconomics Annual* 20: 1–64. https://doi.org/10.1086/ma.20.3585411.

Alesina, Alberto, Elie Murard, and Hillel Rapoport. 2021. "Immigration and Preferences for Redistribution in Europe." *Journal of Economic Geography,* 1–30. https://doi.org/10.1093/jeg/lbab002.

Allison, Graham. 2017. *Destined for War: Can America and China Escape Thucydides's Trap?* Boston: Houghton Mifflin Harcourt.

Alstadsæter, Annette, Niels Johannesen, and Gabriel Zucman. 2019. "Tax Evasion and Inequality." *American Economic Review* 109 (6): 2073–2103. https://doi.org/10.1257/aer.20172043.

Alston, Philip. 2020. *The Parlous State of Poverty Eradication.* A / HRC/44/40, Report of the Special Rapporteur on extreme poverty and human rights to the Human Rights Council, 44th Session, July 7. https://www.ohchr.org/EN /Issues/Poverty/Pages/parlous.aspx.

Amaldoss, Wilfred, and Sanjay Jain. 2005. "Pricing of Conspicuous Goods: A Competitive Analysis of Social Effects." *Journal of Marketing Research* 42 (1): 30–42. https://doi.org/10.1509/jmkr.42.1.30.56883.

Anderson, Benedict. 1983. *Imagined Communities: Reflections on the Origin and Spread of Nationalism.* London: Verso.

Anderson, Robert Warren, Noel D. Johnson, and Mark Koyama. 2017. "Jewish Persecutions and Weather Shocks: 1100–1800." *Economic Journal* 127 (602): 924–958. https://doi.org/10.1111/ecoj.12331.

Andrews, Dan, Chiara Criscuolo, and Peter N. Gal. 2019. "The Best versus the Rest: Divergence across Firms during the Global Productivity Slow-down." CEP Discussion Paper 1645, Centre for Economic Policy, London School of Economics. August. http://cep.lse.ac.uk/pubs/download/dp1645.pdf.

Anelli, Massimo, Gaetano Basso, Giuseppe Ippedico, and Giovanni Peri. 2020. "Does Emigration Drain Entrepreneurs?" CESifo Working Paper 8388,

Center for Economic Studies and ifo Institute (CESifo), Ludwigs-Maximilian University, Munich. June. https://www.cesifo.org/en/publikationen/2020 /working-paper/does-emigration-drain-entrepreneurs.

Angeloni, Ignazio, and Jean Pisani-Ferry. 2012. "The G20: Characters in Search of an Author." Bruegel Working Paper no. 2012 / 04. March 13. https://www .bruegel.org/2012/03/the-g20-characters-in-search-of-an-author/.

Bailey, Daniel. 2015. "The Environmental Paradox of the Welfare State: The Dynamics of Sustainability." *New Political Economy* 20 (6): 793–811. https://doi .org/10.1080/13563467.2015.1079169.

Baldwin, Richard. 2019. *The Globotics Upheaval: Globalisation, Robotics and the Future of Work.* Oxford: Oxford University Press.

Ball, Laurence, Daniel Leigh, and Prakash Loungani. 2017. "Okun's Law: Fit at 50?" *Journal of Money, Credit and Banking* 49 (7): 1413–1441.

Barone, Guglielmo, and Sauro Mocetti. 2021. "Intergenerational Mobility in the Very Long Run: Florence 1427–2011." *Review of Economic Studies* 88: 1863–1891. https://doi.org/ 10.1093 / restud / rdaa075.

Barrett, Scott. 2007. *Why Cooperate? The Incentive to Supply Global Public Goods.* Oxford: Oxford University Press.

Bashmakov, Igor. 2007. "Three Laws of Energy Transitions." *Energy Policy* 35 (7): 3583–3594. https://doi.org/10.1016/j.enpol.2006.12.023.

Baxter, Brian. 1999. *Ecologism: An Introduction.* Washington, DC: Georgetown University Press.

Beckerman, Wilfred. 1974. *In Defence of Economic Growth.* London: Jonathan Cape.

Beckert, Sven. 2015. *Empire of Cotton: A Global History.* New York: Vintage.

Bennis, Amine. 2021. "Power Surge: How the European Green Deal Can Succeed in Morocco and Tunisia." ECFR Policy Brief 366, European Council on Foreign Relations. January. https://ecfr.eu/publication/power-surge-how-the -european-green-deal-can-succeed-in-morocco-and-tunisia/.

Bernard, Andrew B., and Meghan R. Busse. 2004. "Who Wins the Olympic Games: Economic Resources and Medal Totals." *Review of Economics and Statistics* 86 (1): 413–417. https://doi.org/10.1162/003465304774201824.

Bevan, Andrew, and David Wengrow. 2010. *Cultures of Commodity Branding.* New York: Routledge.

Binswanger, Mathias. 2009. "Is There a Growth Imperative in Capitalist Economies? A Circular Flow Perspective." *Journal of Post Keynesian Economics* 31 (4): 707–727. https://doi.org/10.2753/pke0160-3477310410.

Blanchard, Olivier. 2004. "The Economic Future of Europe." *Journal of Economic Perspectives* 18 (4): 3–26. https://doi.org/10.1257/0895330042632735.

Blanchard, Olivier, and Dani Rodrik. 2021. *Combating Inequality: Rethinking Government's Role.* Cambridge, MA: MIT Press.

Bloom, David E., and David Canning. 2007. "Commentary: The Preston Curve 30 Years On: Still Sparking Fires." *International Journal of Epidemiology* 36 (3): 498–499. https://doi.org/10.1093/ije/dym079.

Bloom, Nicholas, Charles I. Jones, John Van Reenen, and Michael Webb. 2020. "Are Ideas Getting Harder to Find?" *American Economic Review* 110 (4): 1104–1144. https://doi.org/10.1257/aer.20180338.

Bloom, Nicholas, John Van Reenen, and Heidi Williams. 2019. "A Toolkit of Policies to Promote Innovation." *Journal of Economic Perspectives* 33 (3): 163–184.

Blyth, Will, Rob Gross, Jamie Speirs, Steve Sorrell, Jack Nicholls, Alex Dorgan, and Nick Hughes. 2014. *Low Carbon Jobs.* UKERC report, UK Energy Resource Centre, London. November. https://ukerc.ac.uk/publications/low-carbon-jobs-the-evidence-for-net-job-creation-from-policy-support-for-energy-efficiency-and-renewable-energy/.

Boccaletti, Giulio. 2021. *Water: A Biography.* London: Pantheon.

Boldizzoni, Francesco. 2020. *Foretelling the End of Capitalism: Intellectual Misadventures since Karl Marx.* Cambridge, MA: Harvard University Press.

Bolt, Jutta, and Jan Luiten van Zanden. 2020. "The Maddison Project: "Maddison Style Estimates of the Evolution of the World Economy: A New 2020 Update." 15. Maddison-Project Working Paper 15, Maddison Project, University of Groningen. October. https://www.rug.nl/ggdc/historicaldevelopment/maddison/publications/wp15.pdf.

Boserup, Ester. 1965. *The Conditions of Agricultural Growth: The Economics of Agrarian Change under Population Pressure.* London: G. Allen and Unwin.

Boushey, Heather. 2019. *Unbound: How Inequality Constricts Our Economy and What We Can Do about It.* Cambridge, MA: Harvard University Press.

Bowen, Alex, Karlygash Kuralbayeva, and Eileen L. Tipoe. 2018. "Characterising Green Employment: The Impacts of 'Greening' on Workforce Composition." *Energy Economics* 72: 263–275. https://doi.org/10.1016/j.eneco.2018.03.015.

Bradford, Anu. 2020. *The Brussels Effect: How the European Union Rules the World.* Oxford: Oxford University Press.

Braudel, Fernand. 1992. *Civilization and Capitalism, 15th–18th Century,* vol. 1: *The Structure of Everyday Life.* Berkeley: University of California Press.

Bregman, Rutger. 2020. *Humankind: A Hopeful History.* Boston: Little, Brown.

Bricker, Jesse, Jacob Krimmel, and Rodney Ramcharan. 2021. "Signaling Status: The Impact of Relative Income on Household Consumption and Financial Decisions." *Management Science* 67 (4): 1993–2009. https://doi.org/10.1287/mnsc.2019.3577.

Brown, Chad P., Monica de Bolle, and Maurice Obstfeld. 2021. "The Pandemic Is Not Under Control Anywhere Unless It Is Controlled Everywhere." *Peterson Institute for International Economics* (blog). https://www.piie.com/blogs/realtime-economic-issues-watch/pandemic-not-under-control-anywhere-unless-it-controlled.

Brumm, Adam, Adhi Agus Oktaviana, Basran Burhan, Budianto Hakim, Rustan Lebe, Jian-xin Zhao, Priyatno Hadi Sulistyarto, et al. 2021. "Oldest Cave Art Found in Sulawesi." *Science Advances* 7 (January): 1–12. https://doi.org/10.1126/sciadv.abd4648.

Brynjolfsson, Erik, and Lorin M. Hitt. 2000. "Beyond Computation: Information Technology, Organizational Transformation and Business Performance." *Journal of Economic Perspectives* 14 (4): 23–48. https://doi.org/10.5951/at.22.1.0022.

Buchanan, Mark. 2019. "Energy Costs." *Nature Physics* 15 (6): 520. https://doi.org/10.1038/s41567-019-0549-x.

Buch-Hansen, Hubert. 2018. "The Prerequisites for a Degrowth Paradigm Shift: Insights from Critical Political Economy." *Ecological Economics* 146 (April): 157–163. https://doi.org/10.1016/j.ecolecon.2017.10.021.

Büchs, Milena, and Max Koch. 2019. "Challenges for the Degrowth Transition: The Debate about Wellbeing." *Futures* 105 (February 2018): 155–165. https://doi.org/10.1016/j.futures.2018.09.002.

Bundick, Brent, and Emily Pollard. 2019. "The Rise and Fall of College Tuition Inflation." *Economic Review,* Federal Reserve Bank of Kansas City. April 11. https://doi.org/10.18651/er/1q19bundickpollard.

Burke, Marshall, Solomon M. Hsiang, and Edward Miguel. 2015a. "Climate and Conflict." *Annual Review of Economics* 7 (1): 577–617. https://doi.org/10.1146/annurev-economics-080614-115430.

———. 2015b. "Global Non-Linear Effect of Temperature on Economic Production." *Nature* 527 (7577): 235–239. https://doi.org/10.1038/nature15725.

Burzyńskia, Michał, Jaime de Melo, Christoph Deuster, and Frédéric Docquier. 2019. "Climate Change, Inequality, and Human Migration." IZA DP no. 12623, Institute of Labor Economics (IZA), Bonn. September. https://www.iza.org/publications/dp/12623/climate-change-inequality-and-human-migration.

Buss, David M. 2001. "Human Nature and Culture: An Evolutionary Psychological Perspective." *Journal of Personality* 69 (6): 955–978. https://doi.org/10.1111/1467-6494.696171.

Butzer, Karl W. 2012. "Collapse, Environment, and Society." *Proceedings of the National Academy of Sciences of the United States of America* 109 (10): 3632–3639. https://doi.org/10.1073/pnas.1114845109.

Butzer, Karl W., and Georgina H. Endfield. 2012. "Critical Perspectives on Historical Collapse." *Proceedings of the National Academy of Sciences of the United States of America* 109 (10): 3628–3631. https://doi.org/10.1073/pnas.1114772109.

Çakmaklı, Cem, Selva Demiralp, Şebnem Kalemli-Özcan, Sevcan Yeşiltaş, and Muhammed A. Yıldırım. 2021. "The Economic Case for Global Vaccinations:

An Epidemiological Model with International Production Networks." NBER Working Paper no. 28395, National Bureau of Economic Research, Cambridge, MA. April.

Calligaris, Sara, Massimo Del Gatto, Fadi Hassan, Gianmarco I. P. Ottaviano, and Fabiano Schivardi. 2016. "Italy's Productivity Conundrum: A Study on Resource Misallocation." European Economy Discussion Paper 030, Directorate-General for Economic and Financial Affairs, European Commission. May. https://doi.org/10.2765/970192.

Caprettini, Bruno, and Hans-Joachim Voth. 2020. "Rage against the Machines: Labor-Saving Technology and Unrest in England, 1830–32." *American Economic Review: Insights* 2 (3): 305–320. https://doi.org/10.1257/aeri.20190385.

Carleton, Tamma, Michael Delgado, Michael Greenstone, Trevor Houser, Solomon Hsiang, Andrew Hultgren, Amir Jina, et al. 2020. "Valuing the Global Mortality Consequences of Climate Change Accounting for Adaptation Costs and Benefits." NBER Working Paper no. 27599, National Bureau of Economic Research, Cambridge, MA. August.

Carleton, Tamma, Solomon M. Hsiang, and Marshall Burke. 2016. "Conflict in a Changing Climate." *European Physical Journal: Special Topics* 225 (3): 489–511. https://doi.org/10.1140/epjst/e2015-50100-5.

Carmichael, Richard. 2019. *Behaviour Change, Public Engagement and Net Zero.* Report for the Committee on Climate Change, Imperial College London. October. https://www.theccc.org.uk/publication/behaviour-change-public -engagement-and-net-zero-imperial-college-london/.

Cavalli-Sforza, Luigi Luca, and Marcus W. Feldman. 1981. *Cultural Transmission and Evolution: A Quantitative Approach.* Princeton, NJ: Princeton University Press.

Cavenaile, Laurent, and Pau Roldan-Blanco. 2021. "Advertising, Innovation, and Economic Growth." *American Economic Journal: Macroeconomics* 13 (3): 251–303. https://doi.org/10.1257/mac.20180461.

Chacon, Richard, and Rubén G. Mendoza. 2012. *The Ethics of Anthropology and Amerindian Research: Reporting on Environmental Degradation and Warfare.* New York: Springer US.

Chancel, Lucas. 2020. *Unsustainable Inequalities: Social Justice and the Environment.* Cambridge, MA: Belknap Press of Harvard University Press.

Chancel, Lucas, Géraldine Thiry, and Damien Demailly. 2014. "Beyond-GDP Indicators: To What End?" IDDRI Study 14. SciencesPo, Paris. September. https://www.iddri.org/sites/default/files/import/publications/sto414en.pdf.

Chaney, Eric. 2013. "Revolt on the Nile: Economic Shocks, Religion, and Political Power." *Econometrica* 81 (5): 2033–2053. https://doi.org/10.3982/ecta10233.

Chemin, Matthieu, and Etienne Wasmer. 2009. "Using Alsace-Moselle Local Laws to Build a Difference-in-Differences Estimation Strategy of the

Employment Effects of the 35-Hour Workweek Regulation in France." *Journal of Labor Economics* 27 (4): 487–524. https://doi.org/10.1086/605426.

"Chemistry Can Help Make Plastics Sustainable—But It Isn't the Whole Solution." 2021. Editorial. *Nature* 590 (7846): 363–364.

Chen, M. Keith, Venkat Lakshminarayanan, and Laurie R. Santos. 2006. "How Basic Are Behavioral Biases? Evidence from Capuchin Monkey Trading Behavior." *Journal of Political Economy* 114 (3): 1–23. https://doi.org/10.1086/503550.

Chen, Tingyun, Jean-Jacques Hallaert, Alexander Pitt, Haonan Qu, Maximilien Queyranne, Alaina Rhee, Anna Shabunina, Jérôme Vandenbussche, and Irene Yackovlev. 2018. "Inequality and Poverty across Generations in the European Union." IMF Staff Discussion Note 18 / 01. International Monetary Fund. January.

Cherif, Reda, and Fuad Hasanov. 2019. "The Return of the Policy That Shall Not Be Named: Principles of Industrial Policy." IMF Working Paper 19 / 74, International Monetary Fund. March. https://doi.org/10.5089/9781498305402 .001.

Chiquiar, Daniel, and Gordon Hanson. 2005. "International Migration, Self-Selection, and the Distribution of Wages: Evidence from Mexico and the United States." *Journal of Political Economy* 113 (2): 239–281.

Christakis, Nicholas A. 2019. *Blueprint: The Evolutionary Origins of a Good Society.* New York: Little, Brown Spark.

———. 2020. *Apollo's Arrow: The Profound and Enduring Impact of Coronavirus on the Way We Live.* New York: Little, Brown Spark.

CIRCaP / LAPS and IAI. 2013. "Italians and Foreign Policy." Laboratory on Social and Political Analysis (CIRCaP / LAPS), University of Siena, and Istituto Affari Internazionale (IAI). https://www.iai.it/sites/default/files/Rapporto _IAI-Circap_en.pdf.

Claeys, Grégory, Simone Tagliapietra, and Georg Zachmann. 2019. "How to Make the European Green Deal Work." Bruegel Policy Contribution 2019 / 13. November. https://www.bruegel.org/2019/11/how-to-make-the-european -green-deal-work/.

Clark, Andrew E., Paul Frijters, and Michael A. Shields. 2008. "Relative Income, Happiness, and Utility: An Explanation for the Easterlin Paradox and Other Puzzles." *Journal of Economic Literature* 46 (1): 95–144. https://doi.org/10.1257 /jel.46.1.95.

Coady, David, Ian Parry, Nghia-Piotr Le, and Baoping Shang. 2019. "Global Fossil Fuel Subsidies Remain Large: An Update Based on Country-Level Estimates." IMF Working Paper 19 / 89, International Monetary Fund. May 2. https://doi.org/10.5089/9781484393178.001.

Cohen, Daniel. 2018. *The Infinite Desire for Growth.* Princeton, NJ: Princeton University Press.

Collier, Paul. 2011. *The Plundered Planet: Why We Must—and How We Can—Manage Nature for Global Prosperity.* New York: Penguin Books.

Comte, August. 1896. *The Positive Philosophy.* Vol. 2. London: George Bell and Sons.

Consoli, Davide, Giovanni Marin, Alberto Marzucchi, and Francesco Vona. 2016. "Do Green Jobs Differ from Non-Green Jobs in Terms of Skills and Human Capital?" *Research Policy* 45 (5): 1046–1060. https://doi.org/10.1016/j.respol.2016.02.007.

Corneo, Giacomo, and Olivier Jeanne. 1997. "Conspicuous Consumption, Snobbism and Conformism." *Journal of Public Economics* 66 (1): 55–71. https://doi.org/10.1016/S0047-2727(97)00016-9.

Cosbey, Aaron, Susanne Droege, Carolyn Fischer, and Clayton Munnings. 2019. "Developing Guidance for Implementing Border Carbon Adjustments: Lessons, Cautions, and Research Needs from the Literature." *Review of Environmental Economics and Policy* 13 (1): 3–22. https://doi.org/10.1093/reep/rey020.

Coyle, Diane. 2014. *GDP: A Brief but Affectionate History.* Princeton, NJ: Princeton University Press.

———. 2019. "Measuring Progress: A Review Essay on the Pricing of Progress: Economic Indicators and the Capitalization of American Life by Eli Cook." *Journal of Economic Literature* 57 (3): 659–677. https://doi.org/10.1257/jel.20181517.

Crépon, Bruno, and Francis Kramarz. 2002. "Employed 40 Hours or Not-Employed 39: Lessons from the 1982 Mandatory Reduction of the Work-week." *Journal of Political Economy* 110 (6): 1355–1389. https://doi.org/10.1086/342807.

Creutzig, Felix. 2020. "Limits to Liberalism: Considerations for the Anthropocene." *Ecological Economics* 177: 10673. https://doi.org/10.1016/j.ecolecon.2020.106763.

Cruz Alvarez, Jose Luis, and Esteban Rossi-Hansberg. 2021. "The Economic Geography of Global Warming." NBER Working Paper no. 28466, National Bureau of Economic Research, Cambridge, MA. February.

Daly, Herman. 2014. *From Uneconomic Growth to a Steady-State Economy.* Cheltenham, UK: Edward Elgar.

Dartnell, Lewis. 2015. *The Knowledge: How to Rebuild Our World from Scratch.* London: Vintage.

Deaton, Angus. 2008. "Income, Health, and Well-Being around the World: Evidence from the Gallup World Poll." *Journal of Economic Perspectives* 22 (2): 53–72. https://doi.org/10.1257/jep.22.2.53.

De Jong, Jasper, Marien Ferdinandusse, Josip Funda, and Igor Vetlov. 2017. "The Effect of Public Investment in Europe: A Model-Based Assessment." ECB

Working Paper no. 2021, European Central Bank. February. https://doi.org /10.2866/139475.

DeLong, J. Bradford. 1998. "Estimates of World GDP, One Million B.C.–Present." Unpublished manuscript. https://delong.typepad.com/print/20061012 _LRWGDP.pdf.

Demaria, Federico, François Schneider, Filka Sekulova, and Joan Martinez-Alier. 2013. "What Is Degrowth? From an Activist Slogan to a Social Movement." *Environmental Values* 22 (2): 191–215. https://doi.org/10.3197/096327113X135815 61725194.

Desmet, Klaus, Robert Kopp, Scott Kulp, Dávid Nagy, Michael Oppenheimer, Esteban Rossi-Hansberg, and Benjamin Strauss. 2021. "Evaluating the Economic Cost of Coastal Flooding." *American Economic Journal: Macroeconomics* 13 (2): 444–486.

Diamandis, Peter H., and Steven Kotler. 2012. *Abundance: The Future Is Better than You Think*. New York: Free Press.

Diamond, Jared. 2004. *Collapse: How Societies Choose to Fail or Succeed*. New York: Viking.

Dickens, Charles. 1854. *Hard Times*. London: Bradbury and Evans.

Directorate-General for Communication. 2019. *Special Eurobarometer 490: Climate Change*. https://data.europa.eu/data/datasets/s2212_91_3_490_eng ?locale=en.

Dorsch, Marcel J., and Christian Flachsland. 2017. "A Polycentric Approach to Global Climate Governance." *Global Environmental Politics* 17 (2): 45–64. https://doi.org/10.1162/GLEP_a_00400.

Douglas, Mary, and Baron Isherwood. 1979. *The World of Goods*. New York: Basic Books.

Dunham, Yarrow, Andrew Scott Baron, and Susan Carey. 2011. "Consequences of 'Minimal' Group Affiliations in Children." *Child Development* 82 (3): 793–811.

Easterlin, Richard A., Laura Angelescu McVey, Malgorzata Switek, Onnicha Sawangfa, and Jacqueline Smith Zweig. 2010. "The Happiness–Income Paradox Revisited." *Proceedings of the National Academy of Sciences of the United States of America* 107 (52): 22463–22468. https://doi.org/10.1073/pnas.1015962107.

Easterly, William, and Stanley Fischer. 1995. "The Soviet Economic Decline." *World Bank Economic Review* 9 (3): 341–371.

———. 2001. "Inflation and the Poor." *Journal of Money, Credit and Banking* 33 (2): 160–178. https://doi.org/10.2307/2673879.

Ehrlich, Paul R. 1968. *The Population Bomb*. San Francisco, CA: Sierra Club Books.

Eichengreen, Barry, and Ugo Panizza. 2014. "A Surplus of Ambition: Can Europe Rely on Large Primary Surpluses to Solve Its Debt Problem?" NBER Working Paper no. 20316, National Bureau of Economic Research, Cambridge, MA. July.

Elgenius, Gabriella. 2011. *Symbols of Nations and Nationalism: Celebrating Nationhood*. London: Palgrave Macmillan.

Esteban, Joan, and Debraj Ray. 2011. "Linking Conflict to Inequality and Polarization." *American Economic Review* 101 (June): 1345–1374. https://doi.org/10.1257/aer.101.4.1345.

Estevão, Marcello, and Filipa Sá. 2008. "The 35-Hour Week in France." *Economic Policy* 23 (55): 418–463. https://doi.org/10.1111/j.1468-0327.2008.00204.x.

European Commission. 2018. "A Clean Planet for All: A European Strategic Long-Term Vision for a Prosperous, Modern Competitive and Climate Neutral Economy." Directorate-General for Climate Action. https://ec.europa.eu/clima/policies/strategies/2050_en.

———. 2019. "Towards a Greener Future: Employment and Social Impacts of Climate Change Policies." Employment and Social Developments in Europe, Annual Review 2019, "Sustainable Growth for All," 166–205. European Commission, EU. https://ec.europa.eu/social/main.jsp?catId=738&langId=en&pubId=8219.

———. 2020a. "Impact Assessment Accompanying the document 'Stepping up Europe's 2030 Climate Ambition: Investing in a Climate-Neutral Future for the Benefit of Our People.'" https://eur-lex.europa.eu/legal-content/EN/TXT/?uri=CELEX:52020SC0176.

———. 2020b. "A New Circular Economy Action Plan for a Cleaner and More Competitive Europe." Communication from the Commission to the European Parliament, the Council, the European Economic and Social Committee of the Regions. Brussels, November 3, 2020. https://eur-lex.europa.eu/legal-content/EN/TXT/?uri=COM:2020:98:FIN.

Fagan, Brian. 2001. *The Little Ice Age: How Climate Made History 1300–1850*. New York: Basic Books.

Fagan, Brian, and Nadia Durrani. 2021. *Climate Chaos: Lessons on Survival from Our Ancestors*. New York: PublicAffairs.

Ferguson, Niall. 2011. *Civilization: The West and the Rest*. London: Penguin Books.

Fiorino, Nadia, Emma Galli, and Ilaria Petrarca. 2012. "Corruption and Growth: Evidence from the Italian Regions." *European Journal of Government and Economics* 1 (2): 126–144. https://doi.org/10.17979/ejge.2012.1.2.4281.

Fitzgerald, F. Scott. 1925. *The Great Gatsby*. Ware, Hertfordshire: Wordsworth Editions, 1993.

Fleurbaey, Marc. 2009. "Beyond GDP: The Quest for a Measure of Social Welfare." *Journal of Economic Literature* 47 (4): 1029–1075. https://doi.org/doi=10.1257/jel.47.4.1029.

Fotopoulos, Takis. 2010. "The De-Growth Utopia: The Incompatibility of De-Growth within an Internationalised Market Economy." *International*

Journal of Inclusive Democracy 4 (4): 103–121. https://doi.org/10.1007/978-90
-481-3745-9.

Fouquet, Roger. 2008. *Heat, Power and Light: Revolutions in Energy Services.*
Cheltenham, UK: Edward Elgar.

———. 2010. "The Slow Search for Solutions: Lessons from Historical Energy
Transitions by Sector and Service." *Energy Policy* 38 (11): 6586–6596.
https://doi.org/10.1016/j.enpol.2010.06.029.

———. 2011. "Divergences in Long-Run Trends in the Prices of Energy and
Energy Services." *Review of Environmental Economics and Policy* 5 (2):
196–218. https://doi.org/10.1093/reep/rer008.

———. 2014. "Long-Run Demand for Energy Services: Income and Price
Elasticities over Two Hundred Years." *Review of Environmental Economics
and Policy* 8 (2): 186–207. https://doi.org/10.1093/reep/reu002.

———. 2016. "Path Dependence in Energy Systems and Economic Development."
Nature Energy 1 (8). https://doi.org/10.1038/nenergy.2016.98.

Fouquet, Roger, and Stephen Broadberry. 2015. "Seven Centuries of European
Economic Growth and Decline." *Journal of Economic Perspectives* 29 (4):
227–244. https://doi.org/10.1257/jep.29.4.227.

Fouquet, Roger, and Peter J. G. Pearson. 2006. "Seven Centuries of Energy
Services: The Price and Use of Light in the United Kingdom (1300–2000)."
Energy Journal 27 (1): 139–177. https://doi.org/10.5547/ISSN0195-6574-EJ
-Vol27-No1-8.

Foxton, Timothy. 2018. *Energy and Economic Growth: Why We Need a New
Pathway to Prosperity.* London: Routledge.

Francis. 2015. *Laudato Si'.* Encyclical Letter. May 24. https://www.vatican.va
/content/francesco/en/encyclicals/documents/papa-francesco_20150524
_enciclica-laudato-si.html.

Frank, Robert H. 1985. "The Demand for Unobservable and Other Nonpositional
Goods." *American Economic Review* 75 (1): 101–116.

———. 2007. *Falling Behind: How Rising Inequality Harms the Middle Class.*
Berkeley: University of California Press.

———. 2011. *The Darwin Economy: Liberty, Competition, and the Common Good.*
Princeton, NJ: Princeton University Press.

Frank, Robert H., Adam Seth Levine, and Oege Dijk. 2014. "Expenditure Cascades." *Review of Behavioral Economics* 1: 55–73. https://doi.org/10.1561/105
.00000003.

Friedman, Benjamin. 2005. *The Moral Consequences of Economic Growth.* New
York: Random House.

Friedman, Milton. 1962. *Capitalism and Freedom.* Chicago: University of Chicago Press.

Fuest, Clemens, and Jean Pisani-Ferry. 2020. "Financing the European Union: New Context, New Responses." Bruegel Policy Contributions 2020/16. September. https://www.bruegel.org/2020/09/financing-the-european-union -new-context-new-responses/.

Fukuyama, Francis. 1992. *The End of History and the Last Man.* New York: Free Press.

Galbraith, John Kenneth. 1958. *The Affluent Society.* London: Penguin Books.

Garrett-Peltier, Heidi. 2017. "Green versus Brown: Comparing the Employment Impacts of Energy Efficiency, Renewable Energy, and Fossil Fuels Using an Input-Output Model." *Economic Modelling* 61 (2017): 439–447. https://doi.org /10.1016/j.econmod.2016.11.012.

Gates, Bill. 2021. *How to Avoid a Climate Disaster: The Solutions We Have and the Breakthroughs We Need.* New York: Random House.

Gavin, Michael, and Ricardo Hausmann. 1998. "Volatility, Uncertainty, Institutional Instability and Growth." In *The Political Dimension of Economic Growth,* edited by S. Borner and M. Paldam, 97–116. London: Palgrave Macmillan.

Gavin, Michael, Ricardo Hausmann, Roberto Perotti, and Ernesto Talvi. 1996. "Managing Fiscal Policy in Latin America and the Caribbean: Volatility, Procyclicality, and Limited Creditworthiness." Working Paper 326, Inter-American Development Bank. March 10. https://publications.iadb.org/en /publication/11943/managing-fiscal-policy-latin-america-and-caribbean -volatility-procyclicality-and.

Gerlagh, Reyer, Snorre Kverndokk, and Knut Einar Rosendahl. 2009. "Optimal Timing of Climate Change Policy: Interaction between Carbon Taxes and Innovation Externalities." *Environmental and Resource Economics* 43 (3): 369–390. https://doi.org/10.1007/s10640-009-9271-y.

Gilens, Martin, and Benjamin I. Page. 2014. "Testing Theories of American Politics: Elites, Interest Groups, and Average Citizens." *Perspectives on Politics* 12 (3): 564–581. https://doi.org/10.1017/S1537592714001595.

Gillingham, Kenneth, Matthew J. Kotchen, David S. Rapson, and Gernot Wagner. 2013. "The Rebound Effect Is Overplayed." *Nature* 493: 475–476. https://doi .org/10.1038/493475a.

Gillingham, Kenneth, David Rapson, and Gernot Wagner. 2016. "The Rebound Effect and Energy Efficiency Policy." *Review of Environmental Economics and Policy* 10 (1): 68–88. https://doi.org/ 10.1093/ reep/ revo17.

Glaeser, Edward. 2012. *Triumph of the City: How Our Greatest Invention Makes Us Richer, Smarter, Greener, Healthier, and Happier.* New York: Penguin.

GMF and Bertelsmann. 2021. *Transatlantic Trends 2021: Transatlantic Opinion on Global Challenges.* German Marshall Fund of the United States and Bertelsmann Foundation. https://www.gmfus.org/sites/default/files/2021-08/TT2021 _Web_Version.pdf.

Godwin, William. 1793. *An Enquiry Concerning Political Justice.* Oxford: Oxford University Press, 2013.

Goldman Sachs. 2020. *Green Hydrogen: The Next Transformational Driver of the Utilities Industry.* Equity Research, Goldman Sachs. Redacted version. September 22. https://www.goldmansachs.com/insights/pages/gs-research /green-hydrogen/report.pdf.

Goldin, Ian. 2021. *Rescue: From Global Crisis to a Better World.* London: Sceptre.

Göpel, Maja. 2016. *The Great Mindshift: How a New Economic Paradigm and Sustainability Transformations Go Hand in Hand.* New York: Springer International.

Gordon, Myron J., and Jeffrey S. Rosenthal. 2003. "Capitalism's Growth Imperative." *Cambridge Journal of Economics* 27 (1): 25–48. https://doi.org/10.1093/cje/27.1.25.

Gordon, Robert. 2016. *The Rise and Fall of American Growth: The U.S. Standard of Living since the Civil War.* Princeton, NJ: Princeton University Press.

Gough, Ian. 2015. "Can Growth Be Green?" *International Journal of Health Services* 45 (3): 443–452. https://doi.org/10.1177/0020731415584555.

Gould, Stephen Jay. 2011. *Full House: The Spread of Excellence from Plato to Darwin.* Cambridge, MA: Belknap Press of Harvard University Press.

Goulder, Lawrence, and Marc Hafstead. 2017. *Confronting the Climate Challenge: U.S. Policy Options.* New York: Columbia University Press.

Goulder, Lawrence H., and Andrew R. Schein. 2013. "Carbon Taxes versus Cap and Trade: A Critical Review." *Climate Change Economics* 4 (3): 1350010. https://doi.org/10.1142/S2010007813500103.

Gowdy, John, J. Barkley Rosser, and Loraine Roy. 2013. "The Evolution of Hyperbolic Discounting: Implications for Truly Social Valuation of the Future." *Journal of Economic Behavior and Organization* 90: S94–104. https://doi.org/10.1016/j.jebo.2012.12.013.

Graeber, David. 2015. *The Utopia of Rules: On Technology, Stupidity, and the Secret Joys of Bureaucracy.* New York: Melville House.

———. 2019. *Bullshit Jobs: A Theory.* London: Penguin Press.

Gramsci, Antonio. 1971. *Selections from the Prison Notebooks.* Translated by Quinton Hoare and Jeffrey Nowell Smith. New York: International Publishers.

Grubb, Michael, Igor Bashmakov, Paul Drummond, Anna Myshak, Nick Hughes, Andrea Biancardi, Paolo Agnolucci, and Robert Lowe. 2018. *An Exploration of Energy Cost, Ranges, Limits and Adjustment Process.* Research Report, UCL Institute for Sustainable Resources, University College London. March. https://www.ucl.ac.uk/bartlett/sustainable/sites /bartlett/files/an_exploration_of_energy_cost_ranges_limits_and _adjustment_process.pdf.

Gustafson, Thane. 2021. *Klimat: Russia in the Age of Climate Change.* Cambridge, MA: Harvard University Press.

Hall, Bronwyn H. 2006. "Innovation and Diffusion." In *The Oxford Handbook of Innovation,* edited by Jan Fagerberg, David C. Mowery, and Richard R. Nelson. Oxford: Oxford University Press.

Hall, Charles A. S., Jessica G. Lambert, and Stephen B. Balogh. 2014. "EROI of Different Fuels and the Implications for Society." *Energy Policy* 64: 141–152. https://doi.org/10.1016/j.enpol.2013.05.049.

Hall, Peter, and David Soskice. 2001. *Varieties of Capitalism: The Institutional Foundations of Comparative Advantage.* Oxford: Oxford University Press.

Hänsel, Martin C., Moritz A. Drupp, Daniel J. A. Johansson, Frikk Nesje, Christian Azar, Mark C. Freeman, Ben Groom, and Thomas Sterner. 2020. "Climate Economics Support for the UN Climate Targets." *Nature Climate Change* 10 (8): 781–789. https://doi.org/10.1038/s41558-020-0833-x.

Hansen, A. H. 1932. *Economic Stabilization in an Unbalanced World.* New York: Harcourt, Brace.

Hardin, Garrett. 1968. "The Tragedy of the Commons." *Science* 162 (3859): 1243–1248.

Hardt, Lukas, and Daniel W. O'Neill. 2017. "Ecological Macroeconomic Models: Assessing Current Developments." *Ecological Economics* 134: 198–211. https://doi.org/10.1016/j.ecolecon.2016.12.027.

Harrison, Kevin, and Tony Boyd. 2018. "Environmentalism and Ecologism." In *Understanding Political Ideas and Movements,* 274–294. Manchester: Manchester University Press. https://doi.org/10.7765/9781526137951.00018.

Hausman, William J. 1996. "Long-Term Trends in Energy Prices." In *The State of Humanity,* edited by Julian Simon. London: Wiley-Blackwell.

Hausmann, Ricardo, and Dani Rodrik. 2003. "Economic Development as Self-Discovery." *Journal of Development Economics* 72 (2): 603–633. https://doi.org/10.1016/S0304-3878(03)00124-X.

Hayek, Friedrich. 1944. *The Road to Serfdom.* London: Routledge.

Helliwell, John F., Richard Layard, Jeffrey Sachs, and Jan-Emmanuel De Neve, eds. 2021. *World Happiness Report 2021.* Sustainable Development Solutions Network, New York. https://worldhappiness.report/ed/2021/.

Henrich, Joseph. 2016. *The Secret of Our Success: How Culture Is Driving Human Evolution, Domesticating Our Species, and Making Us Smarter.* Princeton, NJ: Princeton University Press.

Hepburn, Cameron, Brian O'Callaghan, Nicholas Stern, Joseph Stiglitz, and Dimitri Zenghelis. 2020. "Will COVID-19 Fiscal Recovery Packages Accelerate or Retard Progress on Climate Change?" *Oxford Review of Economic Policy* 36 (20): S359–381. https://doi.org/10.1093/oxrep/graa015.

Hickel, Jason. 2020a. *Less Is More: How Degrowth Will Save the World.* London: Random House.

———. 2020b. "Quantifying National Responsibility for Climate Breakdown: An Equality-Based Attribution Approach for Carbon Dioxide Emissions in

Excess of the Planetary Boundary." *Lancet Planetary Health* 4 (9): 399–404. https://doi.org/10.1016/S2542-5196(20)30196-0.

High-Level Commission on Carbon Prices. 2017. *Report of the High-Level Commission on Carbon Prices.* Washington, DC: World Bank. License: Creative Commons Attribution CC BY 3.0 IGO.

Hirata, Keiko, and Mark Warschauer. 2014. *Japan: The Paradox of Harmony.* New Haven, CT: Yale University Press.

Hirsch, Fred. 1976. *Social Limits to Growth.* Cambridge, MA: Harvard University Press.

Hobbes, Thomas. 1651. *Leviathan Or the Matter, Forme and Power of a Commonwealth, Ecclesiasticall and Civil.* Edited by Richard Tuck. Cambridge: Cambridge University Press, 1991.

Hoffman, Philip T. 2015. *Why Did Europe Conquer the World?* Princeton, NJ: Princeton University Press.

Hoffmann, Ulrich. 2015. "Can Green Growth Really Work and What Are the True (Socio-) Economics of Climate Change?" Discussion Paper No. 222, United Nations Conference on Trade and Development. July. https://unctad.org/system/files/official-document/osgdp2015d4_en.pdf.

Horn, Henrik, and André Sapir. 2013. "Can Border Carbon Taxes Fit into the Global Trade Regime?" Bruegel Policy Brief 2013 / 06. December. https://www.bruegel.org/2013/12/can-border-carbon-taxes-fit-into-the-global-trade-regime/.

Hunt, Jennifer. 1999. "Has Work-Sharing Worked in Germany?" *Quarterly Journal of Economics* 114 (1): 117–148. https://doi.org/10.1162/003355399555963.

Huntington, Samuel P. 1996. *The Clash of Civilizations and the Remaking of World Order.* London: Simon and Schuster.

Iacoboni, Jacopo. 2018. *L'esperimento: Inchiesta Sul Movimento 5 Stelle.* Bari: Laterza.

IEA. 2020. "Renewables 2020: Analysis and Forecasts to 2025." International Energy Agency, Paris. November. https://www.iea.org/reports/renewables-2020.

———. 2021. *Clean Energy Investing: Global Comparison of Investment Returns.* Report of the International Energy Agency, Paris, and the Centre for Climate Finance and Investment, Imperial College Business School. March. https://www.iea.org/reports/clean-energy-investing-global-comparison-of-investment-returns.

ILO. 2018. "Greening with Jobs." World Employment Social Outlook 2018. International Labour Organization, Geneva. May 14. https://www.ilo.org/global/publications/books/WCMS_628654/lang--en/index.htm

IMF. 2019. "How to Mitigate Climate Change." Fiscal Monitor, International Monetary Fund. October. https://www.imf.org/en/Publications/FM/Issues/2019/10/16/Fiscal-Monitor-October-2019-How-to-Mitigate-Climate-Change-47027.

———. 2020. "Mitigating Climate Change: Growth- and Distribution-Friendly Strategies." *World Economic Outlook,* October, 85–114. https://www.imf.org /en/Publications/WEO/Issues/2020/09/30/world-economic-outlook-october -2020.

———. 2021. "A Fair Shot." Fiscal Monitor, International Monetary Fund. April. https://www.imf.org/en/Publications/FM/Issues/2021/03/29/fiscal-monitor -april-2021.

IPCC. 2018. *Global Warming of 1.5°C.* Intergovernmental Panel on Climate Change, Geneva.

IRENA. 2020. "Renewable Power Generation Costs in 2019." International Renewable Energy Agency, Abu Dhabi. https://www.irena.org/publications /2020/Jun/Renewable-Power-Costs-in-2019.

ISTAT. 2019. "La Situazione del Paese." Rapporto Annuale 2019. https://www .istat.it/it/archivio/230897.

Iversen, Torben, and David Soskice. 2019. *Democracy and Prosperity: Reinventing Capitalism through a Turbulent Century.* Princeton, NJ: Princeton University Press.

Iyigun, Murat, Nathan Nunn, and Nancy Qian. 2017a. "The Long-Run Effects of Agricultural Productivity on Conflict, 1400–1900." NBER Working Paper no. 24066, National Bureau of Economic Research, Cambridge, MA. November.

———. 2017b. "Winter Is Coming: The Long-Run Effects of Climate Change on Conflict, 1400–1900." NBER Working Paper no. 23033, National Bureau of Economic Research, Cambridge, MA. January.

Jackson, Tim. 2009. *Prosperity without Growth: Foundations for the Economy of Tomorrow.* London: Earthscan.

———. 2021. *Post Growth: Life after Capitalism.* Cambridge: Polity Press.

Jackson, Tim, and Peter A. Victor. 2015. "Does Credit Create a 'Growth Imperative'? A Quasi-Stationary Economy with Interest-Bearing Debt." *Ecological Economics* 120: 32–48. https://doi.org/10.1016/j.ecolecon.2015.09.009.

Jakob, Michael, and Ottmar Edenhofer. 2014. "Green Growth, Degrowth, and the Commons." *Oxford Review of Economic Policy* 30 (3): 447–468. https://doi .org/10.1093/oxrep/gru026.

Jakob, Michael, William F. Lamb, Jan Christoph Steckel, Christian Flachsland, and Ottmar Edenhofer. 2020. "Understanding Different Perspectives on Economic Growth and Climate Policy." *WIREs Climate Change* 11 (6): 1–17. https://doi.org/10.1002/wcc.677.

Jevons, William S. 1865. *The Coal Question: An Inquiry Concerning the Progress of the Nation, and the Probable Exhaustion of our Coal Mines.* London: Macmillan.

Jia, Ruixue. 2014. "Weather Shocks, Sweet Potatoes and Peasant Revolts in Historical China." *Economic Journal* 124 (575): 92–118.

Jones, Charles I. 1999. "Was the Industrial Revolution Inevitable?" NBER
  Working Paper no. 7375, National Bureau of Economic Research, Cambridge,
  MA. October.
———. 2005. "Growth and Ideas." In *Handbook of Economic Growth*, edited by
  Philippe Aghion and Steven N. Durlauf, 1B: 1063–1111. Amsterdam: Elsevier.
  https://doi.org/10.1016/S1574-0684(05)01016-6.
———. 2016. "The Facts of Economic Growth." In *Handbook of Macroeco-
  nomics*, edited by John B. Taylor and Harald Uhlig, 2A: 3–69. Amsterdam:
  Elsevier.
Jones, Charles I., and Peter J. Klenow. 2016. "Beyond GDP? Welfare across
  Countries and Time." *American Economic Review* 106 (9): 2426–2457.
  https://doi.org/10.1257/aer.20110236.
Juhász, Réka. 2018. "Temporary Protection and Technology Adoption: Evidence
  from the Napoleonic Blockade." *American Economic Review* 108 (11): 3339–
  3376. https://doi.org/10.1257/aer.20151730.
Juhász, Réka, Mara P. Squicciarini, and Nico Voigtländer. 2020. "Technology
  Adoption and Productivity Growth: Evidence from Industrialization in
  France." NBER Working Paper no. 27503, National Bureau of Economic
  Research, Cambridge, MA. October.
Juma, Calestous. 2016. *Innovation and Its Enemies: Why People Resist New Tech-
  nologies.* Oxford: Oxford University Press.
Kahn, Matthew E. 2021. *Adapting to Climate Change: Markets and the Manage-
  ment of an Uncertain Future.* New Haven, CT: Yale University Press.
Kahneman, Daniel. 2013. *Thinking, Fast and Slow.* New York: Farrar, Straus and
  Giroux.
Kallis, Giorgos. 2019. *Limits: Why Malthus Was Wrong and Why Environmental-
  ists Should Care.* Stanford, CA: Stanford University Press.
Kallis, Giorgos, Christian Kerschner, and Joan Martinez-Alier. 2012. "The
  Economics of Degrowth." *Ecological Economics* 84: 172–180. https://doi.org
  /10.1016/j.ecolecon.2012.08.017.
Kallis, Giorgos, Vasilis Kostakis, Steffen Lange, Barbara Muraca, Susan Paulson,
  and Matthias Schmelzer. 2018. "Research on Degrowth." *Annual Review of
  Environment and Resources* 43: 291–316. https://doi.org/10.1146/annurev
  -environ-102017-025941.
Kelton, Stephanie. 2020. *The Deficit Myth: Modern Monetary Theory and the
  Birth of the People's Economy.* New York: PublicAffairs.
Kerr, Sari Pekkala, and William Kerr. 2020. "Immigrant Entrepreneurship in
  America: Evidence from the Survey of Business Owners 2007 & 2012."
  Research Policy 49 (3): 103918. https://doi.org/10.1016/j.respol.2019.103918.
Keynes, John Maynard. 1924. "Alfred Marshall, 1842–1924." *Economic Journal* 34
  (135): 311–372.

———. 1930. "Economic Possibilities for Our Grandchildren." In *Essays in Persuasion*, 358–373. New York: W. W. Norton, 1963. https://doi.org/10.4337/9781788118569.00035.

———. 1936. *The General Theory of Employment, Interest and Money*. London: Palgrave Macmillan, 2007.

Klein, Naomi. 2015. *This Changes Everything: Capitalism vs. the Climate*. New York: Simon and Schuster.

———. 2019. *On Fire: The (Burning) Case for a Green New Deal*. New York: Simon and Schuster.

Klein, Rudolf. 1972. "Growth and Its Enemies." *Commentary*, June.

Klitgaard, Kent A., and Lisi Krall. 2012. "Ecological Economics, Degrowth, and Institutional Change." *Ecological Economics* 84: 247–253. https://doi.org/10.1016/j.ecolecon.2011.11.008.

Knittel, Christopher, and Elizabeth Murphy. 2019. "Generational Trends in Vehicle Ownership and Use: Are Millennials Any Different?" NBER Working Paper no. 25674, National Bureau of Economic Research, Cambridge, MA. March.

Kremer, Michael. 1993. "Population Growth and Technological Change: One Million B.C. to 1990." *Quarterly Journal of Economics* 108 (3): 681–716. https://doi.org/10.2307/2118405.

Krenek, Alexander, Mark Sommer, and Margit Schratzenstaller. 2019. "Sustainability-Oriented Future EU Funding: A European Border Carbon Adjustment." WIFO Working Papers no. 587, Austrian Institute of Economic Research (WIFO), Vienna. August.

Kupers, Roland. 2020. *A Climate Policy Revolution: What the Science of Complexity Reveals about Saving Our Planet*. Cambridge, MA: Harvard University Press.

Kurz, Christopher, Geng Li, and Daniel J. Vine. 2018. "Are Millennials Different?" Finance and Economics Discussion Series 2018-80, Board of Governors of the Federal Reserve System, Washington, DC. November. https://doi.org/10.17016/FEDS.2018.080.

Kuznets, Simon. 1934. "National Income, 1929–1932." NBER Bulletin no. 49, National Bureau of Economic Research, New York.

Lafond, François, Aimee Gotway Bailey, Jan David Bakker, Dylan Rebois, Rubina Zadourian, Patrick McSharry, and J. Doyne Farmer. 2018. "How Well Do Experience Curves Predict Technological Progress? A Method for Making Distributional Forecasts." *Technological Forecasting and Social Change* 128 (1936): 104–117. https://doi.org/10.1016/j.techfore.2017.11.001.

Landes, David. 1998. *The Wealth and Poverty of Nations: Why Some Are So Rich and Some So Poor*. New York: W. W. Norton.

Langgut, Dafna. 2017. "The Citrus Route Revealed: From Southeast Asia into the Mediterranean." *HortScience* 52 (6): 814–822. https://doi.org/10.21273/HORTSCI11023-16.

Latouche, Serge. 1989. *L'Occidentalisation du Monde*. Paris: La Découverte.

———. 2009. *Farewell to Growth*. Cambridge: Polity Press.

Layard, Richard. 2006. "Happiness and Public Policy: A Challenge to the Profession." *Economic Journal* 116 (510): C24–C33. https://doi.org/10.1111/j.1468-0297.2006.01073.x.

Leard, Benjamin, Joshua Linn, and Clayton Munnings. 2019. "Explaining the Evolution of Passenger Vehicle Miles Traveled in the United States." *Energy Journal* 40: 25–54. https://doi.org/10.5547/01956574.40.1.blea.

Leibenstein, Harvey. 1950. "Bandwagon, Snob, and Veblen Effects in the Theory of Consumers' Demand." *Quarterly Journal of Economics* 64 (2): 183–207. https://doi.org/10.2307/1882692.

Leonard, Mark, Jean Pisani-Ferry, Jeremy Shapiro, Simone Tagliapietra, and Guntram B. Wolff. 2021. "The Geopolitics of the European Green Deal." Bruegel Policy Contribution 04 / 21. February. https://www.bruegel.org/2021/02/the-geopolitics-of-the-european-green-deal/.

Lidin, Olof G. 2002. *Tanegashima: The Arrival of Europe in Japan*. Copenhagen: Nordic Institute of Asian Studies (NIAS) Press.

Lipsey, Richard G. 2019. "Policies for Green Growth versus Policies for No Growth: A Matter of Timing." In *Handbook on Green Growth*, edited by Roger Fouquet, 1–21. Cheltenham, UK: Edward Elgar. https://doi.org/10.4337/9781788110686.00007.

Lovelock, James. 2000. *Homage to Gaia: The Life of an Independent Scientist*. Oxford: Oxford University Press.

Luttmer, Erzo. 2001. "Group Loyalty and the Taste for Redistribution." *Journal of Political Economy* 109 (3): 500–528. https://doi.org/10.1086/321019.

———. 2005. "Neighbors as Negatives: Relative Earnings and Well-Being." *Quarterly Journal of Economics* 120 (3): 963–1002. https://doi.org/10.1093/qje/120.3.963.

Lutz, Markus, Markus Flaute, Ulrike Lehr, and Kirsten S. Wiebe. 2015. "Economic Impacts of Renewable Power Generation Technologies and the Role of Endogenous Technological Change." Discussion Paper No. 2015 / 9. Osnabrück: Gesellschaft für Wirtschaftliche Strukturforschung (GWS). http://hdl.handle.net/10419/121455.

Malcolm, Noel. 2016. "Thomas Hobbes: Liberal Illiberal." *Journal of the British Academy* 4 (August): 113–136. https://doi.org/10.5871/jba/004.113.

Malkin, Robert A. 2007. "Barriers for Medical Devices for the Developing World." *Expert Review of Medical Devices* 4 (6): 759–763. https://doi.org/10.1586/17434440.4.6.759.

Malthus, Thomas. 1798. *An Essay on the Principle of Population*. New York: Dover, 2007.

Mann, Michael E. 2019. "Radical Reform and the Green New Deal." *Nature* 573 (7774): 340–341. https://doi.org/10.1038/d41586-019-02738-7.

———. 2021. *The New Climate War: The Fight to Take Back Our Planet.* New York: PublicAffairs.

Marcuse, Herbert. 1964. *One-Dimensional Man.* Boston: Beacon Press.

MarketsandMarkets. 2021. *EV Battery Market by Battery Capacity (<50, 50–110, 111–200, 201–300 and >300), Battery Form (Wire, Laser), Propulsion (BEV, PHEV, PHEV, FCEV), Battery Type, Material Type, Li-ion Battery Component, Method, Vehicle Type & Region—Global Forecast to 2025.* Proprietary report.

Marques, Rui Cunha, and Nuno Ferreira da Cruz. 2015. *Recycling and Extended Producer Responsibility: The European Experience.* London: Routledge.

Marshall, Alfred. 1890. *Principles of Economics.* London: Palgrave Macmillan, 2013.

Marshall, Melanie L. 2015. "Imitating the Rustic and Revealing the Noble: Masculine Power and Music at the Court of Ferrara." In *Eroticism in Early Modern Music,* edited by Bonnie J. Blackburn and Laurie Stras, 83–114. Aldershot, UK: Ashgate.

Martin, Ralf, Sam Unsworth, Anna Valero, and Dennis Verhoeven. 2020. "Innovation for a Strong and Sustainable Recovery." CEPCOVID-19-014. Centre for Economic Performance, London School of Economics. December. https://cep.lse.ac.uk/pubs/download/cepcovid-19-014.pdf.

Marx, Karl. 1867. *Capital: A Critique of Political Economy.* New York: Cosimo Classics, 2007.

Mazzucato, Mariana. 2013. *The Entrepreneurial State: Debunking Public vs. Private Sector Myths.* London: Anthem Press.

———. 2021. *Mission Economy: A Moonshot Guide to Changing Capitalism.* London: Allen Lane.

McAfee, Andrew. 2019. *More from Less: The Surprising Story of How We Learned to Prosper Using Fewer Resources—and What Happens Next.* New York: Simon and Schuster.

McNeill, John Robert, and Peter Engelke. 2014. *The Great Acceleration.* Cambridge, MA: Belknap Press of Harvard University Press.

Meadows, Donella, Dennis Meadows, Jorgen Randers, and William Beherens. 1972. *The Limits to Growth.* New York: Universe Books.

Meadows, Donella, Jorgen Randers, and Dennis Meadows. 2004. *Limits to Growth: The 30-Year Update.* White River Junction, VT: Chelsea Green.

Mearsheimer, John J. 2001. *The Tragedy of Great Power Politics.* New York: W. W. Norton.

Mercure, J.-F., P. Salas, P. Vercoulen, G. Semieniuk, A. Lam, H. Pollitt, P. B. Holden, N. Vakilifard, U. Chewpreecha, N. R. Edwards, and J. E. Vinuales. 2021. "Reframing Incentives for Climate Policy Action." *Nature Energy* 6: 1133–1143. https://doi.org/10.1038/s41560-021-00934-2.

Merler, Silvia. 2020. *La Pecora Nera: L'Italia Di Oggi e l'eurozona.* Milan: Università Bocconi Editore.

Metcalf, Gilbert E., and James H. Stock. 2020. "The Macroeconomic Impact of Europe's Carbon Taxes." NBER Working Paper no. 27488, National Bureau of Economic Research, Cambridge, MA. July.

Miguel, Edward. 2005. "Poverty and Witch Killing." *Review of Economic Studies* 72 (4): 1153–1172. https://doi.org/10.1111/0034-6527.00365.

Miguel, Edward, Shanker Satyanath, and Ernest Sergenti. 2004. "Economic Shocks and Civil Conflict: An Instrumental Variables Approach." *Journal of Political Economy* 112 (4): 725–753.

Milanovic, Branko. 2019. *Capitalism, Alone: The Future of the System That Rules the World.* Cambridge, MA: Belknap Press of Harvard University Press.

Mill, John Stuart. 1849. *Principles of Political Economy, With Some of Their Applications to Social Philosophy.* London: J. W. Parker.

Mokyr, Joel. 1994. "Cardwell's Law and the Political Economy of Technological Progress." *Research Policy* 23 (5): 561–574. https://doi.org/10.1016/0048-7333(94)01006-4.

———. 2016. *A Culture of Growth: The Origins of the Modern Economy.* Princeton, NJ: Princeton University Press.

Molinder, Jakob, Tobias Karlsson, and Kerstin Enflo. 2021. "More Power to the People: Electricity Adoption, Technological Change and Social Conflict." *Journal of Economic History* 81 (2): 481–512. https://doi.org/10.1017/S0022050721000127.

More, Thomas. 1516. *Utopia.* London: Penguin Books, 2003.

Morris, Ian. 2013. *The Measure of Civilization: How Social Development Decides the Fate of Nations.* Princeton, NJ: Princeton University Press.

Murphy, David J., and Charles A. S. Hall. 2011. "Energy Return on Investment, Peak Oil, and the End of Economic Growth." *Annals of the New York Academy of Sciences* 1219 (1): 52–72. https://doi.org/10.1111/j.1749-6632.2010.05940.x.

Neumark, David, and Andrew Postlewaite. 1998. "Relative Income Concerns and the Rise in Married Women's Employment." *Journal of Public Economics* 70 (1): 157–183. https://doi.org/10.1016/S0047-2727(98)00065-6.

Newell, Peter. 2011. "The Elephant in the Room: Capitalism and Global Environmental Change." *Global Environmental Change* 21 (1): 4–6. https://doi.org/10.1016/j.gloenvcha.2010.11.011.

Newell, Peter, and Matthew Paterson. 2010. *Climate Capitalism: Global Warming and the Transformation of the Global Economy.* Cambridge: Cambridge University Press.

———. 2011. "Climate Capitalism." In *After Cancún,* edited by Elmar Altvater and Achim Brunnengräber, 23–44. Wiesbaden: VS Verlag für Sozialwissenschaften. https://doi.org/10.1007/978-3-531-94018-2.

Nordhaus, William D. 1973. "World Dynamics: Measurement without Data." *Economic Journal* 83 (332): 1156–1183. https://doi.org/ 10.2307 / 2230846.

———. 1994. *Managing the Global Commons: The Economics of Climate Change.* Cambridge, MA: MIT Press.

———. 1996. "Historical Reassessments of Economic Progress." In *The Economics of New Goods,* edited by Timothy Bresnahan and Robert Gordon, 27–70. Chicago: University of Chicago Press.

———. 2015. "Climate Clubs: Overcoming Free-Riding in International Climate Policy." *American Economic Review* 105 (4): 1339–1370. https://doi.org/10.1257 /aer.15000001.

———. 2021. "Are We Approaching an Economic Singularity? Information Technology and the Future of Economic Growth." *American Economic Journal: Macroeconomics* 13 (1): 299–332. https://doi.org/10.1257/mac .20170105.

Nordhaus, William D., and James Tobin. 1973. "Is Growth Obsolete?" In *The Measurement of Economic and Social Performance,* edited by Milton Moss, 509–564. Cambridge, MA: National Bureau of Economic Research.

Noy, Ilan, Toshihiro Okubo, Eric Strobl, and Thomas Tveit. 2021. "The Fiscal Costs of Earthquakes in Japan." CESifo Working Paper 9070, Center for Economic Studies and ifo Institute (CESifo), Ludwigs-Maximilian University, Munich. https://www.econstor.eu/bitstream/10419/235440/1/cesifo1 _wp9070.pdf.

OECD. 2012. "OECD Employment Outlook." Paris: OECD Publishing.

———. 2017. *Employment Implications of Green Growth: Linking Jobs, Growth, and Green Policies.* OECD Report for the G7 Environment Ministers. June.

———. 2018. *A Broken Social Elevator? How to Promote Social Mobility.* Paris: Organisation for Economic Co-operation and Development.

———. 2019. *Government at a Glance 2019.* Paris: OECD Publishing. https://doi .org/10.16973/jgs.2011.6.2.010.

———. 2020. "Beyond Growth: Towards a New Economic Approach." New Approaches to Economic Challenges. Paris: OECD Publishing, September. https://doi.org/10.1787/33a25ba3-en.

Ofer, Gur. 1987. "Soviet Economic Growth: 1928–1985." *Journal of Economic Literature* 25 (4): 1767–1833. https://doi.org/10.2307/2009173.

O'Neill, Jim, and Alessio Terzi. 2014. "Changing Trade Patterns, Unchanging European and Global Governance." Bruegel Working Paper 2014/ 02. February. https://www.bruegel.org/2014/02/changing-trade-patterns -unchanging-european-and-global-governance/.

Oreopoulos, Philip, Till von Wachter, and Andrew Heisz. 2012. "The Short- and Long-Term Career Effects of Graduating in a Recession." *American Economic Journal: Applied Economics* 4 (1): 1–29. https://doi.org/10.1257/app.4.1.1.

Oster, Emily. 2004. "Witchcraft, Weather and Economic Growth in Renaissance Europe." *Journal of Economic Perspectives* 18 (1): 215–228. https://doi.org /10.1257/0895330047735635302.

Ostrom, Elinor. 1990. *Governing the Commons: The Evolution of Institutions for Collective Action.* Cambridge: Cambridge University Press.

———. 2009. "A Polycentric Approach for Coping with Climate Change." Policy Research Working Paper 5095, Development Economics Working Group, World Bank.

———. 2012. "Nested Externalities and Polycentric Institutions: Must We Wait for Global Solutions to Climate Change before Taking Actions at Other Scales?" *Economic Theory* 49 (2): 353–369. https://doi.org/10.1007/s00199-010-0558-6.

Ostrom, Vincent. 1991. *The Meaning of American Federalism.* San Francisco: Institute for Contemporary Studies Press.

Ostrom, Vincent, Charles M. Tiebout, and Robert Warren. 1961. "The Organization of Government in Metropolitan Areas: A Theoretical Inquiry." *American Political Science Review* 55 (4): 831–842. https://doi.org/10.1017 /S0003055400125973.

Parey, Matthias, Jens Ruhose, Fabian Waldinger, and Nicolai Netz. 2017. "The Selection of High-Skilled Emigrants." *Review of Economics and Statistics* 99 (5): 776–792. https://doi.org/10.1162/REST_a_00687.

Park, Jonathan. 2015. "Climate Change and Capitalism." *Consilience: The Journal of Sustainable Development* 14 (2): 189–206. https://doi.org/10.7916/ D86H4H4K.

Parker, Geoffrey. 2013. *Global Crisis: War, Climate Change and Catastrophe in the Seventeenth Century.* New Haven, CT: Yale University Press.

Parry, Ian, Simon Black, and Nate Vernon. 2021. "Still Not Getting Energy Prices Right: A Global and Country Update of Fossil Fuel Subsidies." IMF Working Paper 2021 / 236, International Monetary Fund. September. https://www.imf. org/en/Publications/WP/Issues/2021/09/23/Still-Not-Getting-Energy-Prices -Right-A-Global-and-Country-Update-of-Fossil-Fuel-Subsidies-466004.

Pecchi, Lorenzo, and Gustavo Piga, eds. 2008. *Revisiting Keynes: Economic Possibilities for Our Grandchildren.* Cambridge, MA: MIT Press. https://doi .org/10.1215/00182702-2009-042.

Pellegrino, Bruno, and Luigi Zingales. 2017. "Diagnosing the Italian Disease." NBER Working Paper no. 23964, National Bureau of Economic Research, Cambridge, MA. May.

Perez, Carlota. 2019. "Transitioning to Smart Green Growth: Lessons from History." In *Handbook on Green Growth,* edited by Roger Fouquet, 447–463. Cheltenham, UK: Edward Elgar.

Perez-Truglia, Ricardo. 2019. "The Effects of Income Transparency on Well-Being: Evidence from a Natural Experiment." NBER Working Paper no. 25622, National Bureau of Economic Research, Cambridge, MA. September.

Perry, Lora, and Robert Malkin. 2011. "Effectiveness of Medical Equipment Donations to Improve Health Systems: How Much Medical Equipment Is Broken in the Developing World?" *Medical and Biological Engineering and Computing* 49 (7): 719–722. https://doi.org/10.1007/s11517-011-0786-3.

Peruzzi, Michele, and Alessio Terzi. 2021. "Accelerating Economic Growth: The Science beneath the Art." *Economic Modelling* 103, 105593. https://doi.org/10.1016/j.econmod.2021.105593.

Petraglia, Michael D., Huw S. Groucutt, Maria Guagnin, Paul S. Breeze, and Nicole Boivin. 2020. "Human Responses to Climate and Ecosystem Change in Ancient Arabia." *Proceedings of the National Academy of Sciences of the United States of America* 117 (15): 8263–8270. https://doi.org/10.1073/pnas.1920211117.

Pigou, Arthur Cecil. 1920. *The Economics of Welfare.* London: Macmillan.

Piketty, Thomas. 2013. *Capital in the Twenty-First Century.* Cambridge, MA: Harvard University Press.

Piketty, Thomas, and Gabriel Zucman. 2014. "Capital Is Back: Wealth-Income Ratios in Rich Countries 1700–2010." *Quarterly Journal of Economics* 129 (3): 1255–1310. https://doi.org/10.1093/qje/qju018.

Pinker, Steven. 2018. *Enlightenment Now: The Case for Reason, Science, Humanism, and Progress.* London: Penguin Books.

Pollin, Robert. 2018. "De-Growth vs a Green New Deal." *New Left Review,* no. 112: 5–25.

———. 2019. "Advancing a Viable Global Climate Stabilization Project: Degrowth versus the Green New Deal." *Review of Radical Political Economics* 51 (2): 311–319. https://doi.org/10.1177/0486613419833518.

Preston, Samuel H. 1975. "The Changing Relation between Mortality and Level of Economic Development." *Population Studies* 29 (2): 231–248. https://doi.org/10.1093/ije/dym075.

Pritchett, Lant. 2000. "Understanding Patterns of Economic Growth: Searching for Hills among Plateaus, Mountains, and Plains." *World Bank Economic Review* 14 (2): 221–250. https://doi.org/10.1093/wber/14.2.221.

Quah, Danny T. 2019. "The Invisible Hand and the Weightless Economy." In *Handbook on Green Growth,* edited by Roger Fouquet, 464–472. Cheltenham, UK: Edward Elgar.

Rachel, Łukasz, and Lawrence H. Summers. 2019. "On Secular Stagnation in the Industrialized World." *Brookings Papers on Economic Activity* 2019 (1): 1–76. https://doi.org/10.1353/eca.2019.0000.

Raworth, Kate. 2017. *Doughnut Economics: Seven Ways to Think Like a 21st-Century Economist.* London: Random House.

Reinhold, Meyer. 1969. "On Status Symbols in the Ancient World." *The Classical Journal* 64 (7): 300–304.

Ricardo, David. 1817. *The Principles of Political Economy and Taxation.* New York: Dover, 2004.

Richters, Oliver, and Andreas Siemoneit. 2017a. "Fear of Stagnation? A Review on Growth Imperatives." VÖÖ Discussion Paper 6, Vereinigung für Ökologische Ökonomie, Heidelberg. March. http://www.voeoe.de/publikationen /discussion/dp6/.

———. 2017b. "How Imperative Are the Joneses? Economic Growth between Individual Desire and Social Coercion." VÖÖ Discussion Paper 4, Vereinigung für Ökologische Ökonomie, Heidelberg. January. http://www.voeoe.de /publikationen/discussion/dp4/.

Ricke, Katharine, Laurent Drouet, Ken Caldeira, and Massimo Tavoni. 2018. "Country-Level Social Cost of Carbon." *Nature Climate Change* 8 (10): 895–900. https://doi.org/10.1038/s41558-018-0282-y.

Ridley, Matthey, Gautam Rao, Frank Schilbach, and Vikram Patel. 2020. "Poverty, Depression, and Anxiety: Causal Evidence and Mechanisms." *Science* 230 (6522): 1–12. https://doi.org/10.1126/science.aay0214.

Rifkin, Jeremy. 2019. *The Green New Deal: Why the Fossil Fuel Civilization Will Collapse by 2028, and the Bold Economic Plan to Save Life on Earth.* New York: St. Martin's Press.

Ripple, William J., Christopher Wolf, Thomas M. Newsome, Mauro Galetti, Mohammed Alamgir, Eileen Crist, Mahmoud I. Mahmoud, and William F. Laurance. 2017. "World Scientists' Warning to Humanity: A Second Notice." *BioScience* 67 (12): 1026–1028. https://doi.org/10.1093/biosci/bix125.

Robbins, Lionel. 1932. *An Essay on the Nature and Significance of Economic Science.* London: Macmillan.

Rockström, J., W. Steffen, K. Noone, Å. Persson, F. S. Chapin, E. F. Lambin, T. M. Lenton, et al. 2009. "A Safe Operating Space for Humanity." *Nature* 461 (7623): 472–475. https://doi.org/10.1038/461472a.

Rodríguez-Pose, Andrés. 2018. "The Revenge of the Places That Don't Matter (and What to Do about It)." *Cambridge Journal of Regions, Economy and Society* 11 (1): 189–209. https://doi.org/10.1093/cjres/rsx024.

Rodrik, Dani. 1997. *Has Globalization Gone Too Far?* Washington, DC: Institute for International Economics.

———. 2009. *One Economics, Many Recipes: Globalization, Institutions, and Economic Growth.* Princeton, NJ: Princeton University Press.

———. 2011. *The Globalization Paradox: Democracy and the Future of the World Economy.* New York: W. W. Norton.

———. 2014. "Green Industrial Policy." *Oxford Review of Economic Policy* 30 (3): 469–491. https://doi.org/10.1093/oxrep/gru025.

———. 2015. *Economics Rules: The Rights and Wrongs of the Dismal Science.* New York: W. W. Norton.

———. 2018. "Is Populism Necessarily Bad Economics?" *AEA Papers and Proceedings* 108: 196–199. https://doi.org/10.1257/pandp.20181122.

Rodrik, Dani, and Charles Sabel. 2019. "Building a Good Jobs Economy." HKS Faculty Research Working Paper No. RWP20-001, Harvard Kennedy School. January. https://www.hks.harvard.edu/publications/building-good-jobs-economy.

Romer, Paul M. 2016. "Economic Growth." In *The Concise Encyclopedia of Economics,* edited by David R. Henderson. Indianapolis, IN: Liberty Fund.

Rosero-Bixby, Luis, and William H. Dow. 2016. "Exploring Why Costa Rica Outperforms the United States in Life Expectancy: A Tale of Two Inequality Gradients." *Proceedings of the National Academy of Sciences of the United States of America* 113 (5): 1130–1137. https://doi.org/10.1073/pnas.1521917112.

Sachs, Jeffrey D. 2008. *Common Wealth: Economics for a Crowded Planet.* London: Penguin Press.

Saez, Emmanuel, and Gabriel Zucman. 2016. "Wealth Inequality in the US since 1913: Evidence from Capitalized Income Data." *Quarterly Journal of Economics* 131 (2): 519–578. https://doi.org/10.1017/CBO9781107415324.004.

Sala, Enric. 2020. *The Nature of Nature: Why We Need the Wild.* Washington, DC: National Geographic.

Sanchez, Rafael. 2013. "Do Reductions of Standard Hours Affect Employment Transitions? Evidence from Chile." *Labour Economics* 20: 24–37. https://doi.org/10.1016/j.labeco.2012.10.001.

Sandbu, Martin. 2020. *The Economics of Belonging: A Radical Plan to Win Back the Left Behind and Achieve Prosperity for All.* Princeton, NJ: Princeton University Press.

Sapolsky, Robert. 2017. *Behave: The Biology of Humans at Our Best and Worst.* New York: Penguin Press.

Schelling, Thomas C. 1978. *Micromotives and Macrobehavior.* New York: Norton.

———. 1995. "Intergenerational Discounting." *Energy Policy* 23 (4 / 5): 395–401. https://doi.org/10.1016/0301-4215(95)90164-3.

Schor, Juliet. 2010. *Plenitude: The New Economics of True Wealth.* London: Penguin Press.

Schumpeter, Joseph. 1942. *Capitalism, Socialism and Democracy.* New York: Harper and Brothers.

SDSN. 2019. *Roadmap to 2050: A Manual for Nations to Decarbonize by Mid-Century.* Sustainable Development Solutions Network, New York. September. https://roadmap2050.report/.

Sen, Amartya. 1984. "The Living Standard." *Oxford Economic Papers* 36 (November): 74–90. https://doi.org/10.1093/oxfordjournals.oep.a041662.

————. 1985. *The Standard of Living*. The Tanner Lectures on Human Values. Cambridge: Cambridge University Press.

————. 1999. *Development as Freedom*. Oxford: Oxford University Press.

————. 2006. *Identity And Violence: The Illusion of Destiny*. New York: W. W. Norton.

Sharman, J. C. 2019. *Empires of the Weak: The Real Story of European Expansion and the Creation of the New World Order*. Princeton, NJ: Princeton University Press.

Shiller, Robert J. 2019. *Narrative Economics: How Stories Go Viral and Drive Major Economic Events*. Princeton, NJ: Princeton University Press.

Skuterud, Mikal. 2007. "Identifying the Potential of Work-Sharing as a Job-Creation Strategy." *Journal of Labor Economics* 25 (2): 265–287. https://doi.org /10.1086/511379.

Smil, Vaclav. 2017. *Energy and Civilization: A History*. Cambridge, MA: MIT Press.

————. 2019. *Growth: From Microorganisms to Megacities*. Cambridge, MA: MIT Press.

Smith, Adam. 1759. *The Theory of Moral Sentiments*. London: Penguin Classics, 2010.

————. 1776. *An Inquiry into the Nature and Causes of the Wealth of Nations*. London: CreateSpace, 2013.

Smith, Richard. 2016. *Green Capitalism: The God That Failed*. Rickmansworth, UK: College Publications.

Snower, Dennis J. 1994. "The Low-Skill, Bad-Job Trap." IMF Working Paper 94 / 83, International Monetary Fund. July 1.

Solnick, Sara J., and David Hemenway. 1998. "Is More Always Better? A Survey on Positional Concerns." *Journal of Economic Behavior and Organization* 37 (3): 373–383. https://doi.org/10.1016/s0167-2681(98)00089-4.

Standage, Tom. 2021. *A Brief History of Motion: From the Wheel, to the Car, to What Comes Next*. London: Bloomsbury.

Stern, Nicolas, and Joseph E. Stiglitz. 2021. "The Social Cost of Carbon, Risk, Distribution, Market Failures: An Alternative Approach." NBER Working Paper no. 28472, National Bureau of Economic Research, Cambridge, MA. February.

Stock, James H. 2020. "Climate Change, Climate Policy, and Economic Growth." *NBER Macroeconomics Annual* 34 (1): 399–419. https://doi.org/10.1086/707192.

Strassburg, Bernardo B. N., Alvaro Iribarrem, Hawthorne L. Beyer, Carlos Leandro Cordeiro, Renato Crouzeilles, Catarina C. Jakovac, André Braga Junqueira, et al. 2020. "Global Priority Areas for Ecosystem Restoration." *Nature* 586 (7831): 724–729. https://doi.org/10.1038/s41586-020-2784-9.

Streeck, Wolfgang. 2016. *How Will Capitalism End? Essays on a Failing System*. London: Verso.

Strevens, Michael. 2020. *The Knowledge Machine: How an Unreasonable Idea Created Modern Science*. London: Allen Lane.

Suzman, James. 2020. *Work: A History of How We Spend Our Time*. London: Bloomsbury Circus.

Systemiq. 2020. *The Paris Effect: How the Climate Agreement is Reshaping the Global Economy*. Systemiq, London. December. https://www.systemiq.earth/wp-content/uploads/2020/12/The-Paris-Effect_SYSTEMIQ_Full-Report_December-2020.pdf.

Tagliapietra, Simone. 2020a. *Global Energy Fundamentals: Economics, Politics, and Technology*. Cambridge: Cambridge University Press.

———. 2020b. *L'energia Del Mondo*. Bologna: il Mulino.

Tagliapietra, Simone, and Reinhilde Veugelers. 2020. "A Green Industrial Policy for Europe." Bruegel Blueprint Series 31. https://www.bruegel.org/2020/12/a-green-industrial-policy-for-europe/.

Tagliapietra, Simone, and Guntram B. Wolff. 2021. "Form a Climate Club: United States, European Union and China." *Nature* 591: 526–528.

Terzi, Alessio. 2015. "Global Trends in Brain Drain and Likely Scenarios in the Coming Years." In *The Handbook of Global Science, Technology, and Innovation*, edited by Daniele Archibugi and Andrea Filippetti, 411–422. London: Wiley-Blackwell.

———. 2016. "An Italian Job: The Need for Collective Wage Bargaining Reform." Bruegel Policy Contribution 2016/11. July. https://www.bruegel.org/2016/07/an-italian-job-the-need-for-collective-wage-bargaining-reform/.

———. 2017. "The Political Conditions for Economic Reform in Europe's South." In *Economic Crisis and Structural Reforms in Southern Europe*, edited by Paolo Manasse and Dimitris Katsikas. London: Routledge.

———. 2020a. "Crafting an Effective Narrative on the Green Transition." *Energy Policy* 147 (August): 111883. https://doi.org/10.1016/j.enpol.2020.111883.

———. 2020b. "Macroeconomic Adjustment in the Euro Area." *European Economic Review* 128. https://doi.org/10.1016/j.euroecorev.2020.103516.

———. 2021a. "Economic Policy-Making beyond GDP: An Introduction." *European Economy* 142 (June): 1–27. https://doi.org/10.2765/166310.

———. 2021b. "The Roaring Twenties after Covid-19: Revisiting the Evidence for Europe." *Journal of New Finance* 2 (1): 1–14. https://doi.org/10.46671/2521-2486.1013.

Terzi, Alessio, and Pasquale Marco Marrazzo. 2020. "Structural Reforms and Growth: The Elusive Quest for the Silver Bullet." In *Economic Growth and Structural Reforms in Europe*, edited by Nauro F. Campos, Paul De Grauwe, and Yuemei Ji, 37–62. Cambridge: Cambridge University Press.

Trainer, Ted. 2012. "De-Growth: Do You Realise What It Means?" *Futures* 44 (6): 590–599. https://doi.org/10.1016/j.futures.2012.03.020.

Trotsky, Leon. 1918. *The Bolsheviki and World Peace*. Charleston, SC: Nabu Press, 2012.

Twain, Mark, and Charles Dudley Warner. 1873. *The Gilded Age: A Tale of Today*. Chicago: American Publishing Company.

Unsworth, Sam, Pia Andres, Giorgia Cecchinato, Penny Mealy, Charlotte Taylor, and Anna Valero. 2020. "Jobs for a Strong and Sustainable Recovery from Covid-19." CEPCOVID-19-010, Centre for Economic Performance, London School of Economics. October. https://www.lse.ac.uk/granthaminstitute/wp -content/uploads/2020/10/Jobs_for_a_strong_and_sustainable_recovery _from_Covid19.pdf.

Van den Bergh, Jeroen C. J. M., 2017. "Green Agrowth: Removing the GDP-Growth Constraint on Human Progress." In *Handbook on Growth and Sustainability,* edited by Peter A. Victor and Berett Dolter, 181–210. Cheltenham, UK: Edward Elgar. https://doi.org/10.4337/9781783473564.00016.

Van Griethuysen, Pascal. 2012. "Bona Diagnosis, Bona Curatio: How Property Economics Clarifies the Degrowth Debate." *Ecological Economics* 84: 262–269. https://doi.org/10.1016/j.ecolecon.2012.02.018.

Veblen, Thorstein. 1899. *The Theory of the Leisure Class: An Economic Study in the Evolution of Institutions*. New York: Macmillan.

Victor, Peter A. 2012. "Growth, Degrowth and Climate Change: A Scenario Analysis." *Ecological Economics* 84: 206–212. https://doi.org/10.1016/j.ecol econ.2011.04.013.

Victor, Peter A., and Gideon Rosenbluth. 2007. "Managing without Growth." *Ecological Economics* 61 (2–3): 492–504. https://doi.org/10.1016/j.ecolecon .2006.03.022.

Vollrath, Dietrich. 2020. *Fully Grown: Why a Stagnant Economy Is a Sign of Success*. Chicago: University of Chicago Press.

Voltaire. 1764. *Philosophical Dictionary*. London: Penguin Classics, 1984.

Vona, Francesco, Giovanni Marin, Davide Consoli, and David Popp. 2018. "Environmental Regulation and Green Skills: An Empirical Exploration." *Journal of the Association of Environmental and Resource Economists* 5 (4): 713–753. https://doi.org/10.1086/698859.

Wadhams, Peter. 2017. *A Farewell to Ice: A Report from the Arctic*. London: Allen Lane.

Walasek, Lukasz, and Gordon D. A. Brown. 2015. "Income Inequality and Status Seeking: Searching for Positional Goods in Unequal U.S. States." *Psychological Science* 26 (4): 527–533. https://doi.org/10.1177/0956797614567511.

Weber, Max. 1905. *The Protestant Ethic and the Spirit of Capitalism*. London: Routledge, 2001.

———. 1917. "Science as a Vocation." In *The Vocation Lectures,* edited by David Owen and Tracy B. Strong, 1–31. Indianapolis: Hackett, 2004.

Wei, Max, Shana Patadia, and Daniel M. Kammen. 2010. "Putting Renewables and Energy Efficiency to Work: How Many Jobs Can the Clean Energy Industry Generate in the US?" *Energy Policy* 38 (2): 919–931. https://doi.org /10.1016/j.enpol.2009.10.044.

Weiss, Martin, and Claudio Cattaneo. 2017. "Degrowth—Taking Stock and Reviewing an Emerging Academic Paradigm." *Ecological Economics* 137: 220–230. https://doi.org/10.1016/j.ecolecon.2017.01.014.

Weitzman, Martin L. 1998. "Recombinant Growth." *Quarterly Journal of Economics* 63 (2): 331–360.

Westacott, Emrys. 2016. *The Wisdom of Frugality: Why Less Is More—More or Less.* Princeton, NJ: Princeton University Press.

Weyant, John P. 2011. "Accelerating the Development and Diffusion of New Energy Technologies: Beyond the 'Valley of Death.'" *Energy Economics* 33 (4): 674–682. https://doi.org/10.1016/j.eneco.2010.08.008.

Wiedmann, Thomas, Manfred Lenzen, Lorenz T. Keyßer, and Julia K. Steinberger. 2020. "Scientists' Warning on Affluence." *Nature Communications* 11 (1): 1–10. https://doi.org/10.1038/s41467-020-16941-y.

Wilkinson, Richard, and Kate Pickett. 2010. *The Spirit Level: Why Equality Is Better for Everyone.* New York: Penguin Books.

Wilson, David Sloan. 2019. *This View of Life: Completing the Darwinian Revolution.* New York: Vintage.

World Bank. 2008. *The Growth Report: Strategies for Sustained Growth and Inclusive Development.* Commission on Growth and Development, World Bank, Washington DC.

Wrangham, Richard. 2019. *The Goodness Paradox: The Strange Relationship between Virtue and Violence in Human Evolution.* New York: Pantheon.

Wynes, Seth, and Kimberly A Nicholas. 2017. "The Climate Mitigation Gap: Education and Government Recommendations Miss the Most Effective Individual Actions." *Environmental Research Letters* 12: 1–9. https://doi.org /https://doi.org/10.1088/1748-9326/aa7541.

Yergin, Daniel. 2020. *The New Map: Energy, Climate, and the Clash of Nations.* New York: Penguin Press.

Zachmann, Georg, Gustav Fredriksson, and Grégory Claeys. 2018. "The Distributional Effects of Climate Policies." Bruegel Blueprint Series 28. https://www .bruegel.org/2018/11/distributional-effects-of-climate-policies/.

Zhang, David D., C. Y. Jim, George C. S. Lin, Yuan Qing He, James J. Wang, and Harry F. Lee. 2006. "Climatic Change, Wars and Dynastic Cycles in China over the Last Millennium." *Climatic Change* 76 (3–4): 459–477. https://doi.org /10.1007/s10584-005-9024-z.

Acknowledgments

This book contains so much of me and is so shaped by my worldview that the inclination to call it solely my work is strong. The truth is, without the intellectual and moral support of many, it would not have come to fruition, and surely not in its current form.

For discussions and exchanges on topics covered in these pages, and readings of selected parts of the manuscript, I have many to thank: Alessandro Allegra, Liza Archanskaia, Leire Ariz-Sarasketa, Scott Barrett, Paolo Bizzotto, Lorenza Bottacin-Cantoni, Federico Bruni, Francesco Carimi, Tamma Carleton, Orsetta Causa, Valeria Cipollone, Lewis Dartnell, Roberto Defez, Rodrigo Deiana, Viktoria Dendrinou, Anna Dimitrijevics, Björn Döhring, José Dominguez-Abascal, Roger Fouquet, Jiří Friml, Giorgio Gallo, Alicia Garcia-Herrero, Andrea Garnero, Christoph Gran, Ricardo Hausmann, Sarah Mishkin, Carsten Nickel, João Nogueira-Martins, Georgios Petropoulos, Valerio Riavez, André Sapir, Nicoló Spiezia, Michael Strevens, Rosaria Vergara, John Verrinder, Hans-Joachim Voth, Emrys Westacott, and Georg Zachmann. For pointing me to useful data sources I thank Max Roser, and, for their valuable feedback, the participants in three seminars: the "Master per la Pace" session at the University of Pisa; the "Global Energy Fundamentals" session at the Bologna campus of the Johns Hopkins School of Advanced International Studies; and a third gathering hosted by the Directorate-General for Economic and Financial Affairs.

Because some ideas presented in this book build on past papers, special thanks are due to past coauthors Stefano Marcuzzi, Pasquale Marco Marrazzo, Jim O'Neill, and Michele Peruzzi—colleagues and friends from whom I have learned a lot.

Other friends helped throughout the drafting process. These include Natasha Raikhel, who offered constant words of encouragement; Taylor Cade West, who helped polish my language and clear my thoughts on US economic history; and Luca de Angelis, the most thorough reviewer of large parts of the book. Marco Bardicchia was a great source of distraction at times when I needed that most. Filippo Teoldi helped me with data visualization. Omar Barbiero, a wise companion for economic discussions since our Harvard days, played that role again and sharpened my thoughts on the theory of value. Nicoló Melli is one of the few who could stomach reading multiple drafts of the entire manuscript, always providing words of encouragement. Simone Tagliapietra has given me confidence, lending support to the project from the very beginning as well as much-needed expertise on energy and climate policy issues. My colleague and friend Leonor Pires deserves a special mention: the seed for this book was planted by our coffee conversations about downshifting, ecovillages, degrowth, and Extinction Rebellion demands. At Paperò al Glicine, owners Peppe, Paola, and Rosanna (together with Eleonora) earn my praise for the perfect writing atmosphere they created, especially at sunset in the summer.

The staff of Harvard University Press must also be thanked for their support throughout this long process. My editor, Ian Malcolm, much deserves my gratitude for enthusiastically embracing this project right from the start. Olivia Woods provided great assistance, while Julia Kirby much improved my English, while checking the consistency and soundness of my claims with an eagle eye. Likewise, I must thank two anonymous reviewers, whose comments greatly enhanced the quality of the original manuscript, as did the comments of the Harvard University Press Board of Syndics.

Lastly, four people deserve special thanks. Dani Rodrik for believing in me when I needed it the most; Dana Or for being a unique friend, reading through the entire manuscript with great focus; my mother, Fiorella Lo Schiavo, for her unwavering support; and Elinor Wahal. While our romantic relationship has not proven so, my gratitude to you shall be eternal.

Index

Note: Page numbers in *italics* indicate figures and tables.